More Lessons Learned from Research

Helping All
Students
Understand
Important
Mathematics

Volume **2**

Edited by

Edward A. Silver
University of Michigan, Ann Arbor

Patricia Ann Kenney
University of Michigan, Ann Arbor

NCTM | NATIONAL COUNCIL OF
TEACHERS OF MATHEMATICS

Copyright © 2016 by
The National Council of Teachers of Mathematics, Inc.
1906 Association Drive, Reston, VA 20191-1502
(703) 620-9840; (800) 235-7566; www.nctm.org
All rights reserved

ISBN 978-0-87353-725-4

Library of Congress Control Number: 2015413341

The National Council of Teachers of Mathematics is the public voice of mathematics education, supporting teachers to ensure equitable mathematics learning of the highest quality for all students through vision, leadership, professional development, and research.

Printed in the United States of America

Contents

Preface

Over the past seventy years or so a solid body of research evidence has been amassed pointing to the benefits of teaching for understanding in mathematics (e.g., Brownell & Moser, 1949; Brownell & Sims, 1946; Cohen, McLaughlin, & Talbert, 1993; Fuson & Briars, 1990; Hiebert & Carpenter, 1992; Hiebert & Wearne, 1993; Hiebert et al., 1996; Kilpatrick, Swafford, & Findell, 2001; Stein & Lane, 1996). In addition to "teaching for understanding," many other terms (such as *authentic instruction, ambitious instruction, higher-order instruction, meaningful instruction, problem-solving instruction,* and *sense-making instruction*) have been used in roughly synonymous ways to convey that mathematics classrooms should be places where students engage actively, deeply, and intellectually in understanding mathematical ideas rather than being treated as passive recipients of knowledge conveyed by the teacher and textbook.

Although there are many unanswered questions about precisely how specific teaching practices are linked to students' learning mathematics with understanding (see Hiebert & Grouws, 2007), and there remain other questions about how students come to understand important mathematical ideas and the obstacles they may encounter along the way, the mathematics education community has placed increasing emphasis on investigating and using teaching practices that are oriented toward the development of students' conceptual understanding, and also on studying the developmental pathways along which students develop mathematical understanding. The National Council of Teachers of Mathematics (NCTM) has played a key role in promoting the importance of teaching mathematics in ways that foster the learning of mathematics with understanding, and this volume is another resource to support the mathematics education community in this quest.

This volume also builds on the foundation laid by the pioneering work of Judith Sowder and Bonnie Schappelle to promote the integration of educational research with educational practice in mathematics education. In 2002 NCTM published *Lessons Learned from Research*, edited by Sowder and Schappelle. That book was a compendium of adapted or abstracted articles that had originally appeared in the *Journal for Research in Mathematics Education* (*JRME*) during the period of Sowder's editorship (1996–2000). It addressed in a novel way what many view as a longstanding problem in education—namely, the perceived gap between educational practice and policy on the one hand and educational research and scholarship on the other.

Although some may view the gap between research and practice in education as an inevitable consequence of the differences between the professional work of educational researchers and educational practitioners, many others view it as a problem that can and must be solved. The desire to narrow or completely close this gap is often fueled by a belief that educational practice could be more effective if it were informed by the best available knowledge from educational research. Thus, making the findings of educational research available to practitioners is viewed as an important strategy for increasing the quality of education. Yet virtually everyone agrees that the form in which research is disseminated within the research community is unlikely to be immediately comprehensible or useful to those interested in its application in classroom instruction, instructional design, or teacher preparation. As Sowder and Schappelle wrote in the introduction to their book, "Teachers rarely access original research reports, perhaps because researchers tend to write in a style that is often not teacher-friendly. Few teachers ever open an issue of the . . . *Journal for Research in Mathematics Education* or, for that matter, any other research journal, unless they are assigned to do so for professional development or for a graduate class" (2002, p. 1).

To address both the gap between research and practice and the inaccessibility of education research reports to teachers and other mathematics education practitioners, Sowder and Schappelle undertook the creation of a compilation based on original research reports published in *JRME* but rewritten with an audience of teachers in mind. This was a novel approach to bridge the gulf between research and practice, and it was an important first step in contemporary efforts to help researchers and practitioners find common ground and learn from each other. In the decade following the publication of that book, NCTM has sponsored reports, conferences, and awards that encourage two-way traffic across the bridge that was constructed initially by Judy Sowder and Bonnie Schappelle.

With this new publication we take another step in the journey toward bridging the gap between research and practice in mathematics education. Like its companion volume (Silver & Kenney, 2015), which focused on research related to core mathematical processes and practices, such as those delineated in the Standards for Mathematical Practice in the Common Core State Standards (National Governors Association Center for Best Practices & Council of Chief State School Officers [NGA Center & CCSSO], 2010) or the NCTM Standards (NCTM 1989, 2000), this volume contains a collection of adapted or abstracted articles that originally appeared in the *Journal for Research in Mathematics Education*. All contributions to this volume are based on original research reports that appeared in *JRME* during the period 2000–2010, which includes a span of four years (2000–2004) when we edited the journal. (Edward Silver was editor, and Patricia Kenney was assistant editor.) Emulating the approach taken by Sowder and Schappelle (2002), each original article was rewritten to make the content more accessible to and useful for a teacher audience.

The twenty-four articles selected for inclusion in this volume were ones that we judged to be relevant to the theme of helping all students learn mathematics with understanding. The contributions are all based on research articles that examined the learning and teaching of key school mathematics content topics and that investigated the effects of innovative and ambitious teaching and curriculum approaches to mathematics instruction inspired by *Curriculum and Evaluation Standards for School Mathematics* (NCTM, 1989) and *Principles and Standards for School Mathematics* (NCTM, 2000). At its core this collection of research articles addresses whether and how it might be possible to assist all students to gain proficiency with and understanding of important mathematical ideas and processes.

The book is organized into three sections. The first section includes articles related to equity and access in relation to ambitious mathematics expectations for all students, and the second and third sections include articles related to one or more topics that are known to be difficult for many students to learn, important for all students to have a chance to learn, and foundational to success in the further study of mathematics. In the second section we trace ideas and skills associated with the long-standing school mathematics trajectory from arithmetic to algebra and on to calculus. The chapters in the third section are based on research related to the topics of rational numbers and probability, which are well known to be very difficult for many students to learn with understanding.

By organizing the articles into thematically related clusters, we hope to provide readers with an opportunity to think not only about the content of each article individually but also about what the articles in a section might contribute to an understanding of what it might take to achieve the grand challenge of making high-quality mathematics teaching and learning available for all students and having students meet our lofty expectations.

As a general rule we believe that research knowledge accumulates across investigations, including those investigations that build systematically on prior work as well as those that come from different perspectives and converge in some way on a key understanding. Although the research articles in this volume are provided individually, and we think that each has independent merit, it is by looking analytically and synthetically across these articles and other work not included here that the reader is likely to gain the most useful and usable knowledge and valuable insights that can inform and enable improvement of mathematics teaching and learning.

The preparation of this volume has increased our optimism that the gap between research and practice can be bridged. When we contacted the authors of the articles we had identified for this volume, they all agreed to participate. In fact, they agreed with enthusiasm because they saw the value of sharing their work with practitioners. Moreover, working in the spirit of this volume, many of these authors had already produced some version of their work for distribution to practitioners.

We thank the authors for their willingness to prepare adapted versions of their articles or to respond to abstracted versions that we prepared on their behalf. We also want to thank Myrna Jacobs, NCTM's former publications manager, for conveying the request from the NCTM Educational Materials Committee to one of us (Silver) to produce a sequel to Sowder and Schappelle's *Lessons Learned from Research*. We are especially grateful to Julie Schorfheide for her skillful copyediting of manuscripts and to Anita Draper for her assistance in moving the book efficiently through the NCTM production process. Finally, we acknowledge the support of the Usable Scholarship in Education (USE) Initiative at the University of Michigan.

Edward A. Silver
Patricia Ann Kenney

References

Brownell, W. A., & Moser, H. E. (1949). *Meaningful vs. mechanical learning: A study in grade III subtraction* (Duke University Research Studies in Education, No. 8). Durham, NC: Duke University Press.

Brownell, W. A., & Sims, V. M. (1946). The nature of understanding. In N. B. Henry (Ed.), *The measurement of understanding* (pp. 27–43). Forty-fifth yearbook of the National Society for the Study of Education, Part I. Chicago, IL: University of Chicago Press.

Cohen, D. K., McLaughlin, M., & Talbert, J. (Eds.). (1993). *Teaching for understanding: Challenges for policy and practice.* San Francisco, CA: Jossey-Bass.

Fuson, K. C., & Briars, D. J. (1990). Using a base-ten blocks learning/teaching approach for first- and second-grade place-value and multidigit addition and subtraction. *Journal for Research in Mathematics Education, 21,* 180–206.

Hiebert, J., & Carpenter, T. P. (1992). Learning and teaching with understanding. In D. A. Grouws (Ed.), *Handbook of research on mathematics teaching and learning* (pp. 65–97). New York, NY: Macmillan.

Hiebert, J., Carpenter, T. P., Fennema, E., Fuson, K., Human, P., Murray, H., ... Wearne, D. (1996). Problem solving as a basis for reform in curriculum and instruction: The case of mathematics. *Educational Researcher, 25*(4), 12–21.

Hiebert, J., & Grouws, D. A. (2007). The effects of classroom mathematics teaching on students' learning. In F. K. Lester (Ed.), *Second handbook of research on mathematics teaching and learning* (pp. 371–404). Charlotte, NC: Information Age.

Hiebert, J., & Wearne, D. (1993). Instructional tasks, classroom discourse, and students' learning in second-grade arithmetic. *American Educational Research Journal, 30,* 393–425.

Kilpatrick, J., Swafford, J., & Findell, B. (2001). *Adding it up: Helping children learn mathematics.* Washington, DC: National Academy Press.

National Council of Teachers of Mathematics. (1989). *Curriculum and evaluation standards for school mathematics.* Reston, VA: Author.

National Council of Teachers of Mathematics. (2000). *Principles and standards for school mathematics*. Reston, VA: Author.

National Governors Association Center for Best Practices & Council of Chief State School Officers (NGA Center & CCSSO). (2010). *Common core state standards for mathematics*. Washington, DC: Author. Retrieved from http://www.corestandards.org.

Silver, E. A., & Kenney, P. A. (Eds.). (2015). *More lessons learned from research: Volume 1. Useful and usable research related to core mathematical practices*. Reston, VA: National Council of Teachers of Mathematics.

Sowder, J., & Schappelle, B. (Eds.). (2002). *Lessons learned from research*. Reston, VA: National Council of Teachers of Mathematics.

Stein, M. K., & Lane, S. (1996). Instructional tasks and the development of student capacity to think and reason: An analysis of the relationship between teaching and learning in a reform mathematics project. *Educational Research and Evaluation, 2*(1), 50–80.

Perspectives on Equity

Introduction

We begin this volume with a section devoted to the critical issue of equity in the mathematics classroom. The *Principals and Standards for School Mathematics* (National Council of Teachers of Mathematics [NCTM], 2000) includes this succinct summary of the Equity principle: "Excellence in mathematics education requires equity—high expectations and strong support for all students" (p. 11). Each of the six chapters in this section was written by our colleagues who have taken the phrase "mathematics for all" to heart, and each chapter focuses on research efforts that add to our knowledge and understanding of how teachers and other practitioners can create opportunities for each and every learner to achieve in mathematics from kindergarten to high school and beyond.

As editors, we have organized and ordered the chapters so that readers not only will think about the content of each individual article but will also consider how these articles together might contribute to an understanding of equity in the mathematics classroom. A brief summary of each chapter and what we think are some lessons learned from the research follow. We encourage readers to look analytically and synthetically across these articles to generate useful and usable knowledge and valuable insights to inform and enable improved mathematics teaching and learning for all students.

Chapter 1. "The Implementation and Impact of Mathematics Reforms in High-Poverty Middle Schools," by Robert Balfanz, Douglas J. Mac Iver, and Vaughan Byrnes

Robert Balfanz, Douglas Mac Iver, and Vaughan Byrnes take us back to the early 2000s—a time when a number of reform efforts were directed at middle school mathematics. In the chapter, the authors summarize results from the first four years of the Talent Development (TD) Middle School Mathematics Program, designed to develop a comprehensive and sustainable set of evidence-based curricular, professional development, and supportive whole-school reforms aimed at raising mathematical achievement in high-poverty middle schools.

After presenting information on the schools that participated in the project, Balfanz and his colleagues describe important features of the TD Mathematics Program: schoolwide use of coherent and challenging curricular materials, multiple tiers of sustained professional development, teacher participation in program development, and the placement of the program within whole-school reform. To measure the effectiveness of the program, the authors focused their research on answering questions about how successfully the intended reforms were implemented and sustained in the schools, whether the students who experienced the TD program had greater achievement gains, and to what extent the implementation levels and achievement gains were related.

The authors report that the teachers liked the curriculum and felt they were receiving good-quality professional development so they could implement the curriculum successfully. The teachers also were covering more mathematics topics than in the past with less repetition of those topics previously taught and with a focus on problem-solving strategies rather than on rote learning. With respect to mathematics achievement, students experiencing the TD program increased their school average scores and the degree to which they met district- or state-determined benchmarks.

Balfanz, Mac Iver, and Byrnes conclude by highlighting three important findings: (1) it was possible to implement and sustain a comprehensive set of mathematics reforms in high-poverty schools; (2) implementation of the reforms led to significant and sustained achievement gains across multiple classrooms in multiple schools over multiple years; and (3) although achievement results were statistically significant and educationally substantial, they were not of sufficient magnitude to conclude that the enacted reforms *alone* were enough to close the achievement gaps that high-poverty students bring to urban middle schools.

Lessons learned

Through the details the authors present in this chapter (along with a updated reference list of publications about the TD Mathematics project), readers get a glimpse of how complicated reforms directed toward equity can be. The lesson learned here is that implementing reform for equity is not easy, as there are many factors to be considered: students, teachers, administrative structure, and curriculum, to name just a few.

Chapter 2. "The Importance of Teaching in the Promotion of Open and Equitable Mathematics Environments," by Jo Boaler

Continuing the theme of mathematics education reform, Jo Boaler focuses on two research-based programs: her own three-year study of mathematics instruction in secondary schools in England (in particular, a school she called Phoenix Park) and the work of the QUASAR project in a group of U.S. middle schools. The author's focus is on teaching practices that promote equity in student learning of important mathematics, and examples from Phoenix Park and QUASAR show how teachers can engage students both deeply and equitably.

Boaler begins by briefly describing her study of two schools and the work of the QUASAR project and how students in each project made significant gains in achievement, which were distributed equally among the different racial, ethnic, and linguistic groups. The remainder of her chapter is devoted to three teaching practices that helped make mathematics instruction accessible to all students.

The first activity involves *introducing activities through discussion*. Both Phoenix Park teachers and those in the QUASAR classrooms made an effort to participate in the discussion; students were never left to interpret problems alone. The second activity had teachers attending to the *ways in which students communicated their thinking*. Because students were initially not used to explaining their reasoning or talking about mathematics, teachers often had to urge students to do so. The third activity involved *bringing real-world contexts into the mathematics classroom*. A significant problem arises when students are required to engage with contexts as though they are real but ignore factors that would be important in real-life versions of mathematics tasks.

Boaler concludes the chapter by underscoring the importance of teachers paying careful attention to equity issues by making open-ended problems and rich discussions about mathematical situations accessible to all students. She states that it is the *teachers* of mathematics—not the curriculum materials—who are critically important. Empowering teachers with knowledge of equitable teaching practices remains an important goal of mathematics teacher education.

Lessons learned

The three teaching practices that promote equity—introducing activities through discussion, teaching students to justify and reason about mathematics, and making real-world contexts accessible—are just as relevant today as when Jo Boaler first wrote about them more than ten years ago. Making these and other relevant practices part of teacher education programs remains an important goal in mathematics education.

Chapter 3. "Informing Teachers about Identities and Agency Using the Stories of Black Middle School Boys Who Are Successful with School Mathematics," by Robert Q. Berry III

The chapter by Robert Q. Berry III presents stories of several middle school Black boys who are successful in school mathematics and focuses on their experiences of their mathematics identity: how they saw themselves and how they were seen by others as doers of mathematics. Further, Berry explores the connection between identity and agency, with agency being identity in action and the presentation of a person's identity to the world.

Berry begins by showcasing how the boys developed their individual identities and how these identities contributed to success with school mathematics. For example, most of the boys identified career and college goals as motivators to do well in school and mathematics. They also expressed a strong confidence in their mathematics ability, which allowed them to be among the "smart kids." Further, Berry identifies out-of-school contexts that directly related to shaping the boys' positive mathematics identities: co-curricular and special academic programs, religious affiliations, and athletics. The author encourages teachers to leverage students' culture, contexts, and identifies to support and enhance mathematics learning.

The next section of the chapter is devoted to the topic of agency, or how students enact their identities. The boys in Berry's study identified themselves as being smart and good at mathematics; consequently they presented themselves and adopted behaviors of smartness and being good at mathematics. It was important for their teachers to affirm these identities and encourage students to continue to enact them.

Finally, Berry recommends ways that teachers can support students' development of positive mathematics identity and their agencies. First, teachers can promote persistence and reasoning during problem-solving activities. Second, teachers can find ways to capitalize on students' knowledge of mathematics from everyday interaction with their cultural and social backgrounds. Third, teachers who recognize and position students' backgrounds as a resource in their mathematics teaching can connect students' identities with building a sense of agency. Berry closes with a set of questions for teachers to consider as they think about how to encourage the development of identity and agency in the mathematics classroom.

Lessons learned

As illustrated in the stories that Berry includes in the chapter, classroom experiences and school climate can shape student perceptions of being successful learners and doers of mathematics; conversely, these experiences can also discourage students from achieving if teachers have low expectations of them based on race or cultural issues. It is important for practitioners to understand students' identities, know the context of their students' experiences, and foster the development of agency.

Chapter 4. "Teaching Mathematics in a Multilingual Classroom," by Mamokgethi Setati Phakeng

This chapter by Mamokgethi Phakeng focuses on the challenges of teaching mathematics in classrooms in which the languages spoken by students are not the first language of the teacher. The author suggests that mathematics teachers face the challenges of helping their students develop ways of (as she puts is) "acting-interacting-thinking-valuing-talking-writing-reading-listening" that are mathematically appropriate and of reaching all students within the multilingual classroom. Phakeng writes her chapter as a commentary to a journal article she authored with two colleagues (Terence Molefe and Mampho Langa); the article from *Pythagoras* is reprinted as an appendix to the chapter.

Phakeng begins with a summary on what research tells us about teaching and learning in multilingual classrooms. In particular, she notes that language is not only a tool for thinking and communicating, but it can also be political (i.e., people use language to project themselves as certain kinds of people engaged in certain kinds of activities). She focuses on South Africa, where teachers in multilingual mathematics classrooms prefer to teach in English despite their students' limited fluency in that language. Phakeng offers three challenges that teachers in multilingual classrooms face and strategies teachers can use to address them.

The first suggestion is getting to know your students so that you understand the audience for whom you are preparing lessons, and she offers some questions teachers can ask regarding language preferences (e.g., What language do you use to communicate with your parents/ siblings at home?). The second suggestion involves how to create mathematics tasks that are relevant to students' background and experiences. Phakeng includes two tasks used in South African classrooms as examples, and encourages teachers to have copies of the tasks available in whichever home languages students speak. The final suggestion is to consider students' home languages when forming groups in order to foster communication within the groups. Phakeng concludes by acknowledging the challenges of teaching in a multilingual classroom and by encouraging teachers to consider wider social, cultural, and political factors as they plan for mathematics instruction in diverse classrooms.

Lessons learned

Using examples from South African classrooms, Mamokgethi Phakeng raises our awareness about the critical issue of communication in learning and teaching mathematics—especially when the first language of the teacher does not necessarily match those of the students. The challenges identified and the practical suggestions offered are important food for thought for practitioners in light of mathematics lesson planning, task choice, and small-group structure.

Chapter 5. "Building Young Children's Mathematics," by Douglas H. Clements and Julie Sarama

Douglas Clements and Julie Sarama focus their chapter on how to better serve another group of traditionally underserved students—in this case, young children from lower-resourced communities. What these children know when they enter kindergarten predicts their mathematics achievement throughout their school career; moreover, what they know in mathematics predicts their reading achievement even better than early literacy skills do. As a way to provide more experience with mathematics to lower-income children, the authors have developed a curriculum, called *Building Blocks*, that focuses on numeric/quantitative and geometric/spatial skills using children's everyday activities (e.g., block building, art, songs, puzzles) in preschool and the early grades.

A unique feature of this chapter is that the authors wrote it in part as a commentary to an article they published in *Teaching Children Mathematics*; that article is available at nctm.org .more4u. They also include updated information about other research studies on the *Building Blocks* curriculum and how it supports the mathematics learning of young children, including those from traditionally underserved populations. For example, in a study of African American children, Clements and Sarama found that in *Building Blocks* classrooms, these children averaged higher gains than children in more traditional classrooms.

Clements and Sarama include a description of the professional development model used to introduce teachers to the curriculum. They continue by posing and then answering a series of questions about the success of *Building Blocks*, based on their extensive research findings: Why was *Building Blocks* successful? How can the results help teachers? Does "more math" mean "less literacy and language learning"? and Do effects persist?

The authors conclude by underscoring the importance of early mathematics learning—especially to children from low-resourced communities, underrepresented racial/ethnic groups, and who are English language learners. Their model curriculum and the supports for teachers as they implement the model can help teachers of young children establish foundational mathematics skills.

Lessons learned

Clements and Sarama remind us that it is never too early to develop mathematical learning in children. The knowledge that young children bring with them to kindergarten predicts their mathematics achievement throughout their school career and beyond.

Chapter 6. "A Letter to Those Who Dare Teach Mathematics for Social Justice," by Eric "Rico" Gutstein

The final chapter in the section is by Rico Gutstein, who writes a letter to teachers who (as he puts it) dare to implement a mathematics curriculum based on social justice. He begins with a summary of the state of equity and social justice and a description of how factors such as high-stakes testing, the Common Core Standards, and failing economic standards—especially in larger cities—are affecting teachers and teaching. Given the conditions of work and life, Gutstein focuses on mathematics for social justice as a way to allow students to learn and use mathematics to develop a sociopolitical consciousness of the roots of injustice in their lives and in the broader society so that they can eventually act to change the things they believe are wrong.

The author continues by recapping important ideas from his experiences teaching a social justice mathematics curriculum to middle school students at Rivera, a school in a low-income, working-class Mexican immigrant community in Chicago. He gave the students real-world projects in which they used mathematics to investigate social justice issues they cared about: immigration, gentrification, working conditions for tomato pickers, racism in housing prices, and so on. Gutstein goes on to talk about developing and then teaching mathematics for social justice, stating that *teaching* is the key. He writes that a social justice focus cannot be imposed on students; teachers and students have to co-create the classroom space. Politically taboo topics must become the norm; students must learn to question the norm; teachers must hear students about what matters to them, explain how you are teaching and why; and the design of the mathematics curriculum should be based upon student concerns and situations they face in their daily lives. The author continues by giving concrete examples based on his own classroom experiences on how teachers can get started on teaching mathematics for social justice.

Gutstein concludes the letter to teachers by talking about race—his and that of his students. He challenges us to make social justice mathematics teaching an explicitly antiracist pedagogy that is against all forms of racism and oppression.

Lessons learned

Throughout the letter, Rico Gutstein talks about what he learned from his time with the students at Rivera and subsequently with those at the Social Justice High School (aka Sojo). In both cases, he shares with us that teaching for social justice was difficult, messy, and complicated, but it was *possible* for students to begin to read and write the world using mathematics and to learn mathematics using a reform curriculum, teacher-selected real-world projects, and a co-created classroom culture supporting a social justice pedagogy.

References

National Council of Teachers of Mathematics (NCTM). (2000). *Principles and standards for school mathematics.* Reston, VA: Author.

The Implementation and Impact of Mathematics Reforms in High-Poverty Middle Schools

Robert Balfanz, Douglas J. Mac Iver, and Vaughan Byrnes
Johns Hopkins University

We begin this chapter by taking our readers back to the early 2000s. At that time, we noted in our *Journal for Research in Mathematics Education* (*JRME*) article that "middle school mathematics in the United States is in need of reform" (Balfanz, Mac Iver, & Byrnes, 2006, p. 33). We stated that middle school students learned less mathematics than their peers in many other countries (e.g., National Center for Education Statistics, 2000; Schmidt, McKnight, Cogan, Jakwerth, & Houang, 1999) and that the opportunity to acquire a substantial body of mathematical knowledge during middle school was unevenly distributed across the country (e.g., Balfanz, McParland, & Shaw, 2002; Campbell & Silver, 2000). In particular, high-poverty urban middle schools, attended predominately by minority students, appeared to provide fewer of the supports and resources that students need in order to learn a significant amount of mathematics during middle school. Recognition of the increased significance of middle school mathematics led to multiple reform proposals and the consensus that such reform should include several core elements:

- Students need to be provided with a coherent curriculum that is less cursory and repetitive and that systematically develops their intermediate mathematics skills and their mathematical reasoning ability.

- Middle school mathematics teachers, in order to implement a more challenging and comprehensive curriculum, need access to sustained high-quality professional development that is linked to the instructional materials they will be using; that focuses on their classroom activities; and that provides them with the content knowledge, pedagogy, and classroom management skills needed to implement a challenging middle-grades mathematics curriculum.

- Mathematics reforms need to be embedded in state, district, and whole-school reforms that facilitate instructional program coherence by aligning accountability, assessment, and resources; that create teaching and learning environments cognizant of the developmental transitions that occur in the middle grades; and that promote an "Every Child Can Succeed" culture in which students, teachers, and parents do what it takes to provide the supports needed for all students to receive a strong mathematical foundation.

At that time there was evidence that each of these reforms practices altered instructional practice and/or student effort in a productive manner and consequently raised mathematical achievement. However, as we planned our project we found that there had been little research in high-poverty schools, particularly at the middle school level, on the cumulative impact on this set of evidence-based practices, especially across multiple schools and over multiple years. We needed to know more than simply which reform strategies appear to

This chapter is adapted from R. Balfanz, D. J. Mac Iver, & V. Byrnes (2006), The implementation and impact of evidence-based mathematics reforms in high-poverty middle schools: A multi-site, multi-year study, *Journal for Research in Mathematics Education, 37*, 33–64.

raise average state or district levels of achievement; we needed to know the levels of impact these sets of strategies have under different conditions—in particular, (1) the extent to which the combined set of evidence-based mathematics education reforms outlined above can be implemented and sustained in high-poverty middle schools, (2) the level of implementation support needed to overcome existing conditions, (3) the level of impact on achievement that can be expected under different conditions, and (4) if the emerging set of evidence-based reforms in and of itself is sufficient to have a significant impact on improving mathematical achievement.

In our *JRME* article (Balfanz et al., 2006) we reported on results from the first four years of an ongoing effort to develop and implement a comprehensive and sustainable set of evidence-based curricular, professional development, and supportive whole-school reforms aimed at raising mathematical achievement in high-poverty middle schools. In this chapter, we summarize the content of the article; we refer our readers to the article for more complete information. The first section of the chapter provides information on the schools that participated in the study; the second section focuses on factors such as the levels of implementation achieved and the impact of the reforms on multiple measures of mathematics achievement; and the third section contains a discussion of the overall impact of the reforms and explores additional steps needed to achieve high levels of mathematical learning in high-poverty middle schools. Where relevant we include page number references to the *JRME* article that lead the reader to information not included in the chapter.

Designing the Mathematics Reforms and the Context of Our Study

At the start of the project in 1996, one of the authors spent a year observing mathematics instruction and the mathematics program in twelve classrooms within two of the three middle schools that would ultimately participate in the project. The schools were in the School District of Philadelphia and were nonselective neighborhood schools serving low-income minority populations. The observations revealed that these high-poverty schools shared many of the weaknesses reported in the literature about middle school mathematics in general (Balfanz, 1997). For example, the observed mathematics instruction was disorganized and idiosyncratic. Across and within grades, teachers were using different mathematics curricula, partly because of textbook shortages and partly as a reflection of teacher taste. Essentially, each teacher was making individual decisions about the type and level of mathematics needed by his or her classes. One unintended result of this individual decision making was a highly repetitive course of study across the grades, in which less and less grade-level material was introduced each year. For example, during one class day in September, students at all four grade levels (5–8) were learning about place value.

In the case of the teachers, most were unenthusiastic about teaching mathematics but had been assigned to do so, usually in combination with one or two other subjects. With one exception, all of the teachers were elementary certified. Many viewed teaching mathematics as a short-term assignment—a chore they would do until they obtained a better assignment within or outside of the school. There was a high degree of turnover in who taught mathematics and at what grade level.

The overwhelming majority of students in both schools entered the middle grades significantly behind grade level in their mathematics skills and knowledge as measured by scores on the Stanford 9 test. Moreover, because both schools served neighborhoods with high concentrations of poverty (in both schools, more than 80 percent of students were eligible for free or reduced-price lunches), the students brought with them greater levels of exposure to safety risks, unhealthy environments, and high levels of social disorder.

The two schools we observed had unsupported and essentially temporary mathematics teachers using an unorganized curriculum. These teachers faced the difficult task of enabling their students both to master the middle-grades curriculum and to close their elementary skill and knowledge gaps in the context of the myriad challenges endemic to high-poverty schools. This year of observation made it clear to us that, in order to succeed, any attempt to improve mathematics achievement would need curricular, professional development and teacher-support elements, and it would need to take place in the context of schoolwide reform. Based on our observations and an extensive literature review, we formulated a program—the Talent Development (TD) Middle School Mathematics Program—that we describe next.

The TD Middle School Mathematics Program

The TD Mathematics Program was a core component of a larger, whole-school reform design (the Talent Development Middle School) that integrates organizational, curricular, professional-development, school-climate, teacher-student interaction, and student-support reforms into a comprehensive set of reforms for high-poverty middle schools (see Mac Iver, Ruby, Balfanz, & Byrnes, 2003, for more information). The sections that follow briefly describe the essential features of the TD Mathematics Program.

Schoolwide use of coherent and challenging curriculum

To resolve the issue of individually selected mathematics materials, the schools implemented *Everyday Mathematics*, from the University of Chicago School Mathematics Project (UCSMP) elementary curriculum, for grades 5 and 6; in grade 7, UCSMP *Transition Mathematics* was used; and in grade 8, UCSMP *Algebra* was used. These materials were developed on the premise that students should be taught a substantial body of challenging mathematics and that algebra and geometry should be introduced early on and with greater emphasis (UCSMP, 2003). The curriculum also had a strong focus on mathematical reasoning, problem solving, and communication. The UCSMP curriculum received a promising program endorsement from the U.S. Department of Education's Mathematics and Science Expert Panel (1999).

One of the three middle schools participating in the project was attempting to teach all students algebra in eighth grade, so in this school UCSMP *Algebra* was adopted schoolwide during year 1 of implementation of the TD Mathematics Program. The other two schools phased in the *Algebra* text over a three-year period in order to allow time to build both teacher skills and student skills. In these two schools, seventh and eighth graders used *Transition Mathematics* during year 1. During year 2, eighth graders used lessons from *Transition Mathematics* and the *Algebra* text. In year 3, the *Algebra* text was used from the start of eighth grade. All three schools introduced *Everyday Mathematics Grade 5* (in the two grade 5–8 schools) and *Everyday Mathematics Grade 6* in year 1 of program implementation. Thus, by the start of year 3, all three schools were offering all students the same mathematics curriculum and sequence of courses, culminating with all students taking an algebra course in eighth grade.

Multiple tiers of sustained professional development

Teachers in the three schools were offered tiers of professional development linked to the implementation of the new mathematics curriculum. Three days of summer training were followed by monthly three-hour Saturday workshops, with make-up sessions available. Experienced peer teachers and users of the curricula led the workshops. The sessions were grade specific and focused on the unit or lessons the teachers would be using during the following month. (For an example of a professional development session on geometry, we refer the reader to pp. 38–39 of the 2006 *JRME* article.) In all, teachers had access to more than thirty-six hours of professional development per year. Attendance was voluntary, and teachers were paid the district rate (approximately $20 per hour at that time). Beginning in year 2, arrangements were made at a local university to give teachers three graduate credits if they completed thirty-six hours of training and related assignments. The goal here was to provide teachers with more professional development opportunities than had been typically available to U.S. mathematics teachers at that time.

In addition to the professional development sessions, teachers had access to support from a curriculum coach, who spent one or two days per week in each school. The curriculum coach was an experienced district teacher on special assignment to the project. The support was designed to be nonjudgmental, and it varied from classroom to classroom but included modeling, explaining, co-teaching, lesson planning assistance, observing lessons and providing confidential feedback, and making sure teachers had necessary materials for the lessons. During year 3 of the program, we began an initiative to train two to three teacher leaders from each school to be on-site trainers, with the goal of making the schools self-sustaining over time. These teacher leaders received an additional thirty hours of training per year for two years and provided an additional layer of support in the schools.

Teacher participation in program improvement	Each summer, the teachers were invited to take part in working groups to develop supplemental materials to help further customize and localize the instructional materials. The activities varied over the course of the project. For example, during the third summer, September Introductory Units were developed. These units were designed to compensate for the "broken supply lines" found in many urban schools at the start of the school year. At times schools fail to provide teachers with essential supplies and learning materials in a timely fashion. The September Introductory Units made it possible for teachers to begin teaching substantive standards-based lessons right away even if the regular materials their schools were supposed to supply had yet to be ordered, found, or delivered into their hands.
Embedded in whole-school reform	The three schools not only enacted the mathematics education reforms outlined above in the context of implementing the TD model but also adopted schoolwide reforms. The schools were also engaged in reforms in English and science that employed similar approaches to professional development as those outlined for mathematics. In addition, the schools made organizational changes to increase the communal nature of the schooling, including looping (where teachers stay with the same class of students for two years), semi-departmentalizing (teachers taught two subjects to the same class, so they only interacted with sixty to seventy students during a school year), and dividing the school into small learning communities (SLCs). Both survey data and interviews indicated that these reforms led to a greater sense among students that "my teacher cares about me," enabled teachers to adopt riskier but more engaging pedagogy, and helped form a "no excuses" attitude toward student success (e.g., see Mac Iver, Mac Iver, Balfanz, Plank, & Ruby, 2000).
Summary	The TD Mathematics Program was designed to encompass the (at that time) emerging set of evidence-based reforms in mathematics education. It had a high level of instructional program coherence and a challenging mathematics curriculum that culminated with algebra for all in grade 8. In addition, the program provided much more intensive, focused, and sustained professional development than teachers at that time were typically given, was integrated into a set of whole-school reforms, and was accompanied by substantial implementation support.

Program of Research

In any reform effort done at that time (and perhaps even now, about fifteen years later), the questions inevitably arise as to whether the program "worked." To answer this kind of broad question about the TD Mathematics Program, we worked with the School District of Philadelphia, which selected three comparison schools for each participating school that was similar in racial composition, high-poverty status, and past performance during the period before TD began in the district. The characteristics of the three TD schools and the three comparison schools appear in table 1.1.

Table 1.1
Characteristics of participating schools (TD) and comparison schools: Grade span, eighth-grade enrollment, race, and math test scores for school years 1995–1996 through 1996–1997

School	Grade span	Eighth graders enrolled	Black %	White %	Other %	PSSA math score (NCE)	Stanford 9 math total (NCE)
TD A	5–8	258	26.2	12.9	60.9	27.3	35.5
Comparison A	5–8	296	18.9	20.1	61.0	24.1	33.2
TD B	5–8	264	74.8	1.0	24.3	24.9	34.8
Comparison B	5–8	201	65.1	11.2	23.7	27.7	36.2
TD C	6–8	399	98.3	1.2	0.5	30.0	41.1
Comparison C	6–8	210	99.6	0.0	0.4	27.2	38.9

Note: Race and test scores are averages for eighth graders from individual student school records. Thanks to James Kemple and Corinne Herlihy for performing the calculations reported in this table.

Another comparison characteristic important to our research studies was that more than 70 percent of the students in both the TD schools and the comparison schools entered middle school performing below grade level in mathematics; the range was 71 percent to 86 percent. (More complete information about the TD schools, the comparison schools, and their school district can be found on pp. 41–44 in the 2006 *JRME* article.)

We developed an interrelated set of questions about the implementation and impact of the TD mathematics program:

1. How successfully were the intended reforms implemented and sustained in the schools?

2. Did the students who attended the experimental schools have greater achievement gains than students who attended the matched control schools on the high-stakes district assessments and on the lower-stakes state assessments?

3. What was the relationship between implementation levels and achievement gains?

The results are reported in abbreviated form in the next sections. We invite our readers to consult the original *JRME* article and other relevant publications (a list appears as an appendix to this chapter) for a more comprehensive presentation of the statistical models and the findings. In this section of the chapter our primary focus is on the conclusions that we drew from the analyses.

Implementation levels

Multiple methodologies (e.g., student surveys, a curriculum coach's evaluation of classroom activities, teacher focus groups, individual interviews of teachers) were used to analyze the implementation levels of TD and the roadblocks that teachers encountered. Overall, the

11

implementation measures indicated that a moderate-high level of implementation was achieved across the three schools. This implies that despite endemic problems, such as high staff turnover, it is possible to obtain an acceptable level of reform program implementation in high-poverty middle schools.

The sections that follow detail the implementation levels across the schools primarily in the fourth year of the project. This year is highlighted because it represents the most mature year of the project and it is the year for which the largest body of implementation data was available.

Interviews and focus groups

Interviews and focus group results showed that, in general, teachers liked the TD curriculum being implemented, felt that they were receiving good quality professional development, and recognized that the in-classroom implementation was both beneficial and much more intense than teachers commonly received. (See Useem, 1998, 1999, 2000, 2003).

Participation in professional development

Nearly 80 percent of the teachers attended some professional development sessions (participation was voluntary per the local union contract), and about two-thirds achieved the recommended level of thirty-six hours per year for two years. Thus, the majority of TD teachers had achieved a satisfactory level of training in regard to the program's specific curriculum and pedagogy, although a significant portion still lacked adequate training.

Teacher experience and stability

One of our biggest implementation challenges was a high degree of teacher turnover. This meant that each year many of the well-trained teachers left and were replaced by new, untrained teachers. By the fourth year of the study, only 31 percent to 59 percent of the classrooms in the three schools had mathematics from a teacher who had participated in the TD project all four years.

Recommended instructional practices

The survey results from the fourth year of implementation indicated that 71 percent of the classrooms across the three TD schools used five or more of the nine recommended instructional practices (e.g., students explaining how they got their answers; students working with a partner; whole-class work on a challenging problem) versus 51 percent of the classrooms in the control schools.

Curriculum coverage and implementation roadblocks

The initial goal of having TD teachers complete six to eight units of each grade level's instructional materials was by and large *not* achieved. Factors such as pressure to use district test preparation materials, numerous scheduling disruptions, and the wide range of student skills in the heterogeneous classrooms contributed to this situation. The average classroom completed four and a half units from the UCSMP instructional materials.

The curriculum coaches in each school were asked to identify the number and type of roadblocks teachers encountered. Almost all of the identified roadblocks were the result of staff turnover, lack of training, teaching inexperience, and/or poor classroom management. Only one of the three schools identified lack of materials as a major roadblock.

Impact on growth in mathematics problem solving

Extensive achievement data were available from the three TD schools and their three comparison schools across all four years of the study. Most cohorts in these schools took the mathematics problem-solving battery of the Stanford 9 test two or three times during their middle school years. In our preliminary analyses, we found no significant differences between the groups (TD and control) in prior achievement. We used hierarchical linear modeling (HLM) methods to model the achievement of all students who attended one of the six schools at any time between the fall of 1997 and the spring of 2001 *and* took the Stanford 9 at least once during their time of attendance.

The results showed that the TD schools raised their school mean achievement growth in problem solving substantially more than did the comparison schools. The model estimates also revealed a slight decline in the TD advantage over the comparison schools from seventh grade to eighth grade, which may be explained by the shift of the TD eighth-grade curriculum to algebra.

The TD Mathematics Program focused on improving students' mathematical problem-solving skills as opposed to routine mathematics procedures. As a result, we expected TD students to do no better (and no worse) than control students on the Stanford 9's Math Procedures Subtest. As expected, when analyses were conducted on students' growth in procedures scores there were no significant differences between students in the TD schools and in the control schools.

Impact on mathematics achievement growth between fifth and eighth grade on state assessments

To further understand and analyze any achievement benefit incurred by students in the TD Mathematics Program, we examined the extent to which students in both the TD schools and the control schools experienced achievement gains between the end of fifth grade and the end of eighth grade on the Pennsylvania System of State Assessments (PSSA) in mathematics, the state-required exam at that time. The PSSA was viewed as a lower-stakes test than the Stanford 9. Again, we used HLM to analyze the data for this part of the study.

Results showed that TD schools increased their school average scores and the degree to which they met concrete district or state-determined benchmarks. In Pennsylvania at the time of the study, students who scored below the 25th state percentile were considered "below basic." Our findings showed that in the TD schools there was a 10 percent increase in the percentage of students scoring above this critical 25th percentile between the fifth and eighth grades, whereas the control schools had only a 2 percent gain. Thus, although both the TD schools and the control schools had similar percentages of students with below basic skills in the fifth grade (26 percent vs. 23 percent), by the eighth grade more than one-third of the students in the TD schools had crossed this significant threshold, compared to only one-fourth of the control school students. We also found that the TD schools outperformed the control schools in raising the percentage of students scoring above the 10th and 50th percentiles. These results along with others reported in the article (see tables 12 and 13 in Balfanz et al., 2006, p. 55) indicate that the set of evidenced-based mathematics reforms implemented in the TD schools helped students at all levels of the achievement spectrum in educationally significant ways.

Achievement outcomes and implementation levels

Thus far we have reported that students in the TD schools achieved more than their peers in control schools. This final analysis tied the achievement outcomes to variation in implementation and showed that *greater implementation of the reform model produced better outcomes.* For this analysis we used the implementation data for the fourth year of implementation (2000–2001), and we estimated the relationship between different implementation levels and students' achievement in spring 2001 after controlling in the statistical model for prior achievement in spring 2000. An HLM model was used, and details about the model and statistical information can be found on pages 54 and 56–57 of the 2006 *JRME* article. Within the TD schools, the classrooms with higher levels of program implementation averaged higher achievement gains, and we suggest that even larger gains could have been made if more of the implementation roadblocks had been removed.

Conclusion

Here, we highlight three important findings that were reported in the article. First, it is possible to implement and sustain a comprehensive set of mathematics reforms that incorporate evidence-based curriculum, professional development, and whole-school reform practices in high-poverty middle schools. Across the four years of the study, two-thirds to three-fourths of classrooms in the three middle schools obtained at least a medium level of implementation. The support infrastructure put in place (in-class coaching, ongoing professional development, nesting within whole-school reform) was strong enough to withstand the high rates of principal and teacher turnover, shifting district foci, significant rates of student mobility, and dysfunctional responses to scarcity and uncertainty that were (and still probably are) emblematic of high-poverty middle schools.

Second, implementation of the comprehensive set of mathematics reforms led to significant and substantial achievement gains across multiple classrooms in multiple schools over multiple years. Those gains occurred across all levels of the achievement spectrum. All types of students benefited from a richer and more demanding curriculum, better trained and better supported teachers, and an improved teaching and learning environment. Comparing the magnitude of the impact of the TD mathematics reforms on student achievement to prior results for its individual components suggests that benefits were gained by combining evidence-based curricular, professional development, and whole-school reform practices into a coherent and integrated reform effort that was sustained for four years. The strength of our results in the face of challenging implementation conditions (i.e., the roadblocks we mentioned earlier in the chapter) further suggests that a package of evidence-based mathematics reforms should become a standard feature of high-poverty middle schools and perhaps all middle schools with significant achievement gaps. Such reforms include coherent and challenging instructional materials used schoolwide, grade- and curriculum-specific ongoing professional development, in-class coaching, and whole-school restructuring to create improved teaching and learning climates.

Third, although achievement results were statistically significant and educationally substantial, they were not of sufficient magnitude to allow us to conclude that the enacted reforms alone were enough to close all the achievement gaps that high-poverty students bring to urban middle schools. A more detailed analysis (see Balfanz & Byrnes, 2006) revealed that a near majority of students in the TD schools were able to substantially close their achievement gaps, learning on average more than a year's worth of mathematics per year while in middle school. The remaining students, however, did no better than tread water, with their achievement gap remaining constant during the middle grades. This clearly indicates that additional reforms and supports will be needed to provide all middle school students with the mathematics skills and strategies they need to succeed in a rigorous set of high school mathematics courses.

We concluded the article by stating that the results of our study were encouraging. They show that the emerging set of evidence-based reforms in mathematics curriculum, professional development and teacher support, and whole-school reform, when integrated, can significantly raise mathematics achievement in high-poverty middle-grades schools. At the same time, they also show that much work remains to be done before we can reliably provide all middle schools with the tools, technologies, and human resources they need to both effectively teach standards-based mathematics and close the skill and knowledge gaps of all their students.

Suggestions for Practitioner Use of Our Work

We close with a suggestion about how our practitioner readers could use the information about and results from the TD Mathematics Project. It has been almost ten years since the *JRME* article was published, and during that time we have published other articles and books about the project. A list of these publications appears in the appendix. For mathematics educators who with their preservice or in-service teachers are looking carefully at the reform movement in mathematics education, we offer the list as a way to obtain information about one reform effort. The contents of the articles and other publications can foster a discussion of the project in light of recent standards movements such as the Common Core State Standards for Mathematics (National Governors Association Center for Best Practices & Council of Chief State Officers [NGA Center & CCSO], 2010).

References

Balfanz, R. (1997, March). *Mathematics for all in two urban schools: A view from the trenches.* Paper presented at the annual meeting of the American Educational Research Association, Chicago, IL.

Balfanz, R., & Byrnes, V. (2006). Closing the mathematics achievement gap in high poverty middle schools: Enablers and constraints. *Journal of Education for Students Placed at Risk, 11,* 143–159.

Balfanz, R., Mac Iver, D. J., & Byrnes, V. (2006). The implementation and impact of evidence-based mathematics reforms in high-poverty middle schools: A multi-site, multi-year study. *Journal for Research in Mathematics Education, 37,* 33–64.

Balfanz, R., McParland, J., & Shaw, A. (2002). *Re-conceptualizing extra help for high school students in a high standards era.* Washington, DC: Office of Vocational and Adult Education, U.S. Department of Education.

Campbell, P. F., & Silver, E. A. (2000). *Teaching and learning mathematics in poor communities: Report of a task force.* Reston, VA: National Council of Teachers of Mathematics.

Mac Iver, D., Mac Iver, M., Balfanz, R., Plank, S. B., & Ruby, A. (2000). Talent Development Middle Schools: Blueprint and results for a comprehensive whole-school reform model. In M. G. Sanders (Ed.), *Schooling students placed at risk: Research, policy, and practice in the education of poor and minority adolescents* (pp. 292–319). Mahwah, NJ: Erlbaum.

Mac Iver, D. J., Ruby, A., Balfanz, R., & Byrnes, V. (2003). Removed from the list: A comparative longitudinal case study of a reconstitution-eligible school. *Journal of Curriculum and Supervision, 18,* 259-289.

National Center for Education Statistics. (2000). *Pursuing excellence: Comparisons of international eighth-grade mathematics and science achievement from a U.S. perspective, 1995 and 1999.* Washington, DC: U.S. Department of Education.

National Governors Association Center for Best Practices & Council of Chief State School Officers (NGA Center & CCSSO). (2010). *Common core state standards for mathematics.* Washington, DC: Author. Retrieved from http://www.corestandards.org.

Schmidt, W. H., McKnight, C. C., Cogan, L. S., Jakwerth, P. M., & Houang, R. T. (1999). *Facing the consequences: Using TIMSS for a closer look at U.S. mathematics and science education.* Dordrecht, the Netherlands: Kluwer Academic.

University of Chicago School Mathematics Project. (2003, Spring). *USCMP Newsletter, 31.*

U.S. Department of Education, Mathematics and Science Expert Panel. (1999). *Exemplary and promising mathematics programs.* Washington, DC: U.S. Department of Education.

Useem, E. (1998, October). *Teachers' appraisals of Talent Development Middle School training materials and student progress: Results from focus groups* (CRESPAR Report No. 25). Baltimore, MD: Center for Research on the Education of Students Placed at Risk.

Useem, E. (1999, April). *Teachers' appraisals of the Talent Development training model.* Paper presented at the annual meeting of the American Educational Research Association, Montreal, Canada.

Useem, E. (2000). *Interviews with algebra teachers implementing the Talent Development Middle School Model's Mathematics Program.* Philadelphia, PA: Philadelphia Education Fund.

Useem, E. (2003, March). *The retention and qualifications of new teachers in Philadelphia's high-poverty middle schools: A three-year cohort study.* Paper presented at the annual meeting of the Eastern Sociological Society, Philadelphia, PA.

Appendix
References Specific to the TD Mathematics Project

Balfanz, R. (1997, March). *Mathematics for all in two urban schools: A view from the trenches.* Paper presented at the annual meeting of the American Educational Research Association, Chicago, IL.

Balfanz, R. (2009). *Putting middle grades students on the graduation path: A policy and practice brief.* Westerville, OH: National Middle School Association.

Balfanz, R., & Byrnes, V. (2006). Closing the mathematics achievement gap in high poverty middle schools: Enablers and constraints. *Journal of Education for Students Placed at Risk, 11,* 143–159.

Balfanz, R., Herzog, L., & Mac Iver, D. (2007). Preventing student disengagement and keeping students on the graduation path in urban middle grades schools: Early identification and effective interventions. *Educational Psychologist, 42,* 223–235.

Balfanz, R., Legters, N., & Byrnes, V. (in press). Closing achievement gaps in high school mathematics calls for innovation far beyond the Common Core. (Note: Copy available from the Center for Social Organization of Schools, Johns Hopkins University, Baltimore, MD.)

Balfanz, R., Legters, N., & Byrnes, V. (in press). A randomized trial of two approaches to increasing mathematics achievement for underprepared freshmen. (Note: Copy available from the Center for Social Organization of Schools, Johns Hopkins University, Baltimore, MD.)

Balfanz, R., Mac Iver, D. J., & Byrnes, V. (2006). The implementation and impact of evidence-based mathematics reforms in high-poverty middle schools: A multi-site, multi-year study. *Journal for Research in Mathematics Education, 37,* 33–64.

Balfanz, R., McParland, J., & Shaw, A. (2002). *Re-conceptualizing extra help for high school students in a high standards era.* Washington, DC: Office of Vocational and Adult Education, U.S. Department of Education.

Balfanz, R., Ruby, A., & Mac Iver, D. (2002). Essential components and next steps for comprehensive whole-school reform in high-poverty middle schools. In S. Stringfield & D. Lands (Eds.), *Educating at-risk students: One hundred-first yearbook of the National Society for the Study of Education, Part II* (pp. 128–147). Chicago, IL: NSSE.

Mac Iver, D. J., Balfanz, R., & Plank, S. B. (1998). An "elective replacement" approach to providing extra help in math: The Talent Development Middle Schools' Computer- and Team-Assisted Mathematics Acceleration (CATAMA) Program. *Research in Middle Level Education Quarterly, 22*(2), 1–23.

Mac Iver, D., Mac Iver, M., Balfanz, R., Plank, S. B., & Ruby, A. (2000). Talent Development Middle Schools: Blueprint and results for a comprehensive whole-school reform model. In M. G. Sanders (Ed.), *Schooling students placed at risk: Research, policy, and practice in the education of poor and minority adolescents* (pp. 292–319). Mahwah, NJ: Erlbaum.

Mac Iver, D. J., Ruby, A., Balfanz, R., & Byrnes, V. (2003). Removed from the list: A comparative longitudinal case study of a reconstitution-eligible school. *Journal of Curriculum and Supervision, 18,* 259–289.

Ruby, A. (2002). Internal teacher turnover in middle school reform. *Journal of Education for Students Placed at Risk, 7,* 379–406.

Useem, E. (1998, October). *Teachers' appraisals of Talent Development Middle School training materials and student progress: Results from focus groups* (CRESPAR Report No. 25). Baltimore, MD: Center for Research on the Education of Students Placed at Risk.

Useem, E. (1999, April). *Teachers' appraisals of the Talent Development training model.* Paper presented at the annual meeting of the American Educational Research Association, Montreal, Canada.

Useem, E. (2000). *Interviews with algebra teachers implementing the Talent Development Middle School Model's Mathematics Program.* Philadelphia, PA: Philadelphia Education Fund.

Useem, E. (2001). New teacher staffing and comprehensive middle school reform: Philadelphia's experience. In V. A. Anfara, Jr. (Ed.), *The handbook of research in middle level education* (Vol. 1, pp. 143–160). Greenwich, CT: Information Age.

Useem, E. (2003, March). *The retention and qualifications of new teachers in Philadelphia's high-poverty middle schools: A three-year cohort study.* Paper presented at the annual meeting of the Eastern Sociological Society, Philadelphia, PA.

The Importance of Teaching in the Promotion of Open and Equitable Mathematics Environments

Jo Boaler
Stanford University

I begin this chapter with a look back at a very interesting time in the history of mathematics education. The 1980s was a decade in which research on mathematics learning was sufficiently prevalent and established for educators to know that they needed to engage students actively in their mathematics learning in order to erase the widespread math failure and trauma that pervaded the United States (Boaler, 2009). Teachers and curriculum developers began to give students more open problems in which they could think deeply about mathematics, choosing and applying different methods, rather than pages of procedural questions in textbooks. It was an exciting time, and a wide range of high-quality teaching materials were produced.

But the changes, which came to be known as "reforms" in mathematics education, were not well received by all parties, as I set out in a *Journal for Research in Mathematics Education* article in 2002 and later a book (Boaler, 2009). Most of the opposition came from those who had been successful in more traditional teaching approaches, but some researchers also questioned whether more open approaches to mathematics teaching and learning would be accessed equitably by all students (Delpit, 1998; Lubienski, 2000).

The question of whether some students might be disadvantaged by reform-oriented approaches to mathematics remains an important one, especially as teachers in the United States pay attention to the mathematical practices, such as reasoning, problem solving, and connecting ideas, in the Common Core State Standards for Mathematics (National Governors Association Center for Best Practices & Council of Chief State School Officers [NGA Center & CCSSO], 2010). Such practices are highly mathematical, but they require questions and curriculum materials that are sufficiently open for students to problem solve, reason about methods, communicate, and connect ideas. In my 2002 *JRME* article, I considered the first wave of research on reform-oriented curricula and teaching practices and contended that the differences between equitable and inequitable teaching approaches lay *within* teachers' practices. In particular, I argued that we should shift ideas of *students being unable to work in these ways* onto discussions of ways to help schools and teachers make the educational experience more equitable. In this chapter I revisit the school approaches I reviewed in my *JRME* article, from Phoenix Park and QUASAR schools—two approaches in which teachers engaged students both deeply and equitably. Readers may also want to consider the work of teachers in the Railside School in the United States, which promoted high and equitable achievement with impressive results (see Boaler, 2008; Boaler & Staples, 2008; Boaler, 2009).

In my 2002 *JRME* article, I offered a theoretical grounding for questions of teaching and equity; I refer the reader to the article (2002, pp. 241–245) for a complete presentation of these theoretical considerations. In the remainder of this chapter I will focus upon the two examples of teaching highlighted in the article—one from England (Phoenix Park School)

This chapter is adapted from J. Boaler (2002), Learning from teaching: Exploring the relationship between reform curriculum and equity, *Journal for Research in Mathematics Education*, *33*, 239–258.

and another from the United States (the QUASAR project). Both approaches were reform oriented and contributed to a reduction in linguistic, ethnic, and class inequalities. These school approaches, as well as others (see Boaler, 2009), promoted high achievement and equity, casting doubts upon claims that reform-oriented approaches are inaccessible to some students.

Phoenix Park and the QUASAR Project: Evidence of Equitable Teaching

The examples that follow come from two sources. The first source is from my own three-year research study (summarized in Boaler, 1998b, 2002) that compared the experiences and achievements of approximately three hundred students attending two secondary schools in England with vastly different mathematics teaching approaches. Both schools (Phoenix Park and Amber Hill) were situated in low-income areas, and the vast majority of students were white and working class. Phoenix Park used an open-ended approach to mathematics; Amber Hill used a procedural, skills-based approach. I followed a cohort of students at each school from age thirteen to age sixteen (the end of compulsory schooling in England). At the beginning of the study, the cohorts of students at the two schools were matched by gender, race, social class, and prior achievement. Both sets of students had followed the same mathematics approach previously, and there were no significant differences in their levels of mathematics achievement on national tests when the study started. Then their mathematics pathways diverged: One set of students, at Amber Hill, worked in tracks, through procedural textbooks; Phoenix Park students worked on long, open-ended projects through a project-based learning (PBL) approach. Three years later, the Phoenix Park students not only outperformed students on applied assessments but also attained significantly higher grades on the closed, procedural national mathematics examination (GCSE). In addition, although boys at Amber Hill earned significantly higher grades than the girls did, there were no gender disparities at Phoenix Park. More strikingly, at Amber Hill achievement was correlated with social class, as is typical in schools, but at Phoenix Park the social class differences that were evident when students started the school had been erased by the time they left (see Boaler, 1997, 2002, for more details). Thus, the school that used an open-ended approach not only achieved impressive academic results for its students—whose examination results were higher than the national average, despite the students being at significantly lower levels when they entered the school—but also reduced the inequalities that typically exist in schools (Kozol, 2012).

The second example of a reform approach that enhanced equity comes from the QUASAR project (Quantitative Understanding: Amplifying Student Achievement and Reasoning) (see Brown, Stein, & Forman, 1996; Lane & Silver, 1999; Silver, Smith, & Nelson, 1995). This project began with the assumptions that all students could achieve at high levels and that achievement would increase in low-income and minority communities if teachers placed a greater emphasis on problem solving, communication, and conceptual understanding. Teachers in six urban middle schools serving socially and culturally diverse populations of students in the United States spent five years developing and implementing a more open and collaborative mathematics approach. Students learned about facts and algorithms; but they also learned when, how, and why to apply procedures to solve high-level problems, in an approach similar to the one used at Phoenix Park. There were a number of similarities between the QUASAR and Phoenix Park approaches, including mixed-ability classes, a focus on problem solving, high expectations for all students, attention to a broad array of mathematical topics, and the encouragement of discussion with students reasoning about and justifying their work. As measured over time, the QUASAR students' achievement revealed extremely positive results. The students made significant gains in achievement, and they performed at significantly higher levels than comparable groups of students on a range of different assessments. Furthermore, the gains were distributed equally among the different racial, ethnic, and linguistic groups of students.

Results from these two studies cast considerable doubt on claims that open-ended approaches are less suitable for working-class and minority children, but they also raise important questions for us as mathematics educators. Why did the reform approaches of the Phoenix Park and QUASAR teachers promote equity? The answer to this question was important when my *JRME* article was published in 2002, and it remains important today. In the next section of the chapter, I summarize the points I made in the article concerning particular practices of teaching and learning that the Phoenix Park and QUASAR teachers employed and that appeared to have had an impact on the promotion of equitable experiences and achievements.

Teaching and Learning Practices That Promoted Equity

I will begin by describing the curriculum and teaching practices used at Phoenix Park School. The curriculum at Phoenix Park was designed by the teachers, who collected a range of different open-ended projects that generally lasted for two to three weeks of mathematics lessons (see Boaler, 1997, 2002, for more information). The Phoenix Park teachers worked with teachers from five other schools and were part of a working group of teachers in the Association of Teachers of Mathematics (the U.K. equivalent of National Council of Teachers of Mathematics [NCTM]) creating and teaching through open-ended mathematics projects.

When the students arrived at Phoenix Park, most were immediately receptive to the open-ended mathematics approach, despite having spent the previous eight years working on more closed and traditional mathematics questions. However, some of the students—mostly boys—found the openness of the work disconcerting. They acknowledged being uncomfortable with the lack of structure or direction in the problems and indicated they would prefer a more traditional approach. When I interviewed the Phoenix Park students at the beginning of the year, they described their reactions to the mathematics classes clearly. The following dialogues come from two interviews with Shaun and Megan, when they were in grade 8. They were both in the small group of more resistant and less well behaved students. (These names and others that follow are pseudonyms; *JB* is, of course, the author.)

Shaun: When I go into a maths lesson I usually sit down and think—who am I going to throw a rubber [eraser] at today?

JB: Can you think of a maths lesson that you've enjoyed?

Megan: Messing about, that's what I enjoy doing.

JB: What would have made maths better?

Megan: Working from books—you don't mess about if you've got a book there, you know what to do.

Although some students blamed their misbehavior on the openness of the work, the teachers did not give in to this attitude by giving books or structure to the students. The Phoenix Park teachers believed that the open-ended approach was valuable for *all* students and that it was their job to make the work equally accessible to all. They therefore developed a range of practices that served to increase students' access to the problems and the methods they were expected to use.

Their success in doing so was evidenced by assessment results as well as a change in students' behaviors and orientations. By the time students were in the next grade they had learned to expect to problem solve and reason, and even the initially resistant students expressed a preference for this approach, as highlighted by a teaching episode I witnessed when the regular Phoenix Park teacher was absent. The students were being taught by a trainee teacher who decided to explain mean, mode, and median to them and then give them exercises to follow. The students gathered around the board as the teacher lectured and the students took notes. After a while Shaun (quoted above) asked, "When are we going to do some work, Sir?" The teacher looked uncertain and said, "This is work." Shaun resisted, saying, "This isn't work, this is just copying. When are we going to do some work?" At this time he had shifted his idea of mathematics work from passive copying to being asked to think and problem solve.

In the next sections, I recap three teaching practices that helped make mathematics instruction accessible to all students and that illustrate the importance of paying attention to detailed teacher moves. Similar teacher moves were evident in QUASAR project classrooms, and I include examples where appropriate.

Introducing activities through discussion

One practice central to the Phoenix Park teachers' approach was introducing students to activities through discussion in which the teachers themselves participated. Students were never left to interpret the problems alone. Instead, the teachers always spent time with individuals, groups, or the whole class introducing ideas and making sure that all students knew how to start their explorations. Phoenix Park teachers would frequently ask students to gather around the board when new problems were being introduced and when homework was assigned, in order to have some discussion of the problems posed (see Boaler, 2002, for more details about the ways in which teachers introduced activities).

Likewise, teachers in QUASAR classrooms spent time introducing problems. Margaret Schwan Smith, a member of the QUASAR project research team, reported that she observed one urban middle school teacher who would ask her students to read problems aloud in class and then would hold a discussion about the problem's context and any unfamiliar vocabulary (personal communication, January 18, 2001). She would ask students to discuss what they thought the problem called for them to do, and then she would have them work in groups while she circulated to check that individual students understood what they should be doing. In order to make tasks equally accessible to all students, the Phoenix Park and QUASAR teachers held group and class discussions on the aim of activities, the meaning of contexts, and the challenging points within problems. These methods stand in contrast with those used in classrooms where students are left to interpret the meaning of problems from their reading of reform curriculum materials, which were often more wordy and linguistically demanding. The way in which work is introduced to students and the access that students have to the mathematical ideas that they are intended to explore seems to be extremely important.

Teaching students to justify and reason

Phoenix Park teachers paid attention to the ways in which students communicated their mathematical thinking as well as students' understanding of why this was an important aspect of their work. They frequently urged individual students to explain their reasoning and to communicate in more detail because the students were not used to doing so, based on their prior mathematics classroom experiences. In one of the lessons I observed, a student gave the teacher his solution to a problem on which he had been working. His paper showed some of his methods and a correct answer. The teacher (whom I called Rosie Thomas) studied the paper and then said, "Brilliant work, John, but you can't just write it down, there must be some sense to why you've done it, some logic. Why did you do it that way? Explain it."

Rosie's comment that "there must be some sense to why you've done it" typifies the sort of encouragement given to the students at Phoenix Park. The teachers strove to expand the way in which the students thought about mathematics, extending the students' value systems to incorporate more than the desire to get correct answers. There was considerable evidence that teachers were successful in that regard, as illustrated in this excerpt from my interview with Ian:

Ian: It's an easier way to learn, because you're actually finding things out for yourself, not looking for things in a textbook.

JB: Was that the same in your last school, do you think?

Ian: No, like if we got an answer, they would say, "You got it right." Here you have to explain how you got it.

JB: What do you think about that?

Ian: I think it helps you.

In one of the lessons I observed, the teacher posed the following question: "[What] if someone new came into class and they asked you what makes a good piece of work? What does Ms. Thomas like? What would you say?" Students offered suggestions such as "lots of writing," "have an aim," "draw a plan," and "write about patterns." Each time, the teacher came back with further questions: "Is the amount of writing important?" "What does that mean?" "Why is a plan important?" "What does a good plan look like?" The students struggled with many of their explanations, but they were engrossed in the discussion for some time. They were clearly appreciative of the opportunity to learn about valued ways of working. It is important to note that this lesson focused *explicitly* on the mathematical practices of reasoning and justification—practices that are valued in the reform-oriented classroom but that are not always given explicit attention by teachers.

Making real-world contexts accessible	Many reform curricula are replete with contexts that are intended to bring the "real world" into the mathematics classroom. One of the problems presented by real-world contexts is that they often require familiarity with a situation that is described, when many students are unfamiliar with the situation or example. In my analysis of the inequities posed by state assessments in California (Boaler, 2003) I recalled a language learner emerging from the test and asking his teachers, "What is a soufflé?" A significant problem also arises when students are required to engage with the contexts as though they are real but ignore factors that would be important in real-life versions of the tasks. I have written elsewhere about the strangeness of school mathematics contexts (Boaler, 1993, 2009), suggesting that for many students, walking into a math classroom is like walking into *MathLand*, a world in which trains speed toward each other on the same tracks, baths run at the same rate each second, and friends divide pizzas into perfectly equal slices. Students know they need to suspend common sense to engage in these math tasks. Knowing how much consideration to give to the real-world factors presented in tasks has become a form of school knowledge that is often inequitably distributed (see Boaler, 1998a, 2003, 2009) but that all students need.

The QUASAR teachers addressed this issue by engaging students in conversations about the meaning of different contexts they encountered in tasks. In one example (Silver et al., 1995), a teacher introduced a task about the most economical way to buy bus tickets—as a weekly pass or as daily tickets. The textbook authors of the task intended students to calculate the most cost-effective tickets, but the students considered other variables—such as how often they would use a weekly pass and the different family members who could also use it. When a QUASAR teacher realized that students were situating their reasoning in the context of their daily lives and that there was more than one correct answer to such problems, she changed her expectations and provided students with opportunities to explain their reasoning. As

Silver, Smith, and Nelson concluded, "increasing the relevance of school mathematics to the lives of children involves more than merely providing 'real world' contexts for mathematics problems; real-world solutions for these problems must also be considered" (p. 41).

Phoenix Park teachers attempted to make the different mathematical explorations relevant to the students by relating them to their lives. For example, when introducing work on statistics, teachers asked their students to collect data that was of interest to them from newspapers and magazines. When the problems involved patterns and tessellations, the teachers asked students to bring in examples of patterns they liked. They, like the QUASAR teachers, did not expect students to interpret contextualized questions exactly as the writer of the questions intended.

Both the Phoenix Park and the QUASAR teachers encouraged students to interpret mathematical and real-world variables and their relationship with one another. In doing so, teachers paid careful attention to the access students had to problems and to the promotion of equity, also helping students understand mathematics as a flexible means by which to interpret and make sense of the world.

Summary

My description of practices that the Phoenix Park and QUASAR teachers used to enhance students' access to reform approaches—helping them understand the questions posed to them, teaching them to appreciate the need for written communication and justification, and discussing with them ways of interpreting contextualized questions—includes only a small part of teachers' repertoire of practices. Nevertheless, these examples provide some indication of the complex support that teachers using reform-oriented approaches may need to provide to students. Ball and Bass (2000a, 2000b) offer a careful analysis of the mathematical understandings that teachers need when they engage students in collaborative explorations. Their analysis shows that teaching approaches based on student investigations, exploration, and discussion require considerable teacher knowledge and expertise. The work of teachers such as those in Phoenix Park, QUASAR, and Railside (Boaler, 2008; Boaler & Staples, 2008) classrooms provide the field with rich examples of successful and equitable teaching from which we may all learn.

Conclusion and Final Thoughts on Equitable Teaching

We have reached a significant and critical time in mathematics education, one in which the needs of the economy and of society more generally converge with the results of research on learning in highlighting the importance of students engaging actively with mathematics, problem solving, reasoning, and communicating. Mathematical reasoning is at the core of twenty-first-century work and is also an inherently mathematical practice that is central to the discipline of mathematics (see Boaler, 2013, 2016). But as we move to mathematics environments in which students are engaged in problem solving and reasoning, it is as important now as it was in 2002 to pay careful attention to equity. Students of color and students from low-income homes are often the students who have had least access to high-quality teaching and instead are presented with worksheets and low-level remedial work (see also Anyon, 1980). This situation makes it particularly important to pay careful attention to the ways in which teachers make open work accessible to a range of students.

As we consider the important work that has emerged from studies of mindset in recent years (Boaler, 2016; Dweck, 2006) it becomes clear that equitable teaching starts with the understanding that every student can reach high levels, and with teachers who are committed to that goal. The teachers at Phoenix Park, QUASAR, and Railside schools all shared that important understanding and commitment, and their teaching practices built on the knowledge that all students can reach high levels in mathematics (see also Boaler, 2016). As more and more teachers engage in open teaching, offering students the opportunity to problem solve and reason, the field should focus on the collection of rich examples of teachers' practices such as those used by the teachers in these two approaches. For it is the teachers of mathematics, not the curriculum materials, who hold students' futures in their hands, and empowering teachers with knowledge of equitable teaching practices should remain an important goal for our field.

References

Anyon, J. (1980). Social class and the hidden curriculum of work. *Journal of Education, 162*, 67–92.

Ball, D. L., & Bass, H. (2000a). Bridging practices: Intertwining content and pedagogy in teaching and learning to teach. In J. Boaler (Ed.), *Multiple perspectives on mathematics teaching and learning* (pp. 83–104). Westport, CT: Ablex.

Ball, D. L., & Bass, H. (2000b). Making believe: The collective construction of public mathematical knowledge in the elementary classroom. In D. Phillips (Ed.), *Yearbook of the National Society for the Study of Education: Constructivism in education* (pp. 193–224). Chicago, IL: University of Chicago Press.

Boaler, J. (1993). The role of contexts in the mathematics classroom: Do they make mathematics more real? *For the Learning of Mathematics, 13*, 12–17.

Boaler, J. (1997). *Experiencing school mathematics: Teaching styles, sex, and setting.* Buckingham, England: Open University Press.

Boaler, J. (1998a). Alternative approaches to teaching, learning and assessing mathematics. *Evaluation and Program Planning, 21*, 129–141.

Boaler, J. (1998b) Open and closed mathematics approaches: Student experiences and understandings. *Journal for Research in Mathematics Education. 29*, 41–62.

Boaler, J. (2002). Learning from teaching: Exploring the relationship between reform curriculum and equity. *Journal for Research in Mathematics Education, 33*, 239–258.

Boaler, J. (2003). When learning no longer matters: Standardized testing and the creation of inequality. *Phi Delta Kappan, 84,* 502–506.

Boaler, J. (2008). Promoting "relational equity" and high mathematics achievement through an innovative mixed ability approach. *British Educational Research Journal, 34,* 167–194.

Boaler, J. (2009). *What's math got to do with it? How parents and teachers can help children learn to love their least favorite subject.* New York, NY: Penguin.

Boaler, J. (2013). Ability and mathematics: The mindset revolution that is reshaping education. *FORUM, 55,* 143–152.

Boaler, J. (2016). *Mathematical mindsets: Unleashing students' potential through creative math, inspiring messages and innovative teaching.* San Francisco, CA: Jossey-Bass.

Boaler, J., & Staples, M. (2008). Creating mathematical futures through an equitable teaching approach: The case of Railside School. *Teachers' College Record, 110,* 608–645.

Brown, C. A., Stein, M. K., & Forman, E. A. (1996). Assisting teachers and students to reform their mathematics classroom. *Educational Studies in Mathematics, 31,* 63–93.

Delpit, L. (1988). The silenced dialogue: Power and pedagogy in educating other people's children. *Harvard Educational Review, 58,* 280–298.

Dweck, C. S. (2006). *Mindset: The new psychology of success.* New York, NY: Ballantine Books.

Kozol, J. (2012). *Savage inequalities: Children in America's schools.* New York, NY: HarperPerennial.

Lane, S., & Silver, E. A. (1999). Fairness and equity in measuring student learning using a mathematics performance assessment: Results from the QUASAR project. In A. L. Nettles & M. Nettles (Eds.), *Measuring up: Challenges minorities face in educational assessment* (pp. 97–120). Boston, MA: Kluwer Academic.

Lubienski, S. (2000). Problem solving as a means towards mathematics for all: An exploratory look through the class lens. *Journal for Research in Mathematics Education, 31,* 454–482.

National Governors Association Center for Best Practices & Council of Chief State School Officers (NGA Center & CCSSO). (2010). *Common core state standards for mathematics*. Washington, DC: Author. Retrieved from http://www.corestandards.org.

Silver, E. A., Smith, M. S., & Nelson, B. S. (1995). The QUASAR project: Equity concerns meet mathematics reforms in the middle school. In W. G. Secada, E. Fennema, & L. B. Adajian (Eds.), *New directions in equity in mathematics education* (pp. 9–56). New York, NY: Cambridge University Press.

Informing Teachers about Identities and Agency Using the Stories of Black[1] Middle School Boys Who Are Successful with School Mathematics

Robert Q. Berry III
University of Virginia

Bilal is an eighth grade Black male who has been successful with school mathematics and school in general. Bilal stated that mathematics is an easy subject for him to learn because he likes mathematics and he loves the challenge of problem solving. In fact, he credits his father for helping him develop a love for mathematics. When Bilal was younger, his father would play mathematics games, do mathematics puzzles, and teach him mathematics tricks. Because of Bilal's early experiences, he has always done well in mathematics. In fact, Bilal was labeled Academically Gifted (AG) in the fourth grade.

When Bilal was in sixth grade his achievement in mathematics fell from his typical performance of earning A's to that of C's. Bilal stated that his underperformance was due to his mathematics teacher's inability to convey mathematics concepts in an understanding manner. When Bilal's parents met with the teacher, during the first semester, they were told that the advanced course for sixth graders was too challenging for Bilal and that he should be placed in a lower level mathematics course. Bilal's parents found this recommendation to be odd because while Bilal was not doing terribly bad in the mathematics class, the teacher did not offer any suggestions nor advice on how they can help Bilal with his mathematics. Furthermore, Bilal was the only Black male in this course. From Bilal's parents perspective the teacher was more focused on removing Bilal from the class rather than helping him achieve. Bilal worked hard with the help of his parents and earned a letter grade of B in advanced sixth grade mathematics course. He went on to earn an A in seventh grade pre-Algebra and has done well in Algebra 1 as an eighth grader. (Berry, 2008, p. 464)

I introduced Bilal's story in my *Journal for Research in Mathematics Education* (*JRME*) article (Berry, 2008). His story reveals that he sees himself as being a good student who is good at mathematics. When considering the context and the perspectives of Bilal's parents, we see Bilal's interwoven identities: being a middle school Black boy who is good at mathematics

---

[1] Following U.S. federal government practice, I use the term *Black* to encompass a wide range of students for whom this term represents one aspect of their racial identity. While there is much diversity within this category, for the purposes of this chapter I use the term in its broadest sense. In the 2008 *JRME* article, I used *African American;* since then I have shifted my thinking to using the term *Black* to represent the diversity within the Black community.

This chapter is adapted from R. Q. Berry III (2008), Access to upper-level mathematics: The stories of successful African-American middle school boys, *Journal for Research in Mathematics Education, 39,* 464–488.

and is identified as gifted. This short excerpt along with other interviews and observations reveal that Bilal, like the other Black boys in my study, negotiated several interwoven identities that were important to them. Further, these boys' identities were supported by their parents and teachers. Consequently, when Bilal was confronted with being placed in a lower mathematics course in middle school, we can imagine that he began questioning his interwoven identities:

- Am I being recommended for placement in a lower course because I am is no longer a good student who is good at mathematics?

- Am I being recommended for placement in a lower course because I am no longer identified as gifted?

- Am I being recommended for placement in a lower course because middle school is different from elementary?

- Am I being recommended for placement in a lower course because I am a Black boy?

I do not know whether Bilal's sixth-grade mathematics teacher knew his story or understood the complexities of a Black boy negotiating these interwoven identities. From Bilal's perspective, he was juggling with the transition from elementary school to middle school and focused on his social identity. He stated, "That year, I performed way below my potential because I was not concentrating on school and I was trying to fit in with new friends at school. I went through an academic slump; I was making C's and D's." Imagine the impact on Bilal's identities and how different his experiences would have been had he been placed in a lower mathematics course. Would Bilal still see himself as a good student who is good at mathematics? Would his peers see him differently? If so, what would that mean to Bilal? Would other teachers see him as being a student and good at mathematics?

Table 3.1
Cross-subject characteristics

Characteristics	Participants							
	Cordell	Clayton	Jabari	Darren	Phillip	Akil	Bilal	Andre
Strong academic identity x	x	x	x		x	x	x	x
Likes mathematics	x	x	x	x	x	x	x	
Religious identity	x	x	x		x	x	x	
Co-curricular identity	x	x	x		x	x	x	
Athletic identity	x	x	x	x	x	x	x	
Positive preschool experiences	x	x	x		x	x	x	x
AG placement	x	x	x		x		x	
Not recognized as AG by teacher	x	x			x		x	
Parents discussed race as factor in experiences	x	x	x	x	x	x	x	x
Parents as guardian of opportunities	x	x	x	x	x	x	x	

This chapter is about using the stories of Bilal and seven other middle school Black boys who are successful with school mathematics to focus on the experiences that shaped these boys' interwoven identities, with specific focus on their mathematics identity—how these boys see themselves mathematically and how they are seen by others (teachers, parents, and peers) as doers of mathematics. Success was defined as being enrolled in algebra 1 or geometry in middle school. The information in table 3.1 (an edited version of one that appeared in the *JRME* article, p. 478) focuses on the characteristics relevant for this chapter with specific focus on those items related to identity and agency. The table shows ten characteristics related to the findings from the 2008 article across the eight boys. The first five characteristics focus on identities and the last five are characteristics related to the connection between agency and identity. Below, I discuss identity with a specific focus on how interwoven identities support and connect to mathematics identity. I then connect identities to agency. I conclude the chapter by discussing the teaching practices that were used by the teachers of the boys to cultivate the boys' identities as capable of participating in and being doers of mathematics.

Identities: Supporting Mathematics Teaching and Learning

The stories of the eight boys in my 2008 article suggest that their interwoven identities contributed to their success with school mathematics. Understanding the stories of these boys is appropriate because stories provide the context for understanding, feeling, and interpreting their identities to which they give *voice* in announcing to the world who they think they are, who they see themselves becoming, how others see them, and how they act as a result of these understandings, feelings, and interpretations. Voice is identity—having a sense of self, a sense of purpose, and a sense of relationship to others. Identity is a dynamic and context-driven construct that changes, grows, and evolves over time. Identities of students include early identifications with college attendance and careers such as doctor, scientist, teacher, engineer, or sports professional (Aguirre, Mayfield-Ingram, & Martin, 2013). This early identification is important because it serves as a source of strength and motivation to do well in school and specifically in mathematics (Martin, 2003). Of the eight boys I studied, seven identified career and college goals as motivators to do well in school and in mathematics. For example, Andre stated, "I want to go to the Air Force Academy and become a pilot. You have to be good at math to get into the Academy." Clayton stated, "Good math grades will get me into college."

Understanding the strengths and motivations that serve to develop students' identities should be embedded in the daily work of all teachers (Aguirre et al., 2013). Mathematics teaching involves not only helping students develop mathematical skills and mathematical understanding but also empowering students to seeing themselves as capable of participating in and being doers of mathematics. When students identify themselves in this way, they make positive connections and are motivated to achieve at high levels. This understanding of students' identities gives teachers insights into how and why some students make positive connections with mathematics and others do not (Aguirre et al., 2013). Teachers can use this understanding to provide opportunities for students to use mathematics to examine personal, communal, and social contexts. In providing these opportunities, students may find the motivation and connections with mathematics to see its relevance for their future, thus developing a mathematics identity.

Mathematics identity includes beliefs about one's self as a mathematics learner, one's perceptions of being seen by others as a mathematics learner, beliefs about the nature of mathematics, engagement in mathematics, and perception of oneself as a potential participant in mathematics (Solomon, 2009). Seven boys expressed a strong belief in themselves as mathematics learners. They were confident about their mathematics ability, and they perceived that their abilities allowed them to be among the "smart kids." By identifying themselves as "smart kids," these boys positioned themselves as members of a particular group with certain behavioral and social expectations. From their perceptions, smart kids do their work, answer questions and participate in class, and are good at mathematics. These identity-affirming behaviors influenced the ways in which the boys participated in mathematics and how they saw themselves as doers of mathematics. For example, Cordell stated, "The smart kids make me work harder." Clayton stated, "I like being with the smart kids because that means I am one of the smartest . . . [S]ince I answer all of the questions in math that means I am the smartest of the smart kids" (Berry, 2008, p. 482). A student who identifies as being good at mathematics exhibits behaviors similar to Clayton's description by answering questions and perhaps being an overly active participant to maintain his status. Conversely, students who identify as not being good at mathematics may remain silent and limit their participation because they fear being judged.

The identity-affirming behaviors described above have implications for mathematics teaching. Teachers can cultivate mathematical abilities by providing opportunities for students to make sense of and persevere in challenging mathematics. That is, students should be engaged with mathematics that requires active participation, asking questions, problem posing, and reasoning. Students can see others and model identity-affirming behaviors that support the development of mathematics identity. Consequently, this kind of teaching values all students' thinking and uses pedagogical practices, such as differentiated tasks, mixed ability groupings, and publicly praising contributions and perseverance, to cultivate and affirm mathematical participation and behaviors (National Council of Teachers of Mathematics [NCTM], 2014).

Many influences shape students' mathematical identity; some are directly related to in-school activities as described above and others are in out-of- school contexts. In my 2008 *JRME* article, I discussed what I termed *alternative identities*—(a) co-curricular and special academic program identity, (b) religious identity, and (c) athletic identity—as providing these boys a mechanism that helped them deal with peer pressure and foster resiliency. These boys used these identities to engage in additional academic and mathematics opportunities. Seven of the boys were involved in co-curricular and special summer and after-school programs. These special programs provided the boys with access to mathematics and science resources, hands-on activities, and additional learning experiences. All the boys in these special programs cited that the programs helped them develop positive habits, provided them with additional learning opportunities, and helped them focus on future goals. Jabari stated, "I participated in the Pre-College Program's Summer Scholars program. . . . We went to a chemistry lab and did experiments, and we had a mathematics bowl competition. . . . I think I am going to do something with math and science in my future" (Berry, 2008, p. 483). Seven boys and their parents identified themselves as spiritual people, and their stories reflected their participation in church activities. For these boys, church was not just the location for religious services, but an integral part of their daily lives. They were involved with church athletic teams, youth ministries, church choirs, church tutorial programs, and other church functions. Church was the context in which these boys participated in their communities. Sports provided seven of the boys with an athletic identity (five boys participated in athletics year-round). Participation in athletics was a motivating factor for the boys to maintain good grades. Three parents mentioned that their requirement for athletic participation was higher than the school's requirement. Bilal's mother stated, "No grades, no play. . . . He better bring home all A's and B's" (Berry, 2008, p. 483).

Mathematics teaching should leverage students' culture, contexts, and identities to support and enhance mathematics learning (NCTM, 2014). By understanding the significance and relevance of the alternative identities and contexts, mathematics teachers can draw on community resources to understand how they can use contexts, culture, conditions, and language to support mathematics teaching and learning (Berry & Ellis, 2013; Planas & Civil, 2013). As a result, learning mathematics becomes a part of students' mathematics identity, leading to increased engagement and motivation in mathematics (NCTM, 2014). Imagine mathematics teaching using Jabari's experience with the Pre-College Program's Summer Scholars program as a resource both for positively influencing students' dispositions toward mathematics and for mathematics teaching and learning to help all students. Likewise, imagine how schools could make connections to communities through churches to motivate students' academic participation.

Agency: Enacting Our Identities

Agency is the behavioral aspect of identity focusing on participation and performing effectively in contexts (Aguirre et al., 2013). Simply put, agency is our identity in action and the presentation of our identity to the world (Murrell, 2007). Mathematical agency is about participating in mathematics in personally and socially meaningful ways. As stated earlier, almost all of the boys identified themselves as being smart and good at mathematics; consequently, they presented themselves and adopted behaviors and actions of smartness and of being good at mathematics. Once the presentation of smart and being good at mathematics was affirmed by teachers and others, these boys saw themselves as active participants and doers of mathematics.

Agency can be conceptualized in two ways: high sense of agency and low sense of agency. A high sense of agency is having a high degree of self-exploration that is associated with a high degree of self-direction in determining one's life course (Côté & Schwartz, 2002). Students with a high sense of agency make decisions about their participation in mathematics in pursuing mathematical experiences that provide them with the broadest academic options, such as being in the high mathematics group, seeking additional co-curricular and special programs, choosing to participate in positive community activities, or being associated with the smart kids. These boys chose actions and behaviors that supported success and participation with mathematics. For example, Bilal show a high sense of agency by stating, "I gotta excel in everything I do. Be the best that I can be . . . [B]eing the best means doing your work, asking questions, and being involved in class." Echoing Bilal, Andre's sense of agency is summed up in this statement: "Good math students are focused, do their work, and want to make A's all the time . . . I am a good math student" (Berry, 2008, p. 482). All the boys in my study exercised a high sense of agency to resist the negative identities imposed on them by having a sense of control over their academic success. They knew that, in order to maintain their good standing in mathematics, they needed to be participants in the mathematics classroom.

Conversely, a low sense of agency is a low degree of self-exploration that is translated into a low level of control over one's life course. Students with a low sense of agency have limited participation in their mathematics experiences and may perceive that participation will not change their situation. These students are passive recipients in their mathematical experiences, perhaps because of negative structural and institutional forces. Consider Bilal's case: If he did not have a high sense of agency and parental support, the default option for him would have been placement in a lower mathematics course that would have provided him with a different mathematics experience. Students with a low sense of agency might make a statement such as "Why should I do my work? It won't matter anyway."

For the Black boys in my study, identification of giftedness appeared to have an impact on their identities and agency. This was critical, because the boys and their parents perceived them as having to contest negative discourses and images directed at them. Five boys were identified as gifted by the fourth grade; in regard to identity and agency, this identification gave them the perception of receiving richer mathematics instruction in elementary school than other students. Four boys' identification came in spite of teachers' failure to recognize their intellectual potential. Without advocacy from parents and caring adults, these four boys would have had a different mathematical experience. A pattern that emerges from the lack of identification was the focus on the boys' behavior in evaluations of their academic potential. Cordell, for example, was bored with school and started to have behavioral problems. His teacher and principal did not want him tested for the gifted program because they thought there was no evidence to indicate that he was academically gifted. Cordell's mother saw race as a factor in the school's decision not to test Cordell because both the teacher and the principal were white and, according to her perception, they did not want to give Cordell an opportunity. The teacher and the principal were focused on his behavior rather than his academic potential. Although we cannot generalize the stories of these boys, they raise concerns about the potential number of Black boys in general who are not given the opportunity to gain access to mathematical experiences that allow them to participate and engage in positive, identify-affirming mathematics behaviors.

How Teacher Support Identities and Agency

Teachers support students' sense of agency in the values they communicate through their words and actions. In the mathematics classroom, values are often expressed as teachers having high expectations and showing care for learners. For many students, expectations and care are a proxy for how their teachers value them as people and as learners. When asked "How does a mathematics teacher show that he or she cares about you?" Darren responded with comments about Ms. Blaine that show how high expectations are an indicator of how this teacher demonstrated values. Darren expressed these thoughts about his teacher:

> My teacher, Ms. Blaine, cared about all of us. She would bend over backwards to help us when we needed it. She really helped me. She talked to me and told me that I had a lot of potential in math and that I should use it to get ahead in life. [She thought] I was capable of doing a lot in math. That's what really motivated me . . . [S]he understands that there is a lot of pressure put on African American males. . . . She lets me know I can be cool and smart at the same time. (Berry, 2004, p. 101)

It is clear that Ms. Blaine affirms Darren's mathematics, racial, and social identities by telling him he is good at mathematics and acknowledging the pressure of being a Black male. We can imagine that a teacher like Ms. Blaine might be part of Darren's consciousness as he negotiates his way among being cool, being smart, participation and nonparticipation in school, and issues of race. Ms. Blaine provides Darren with an academically and socially safe space for him to negotiate his identities. From Darren's perspective, Ms. Blaine demonstrates caring not only as an effect but as a means for shaping his disposition toward mathematics, molding his interwoven identities, and developing his sense of agency by helping him believe that he is capable of doing mathematics and affirming his behaviors of being a doer of mathematics.

Teachers can affirm students' mathematics identity and help them develop a sense of agency by promoting persistence and reasoning during problem solving (Aguirre et al., 2013). Additionally, teachers can encourage students to see themselves as confident problem solvers and as active participants in mathematics. Phillip described his teacher, Mr. Wallace, as an authority figure who expects and demands the best from his students:

> My algebra teacher, Mr. Wallace, is one of them [a role model]. I see him as a father figure because he is strict, friendly, and spiritual. I like the fact that he is strict because this shows that he expects the best from his students. In Mr. Wallace's class, "can't" is like a curse word, so when he hears that word, he will work with you until you believe you "can." (Berry, 2003, p. 97)

Phillip's description of Mr. Wallace is significant because it suggests that Mr. Wallace is authoritative while also caring enough to promote persistence and invest in students to demand that they do their best work. We see identity-affirmation and supporting agency in Mr. Wallace in the ways he work with students to help them believe they "can" do mathematics. It appears that nonsuccess and nonparticipation are not options for Mr. Wallace's students. Phillip perceived that the strict context of Mr. Wallace's classroom should be understood as an effort to push students to attain success and as a way to manage the class environment that supported and fostered success.

Observations of Mr. Wallace's class over a four-week period showed that his instructional style consisted of starting with direct instruction, then moving students to small group activity and back to whole-class instruction. Mr. Wallace expected students to share their problem solving and thinking in small groups and with the whole class. Additionally, he had the class respond to questions by using a call-and-response style. This style of instruction is consistent with the preaching traditions of call and response in Black churches (Hollins, 1982). By conforming to behaviors found in the church and using dyads, Mr. Wallace made it clear that nonparticipation was not an option in his class. Mr. Wallace used dyads in which students worked in pairs; one partner was the talker and the other was the listener. Mr. Wallace posed a question or problem, and the talker shared his thinking with his listening partner. Mr. Wallace affirms one student in a dyad by stating, "I like the way you explained that . . . [N]ow show me how your picture is like your equation" (Berry, 2003, n.p.). In this statement, Mr. Wallace affirms the explanation but also pushes the student to make a deeper mathematics connection.

Students bring with them knowledge of mathematics through everyday interaction with their cultural and social backgrounds. The use of context in mathematics teaching and learning can help teachers better recognize and build upon the cultural and social resources that students bring to the mathematics classroom. Context in mathematics teaching and learning helps students better make sense of decisions about the choice of mathematical procedures when problem solving (Boaler, 1998). Phillip described how Mr. Wallace made mathematics relevant for students:

> Mr. Wallace, he basically explains things so that you can understand it and he give you lots of examples and . . . he used everyday life . . . the newspaper, anything you can find around the house, and sports. Most of the time, he likes to use basketball. (Berry, 2003, p. 97)

Mr. Wallace used multiple resources to make intentional connections as a means to support students' understanding by selecting mathematics tasks that were engaging and relevant for the students while balancing the use of direct instruction. He took care to make sure students knew, understood, and had the readiness for continued studies in mathematics.

Teachers who recognize and position students' backgrounds as a resource in their mathematics teaching can connect students' identities with building a sense of agency. By using students' background and competencies to plan and anticipate students' thinking, teachers can support students' participation. When teachers anticipated the boys' thinking, actions, and problem solving, the boys perceived this as knowing and understanding their identities. Jabari's description of Ms. Jackson's anticipation of his thinking and action demonstrates this dynamic:

> I don't know how she does it, but sometimes she know what we are going to do and say before we do and say anything . . . [S]he knows us so well that she gets us out of trouble before we get in trouble. . . . In math, she know the right thing to say to help us with our work. (Berry, 2003, n.p.)

Teachers' anticipation of students' thinking and action is a form of helping students develop their identities because it conveys to students that their teachers care enough about them to understand how they think and see the world. Because of the forward thinking by teachers, students may perceive that they cannot get by with minimal efforts: Their teachers know them too well.

Final Thoughts

In the final section of my 2008 article, I wrote:

> One might merely conclude that because these boys encountered educational gatekeepers who could potentially limit their access to rigorous mathematics, parental involvement and advocacy is necessary if African American boys are to be successful with school mathematics. I contend that such a conclusion is shortsighted and does little to promote changes in policies and practices as they relate to access and opportunity. Nor does such shortsightedness promote the understanding of the complexities of the racialized experiences, stereotypes, lowered expectations, and challenges faced by African American boys. (p. 484)

Based on my research, I recognized that practices needed to be changed and that more understanding of the complexities faced by Black boys is necessary. Building from that conclusion, I think it is important that we make connections between teaching practices and students' identities and agency. As related to Black boys specifically, we must develop an appreciation for the complexities faced by Black boys by framing identity and agency as the constructs to reflect on our teaching practices to support Black boys as doers of mathematics. As teachers, we should consider focusing not only on teaching mathematics content but also on the several negotiated, interwoven identities that are important to students. Identities are dynamic and develop over contexts and time. With this understanding, teachers play a significant role in helping and contributing to the development of students' identities. Teachers can support and help the development of students' identities across contexts and over time. Mathematics teaching practices can shape identify and foster agency. For this to happen, teachers should understand students' interwoven identities, know the context of students' realities, and appreciate how these identities and realities interact to support agency. One way to understand students' identities is to gather students' stories through interviews, autobiographies, and observations. Consider the questions below as you think about your identities and agency as they relate to your teaching:

1. What are the interwoven identities expressed by my students?

2. In what ways, if any, do I affirm the identities of my students?

3. What teaching practice do I use to support and affirm students' mathematics identity development?

4. How do students demonstrate high and low sense of agency in my mathematics classroom?

5. For students' demonstrating low sense of agency, how can I help these students develop a positive mathematics identity to support a high sense of agency?

6. How do I model a high sense of agency and how do I provide opportunities for students to demonstrate a high sense of agency?

References

Aguirre, J. M., Mayfield-Ingram, K., & Martin, D. B. (2013). *The impact of identity in K–8 mathematics learning and teaching: Rethinking equity-based practices.* Reston, VA: National Council of Teachers of Mathematics.

Berry, R. Q., III. (2003). *Voices of African American male middle school students: A portrait of successful middle school mathematics students* (Unpublished doctoral dissertation). University of North Carolina, Chapel Hill.

Berry, R. Q., III. (2004). The equity principle through the voices of African American males. *Mathematics Teaching in the Middle School, 10*(2), 100–103.

Berry, R. Q., III. (2008). Access to upper-level mathematics: The stories of successful African American middle school boys. *Journal for Research in Mathematics Education, 39*, 464–488.

Berry, R. Q., III, & Ellis, M. W. (2013). Multidimensional teaching. *Mathematics Teaching in the Middle School, 19*(3), 172–178.

Boaler, J. (1998). Open and closed mathematics: Student experiences and understandings. *Journal for Research in Mathematics Education, 29*, 41–62.

Côté, J. E., & Schwartz, S. J. (2002). Comparing psychological and sociological approaches to identity: Identity status, identity capital, and the individualization process. *Journal of Adolescence, 25*(6), 571. doi:10.1006/jado.2002.0511

Hollins, E. R. (1982). The Marva Collins story revisited: Implications for regular classroom instruction. *Journal of Teacher Education, 33*(1), 37–40.

Martin, D. B. (2003). Hidden assumptions and unaddressed questions in mathematics for all rhetoric. *Mathematics Educator, 13*(2), 7–21.

Murrell, P. C. (2007). *Race, culture, and schooling: Identities of achievement in multicultural urban schools.* New York, NY: Lawrence Erlbaum.

National Council of Teachers of Mathematics (NCTM). (2014). *Principles to actions: Ensuring mathematical success for all.* Reston, VA: Author.

Planas, N., & Civil, M. (2013). Language-as-resource and language-as-political: Tensions in the bilingual mathematics classroom. *Mathematics Education Research Journal, 25*(3), 361–378.

Solomon, Y. (2009). *Mathematical literacy: Developing identities of inclusion.* New York, NY: Routledge.

Chapter 4

Teaching Mathematics in a Multilingual Classroom

Mamokgethi Setati Phakeng
University of South Africa

Learning mathematics has elements that are similar to learning a language. Just as in learning a language, in mathematics students have to learn new terminology and symbols, how to use them in conversations, and the various ways in which that vocabulary is used in different mathematical contexts. Halliday (1978) refers to this as the "mathematics register." An essential part of learning mathematics therefore is learning how to communicate mathematically, or learning how to use the mathematics register.

Communication, however, involves a great deal more than just language. It involves acting-interacting-thinking-valuing-talking-writing-reading-listening in the *appropriate way* with the *appropriate props* at the *appropriate times* (Gee, 1999). Mathematics teachers are therefore expected to help their students develop or learn ways of acting-interacting-thinking-valuing-talking-writing-reading-listening in ways that are mathematically appropriate so that they can understand and be understood by other members of the wider mathematics community. The question we must ask is this: How can teachers do this in multilingual mathematics classrooms, where students are still learning the language of learning and teaching (LoLT)?

This chapter presents three practical ideas that teachers in multilingual mathematics classrooms can use to support mathematics learning in their classrooms. I begin the chapter with an overview of findings from research on teaching and learning mathematics, drawing mainly from two journal articles: Setati (2005) and Setati, Molefe, and Langa (2008). In so doing I create a foundation for a discussion on how the findings from research can be used in mathematics classrooms. Here I will include as many practical suggestions as possible.

What Research on Teaching and Learning Mathematics in Multilingual Classrooms Tells Us

Most research on mathematics education in bilingual and multilingual classrooms has argued for the use of the learners' home language(s) as resources for learning and teaching mathematics. While this argument makes sense, I have noted that most of this research was framed by a conception of mediated learning, where language is seen only as a tool for thinking and communicating (see Setati, 2005). The argument appropriately foregrounds the mathematics, but it does not consider the political role of language. Thus, mathematics teachers in multilingual classrooms need to recognize and acknowledge language as political, because without such we will fail to understand and work with the demands that learners in multilingual classrooms face. As Gee (1999) aptly puts it, language is political: It always has implications for how social goods are or ought to be distributed. Whenever people speak or write they create a political perspective; they use language to project themselves as certain kinds of people engaged in certain kinds of activities. Words are thus never just words; and

This commentary chapter is adapted from two sources: M. Setati (2005), Teaching mathematics in a primary multilingual classroom, *Journal for Research in Mathematics Education, 36*, 447–477, and M. Setati, T. Molefe, & M. Langa (2008), Using language as a transparent resource in the teaching and learning of mathematics in a grade 11 multilingual classroom, *Pythagoras, 67*, 14–25. The *Pythagoras* article is reprinted as an appendix to this chapter and is used with permission.

language is not just a vehicle to express ideas (a cultural tool), but also is a political tool that we use to enact (i.e., to be recognized as) a particular *who* (identity) engaged in a particular *what* (situated activity). Such is the situation with teachers in multilingual mathematics classrooms: Their language choices and preferences are not trivial.

Analysis of the lesson observation data collected in a study with the purpose of understanding the language practices of teachers in multilingual mathematics classrooms showed a dominance of English and the use of procedural discourse (Setati, 2005). The dominance of English and procedural discourse was also evident in the mathematics tasks and tests that the teachers gave learners as well as in the interactions that occurred during the observed lessons, which suggested that conceptual discourse was not seen as valued mathematical knowledge. Other researchers have interpreted this dominance of procedural teaching as a function of the teachers' lack of or limited knowledge of mathematics (Taylor & Vinjevold, 1999). My research suggests that the problem is much more complex. A tension seems to exist between the need to ensure that students gain access to English and the important but not always recognised and acknowledged need to also ensure access to mathematical knowledge.

In a follow-up study (Setati, 2008) I found that teachers in multilingual mathematics classrooms in South Africa prefer English as the language of learning and teaching (LoLT). The teachers in the study stated ideological and pragmatic reasons for their preference to teach mathematics in English, despite their students' limited fluency in it. These teachers are aware of the power of English and the status it bestows on those who can communicate in it, and so they referred to it as an international or universal language. Awarding such a status to English suggests that they see English as being *bigger than* themselves, as they do not have any control over the international nature of English. All they can do is prepare their learners for participation in the international world; and according to them, teaching mathematics in English is an important part of this preparation. As indicated earlier, teachers' language choices are not trivial; their decisions about which language to use and how and when to do so reflect not only curriculum and pedagogic decisions, but also the political context of their practice together with the identities and activities they are enacting (Setati, 2005). This situation explains why recommendations by research to use the students' home languages as valuable resources for teaching and learning mathematics are hard to translate into practice in multilingual contexts: The need for social access predominates over that for cognitive access. Elsewhere I have argued that in a context where the hegemony of English is so prevalent and students have limited fluency, regarding the students' home languages as a resource tends to be seen as a threat to their development of fluency in English. The challenge for teachers in multilingual mathematics classrooms is how to ensure that students gain access to mathematics without losing an opportunity to gain fluency in English. In the section that follows I will share a strategy that teachers in multilingual classrooms can use to deal with this challenge.

Another challenge that teachers in multilingual classroom face is that of ensuring that language becomes a transparent resource, which means that it must be both visible and invisible. Language must be visible so that it is clearly seen and understood by all, and invisible in that, when students interact with written texts and discuss mathematics, it does not distract their attention from the mathematical task under discussion but facilitates their mathematics learning (Setati, Molefe, & Langa, 2008).

The third challenge for teachers in multilingual classrooms is that of selecting tasks that are relevant and challenging for the students. The challenge of task selection is not unique to multilingual mathematics classrooms; but in these classrooms, the challenge is made more complex because many of the tasks in mathematics textbooks are mainly relevant for students who belong to the dominant culture and race. So in selecting or designing tasks it is important that the teacher in a multilingual classroom pays attention to the context.

Strategies That Teachers Can Use to Address the Three Challenges

I begin with the strategy of getting to know your students. Knowing for whom you are preparing your lessons will make you an effective teacher. In order to plan how to teach your students or how to present mathematics in a manner that will motivate students to study and use it, a teacher needs to know what motivates the students, what language and cultural background the students are bringing to the classroom, and what interests the students. One strategy that I have found useful and efficient to gain this knowledge is to give the students a survey that asks questions about themselves. A teacher could hand it out the first or second week of school in order to find out what the students are interested in, their background, and other things that could be incorporated into lessons. This is also a way for shy and introverted students to express who they are in writing. The survey can include questions about language, such as these:

- How many languages do you speak?

- What language do you use to communicate with your parents at home?

- What language do you use to communicate with your siblings?

- What language do you use to communicate with your friends?

- What language do you prefer to learn mathematics in?

- Why do you prefer that language to learn mathematics in?

Knowing a little about the students' language, culture, and socioeconomic background helps the teacher understand students and in turn to be able to answer questions such as How can I help this student learn better? or What in the students' life can I relate this topic to so that it is interesting?

Based on knowledge about students, teachers can better select mathematics tasks that are relevant to students' background. Below are examples of two tasks that present real-world problems that we used in multilingual classrooms we worked with in South African township schools. We selected these tasks based on our knowledge of the students. We thought the students would find these tasks interesting and useful to engage with, not only because of the familiar contexts but mainly because they deal with pressing situations in their communities.

The first task, shown in figure 4.1, takes place in a hair salon, where Derrick must ensure that he makes the maximum amount of money in the limited time (eight hours) he has to work in the salon each day. Many students in these communities work as hairdressers on weekends in order to make money, and so their interest in this problem was based on their own desire to make the maximum profit within eight hours.

Derrick's Salon

Derrick owns a hair salon. He specializes in two types of hairstyles: the dreadlocks and singles. Let's assume that x represents the number of dreadlocks hairstyles and y represents the number of singles hairstyles. It takes at least 1 hour to do a dreadlocks hairstyle and at least 5 hours to do a singles hairstyle, and Derrick can only work up to 8 hours per day. It costs at most R40 to do dreadlocks and a maximum of R60 to do singles.

Derrick has a problem: He makes a profit of R20 on dreadlocks and a profit of R35 on singles. Because of his other commitments Derrick still wishes to work for a maximum of 8 hours a day only. How many dreadlocks hairstyles and singles hairstyles will give Derrick maximum profit?

Fig. 4.1. Hair Salon task

The second task, shown in figure 4.2, presents a real situation in South Africa, where families had to choose their preferred costing model for electricity consumption. This matter was controversial in townships, as families were struggling to figure out which costing method would be cheaper. The students were thus interested in the problem because they felt that they could use the knowledge gained from engaging with it to help their families choose a costing model. For a detailed discussion on how students in one of the classrooms dealt with the task, see Setati et al. (2008), which is reprinted as an appendix to this chapter.

Cost of Electricity

The Brahm Park electricity department charges R40,00 monthly service fees then an additional 20c per kilowatt-hour (kWh). A kilowatt-hour is the amount of electricity used in one hour at a constant power of one kilowatt.

1. The estimated monthly electricity consumption of a family home is 560 kWh. Predict what the monthly account would be for electricity.

2. Three people live in a townhouse. Their monthly electricity account is approximately R180,000. How many kilowatt-hours per month do they usually use?

3. In winter the average electricity consumption increases by 20%. What would the monthly account be for the family home in (1) above and for the townhouse?

4. In your opinion, what may be the reason for the increase in the average electricity consumption in (3) above?

5. Determine a formula to assist the electricity department to calculate the monthly electricity bill for any household. State clearly what your variables represent and the units used.

6. a) Complete the following table showing the cost of electricity in Rand for differing amounts of electricity used:

Consumption (kWh)	0	100	200	300	400	500	600	700	800	900
Cost (in Rand)										

b) Draw a graph on the set of axes below to illustrate the cost of different units of electricity at the rate charged by the Brahm Park electricity department.

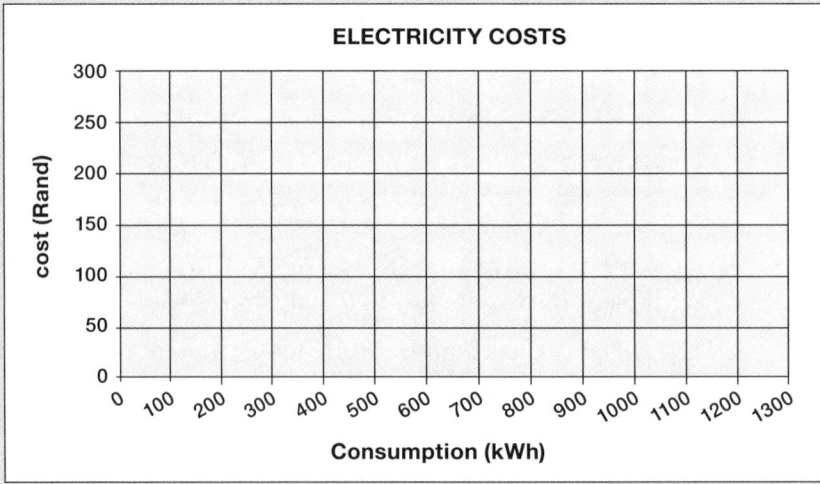

Fig. 4.2. Cost of Electricity task

After careful consideration, the electricity department decided to alter their costing structure. They decided that there will no longer be a monthly service fee of R40,00, but now each kilowatt-hour will cost 25c.

7. What would be the new monthly electricity accounts for the family home and the townhouse?

8. a) Complete the following table showing the cost of electricity in Rand for differing amounts of electricity used, using the new costing structure:

Consumption (kWh)	0	100	200	300	400	500	600	700	800	900
Cost (in Rand)										

b) Draw a graph on the same set of axes in question 6(b) to illustrate the cost of electricity for different units of electricity using the new costing structure.

9. Do both the family home and the townhouse benefit from this new costing structure? Explain.

10. If people using the electricity had the option of choosing either of the two costing structures, which would you recommend? Clearly explain your answer using tables you have completed and graphs drawn in questions 6(a) and 6(b) and 8(a) and 8(b) above.

Malati Draft Materials (30-04-2005)

Fig. 4.2. *Continued*

The next suggestion is to present all mathematics tasks you give to students in two languages: their home language and the language of learning and teaching (e.g., English). I am aware that this means that you as a teacher have to translate tasks into several home languages, some of which you are not familiar with or fluent in. While this may be a challenge, it is a much better one to deal with than the challenge of students who are not interested in mathematics or the lessons you are presenting. Share the responsibility of translating the tasks with other mathematics teachers who teach in multilingual classrooms. If none of you can speak the languages, then get parents involved in assisting with the translations. In this way the teacher can share the challenge with others who are also affected by it. Presenting tasks in this manner will communicate to the students the important and most affirming message that their languages are valid and legitimate languages of learning.

Another idea is to let students sit according to their home language groups during the lessons to allow greater possibility of interaction in their languages of choice. Make it explicit to students that they are allowed to communicate in any language with which they feel comfortable. Avoid correcting your students' grammatical errors during the lessons; ignore all grammatical errors and focus only on mathematics fluency. Even if the students mix languages, just focus on what they are saying rather than on how they are saying it, even if this approach means revoicing what they have just said in mathematically correct terms. In this manner you will render language a transparent resource—a resource that everyone knows is there and is in use, but it is not in focus.

Concluding Remarks

The challenges of teaching mathematics in multilingual classrooms are at once about the nature of mathematics, who the students are and their fluency in the language of learning and teaching, and the quality of mathematics teaching. Therefore ideas about how to attend to the challenges should be not only about language, as if to suggest that that fluency in the language of learning and teaching will solve all problems. Mathematics teaching and learning is complex and therefore cannot be understood in isolation from the pedagogic issues specific to mathematics as well as the wider social, cultural, and political factors that infuse schooling.

References

Gee, J. P. (1999). *An introduction to discourse analysis: Theory and method.* London, England: Routledge.

Halliday, M. A. K. (1978). *Language as social semiotic.* London, England: Edward Arnold.

Setati, M. (2005). Teaching mathematics in a primary multilingual classroom. *Journal for Research in Mathematics Education, 36,* 447–477.

Setati, M. (2008). Access to mathematics versus access to the language of power: The struggle in multilingual classrooms. *South African Journal of Education, 28,* 103–116.

Setati, M., Molefe, T., & Langa, M. (2008). Using language as a transparent resource in the teaching and learning of mathematics in a grade 11 multilingual classroom. *Pythagoras, 67,* 14–25.

Taylor, N., & Vinjevold, P. (1999). *Getting learning right.* Johannesburg, South Africa: Joint Education Trust.

Appendix

Using Language as a Transparent Resource in the Teaching and Learning of Mathematics in a Grade 11 Multilingual Classroom

Mamokgethi Setati

Terence Molefe

Mampho Langa

University of South Africa
setatrm@unisa.ac.za

Fons Luminus High School
molefebt@yahoo.com

Oprah Winfrey Leadership Academy for Girls
Mampho.Langa@owla.co.za

In this paper, we draw on a study conducted in Grade 11 classrooms to explore how the learners' home languages can be used for teaching and learning mathematics in multilingual classrooms in South Africa. This report is part of a wider study that is still in progress. Based on an analysis of data collected through lesson observations in a Grade 11 class and learner interviews we argue for the deliberate, proactive and strategic use of the learners' home languages as a transparent resource in the teaching and learning of mathematics in multilingual classrooms. Such use of the languages will ensure that learners gain access to mathematical knowledge without losing access to English, which many parents, teachers and learners presently see as a necessary condition for gaining access to social goods such as higher education and employment.

What does it mean to teach or learn mathematics in a language that is not your home, first or main language? This is the situation in the majority of classrooms in South Africa. In these classrooms the language of learning and teaching (LoLT) is English – one of the eleven official languages; however, neither the teacher nor the learners have English as their main, home or first language. Research shows that teachers and learners in these classrooms prefer that English be used as the LoLT (Setati, in press). In addition to this, anecdotal evidence shows that a majority of learners in these classrooms are not motivated to study mathematics, and they are doing it either because they have to or because they know that it is needed for any higher education study related to science, technology or commerce. What strategies are appropriate for use in these classrooms? Embedded in this question are issues about language and learning and also about motivating learners' interest in mathematics.

In this paper we draw on a wider study exploring relevant pedagogies for teaching and learning mathematics in multilingual classrooms. We specifically focus on data collected in one Grade 11 classroom to explore how the learner's home languages can be used as a transparent resource in the teaching and learning of mathematics in multilingual classrooms.

We begin the paper with a review of research on teaching and learning mathematics in multilingual classrooms. Through this we expose three prevalent dichotomies in research on teaching and learning mathematics in multilingual classrooms. First, is the dichotomy between using English as LoLT as opposed to using the learners' home language(s) as LoLT. Second, is the dichotomy about drawing on socio-political perspectives when analysing interactions in multilingual mathematics classrooms as opposed to drawing on cognitive perspectives. The third dichotomy is about gaining access to mathematical knowledge as opposed to access to English.

We then discuss the theory that informed the analysis we present in the paper. This discussion provides a theoretical background for the argument for the deliberate, proactive and strategic use of the learners' home languages as a transparent resource in the teaching and learning of mathematics.

Previously published as Setati, M., Molefe, T., & Langa, M. (2008). Using language as a transparent resource in the teaching and learning of mathematics in a Grade 11 multilanguage classroom. *Pythagoras, 67*, 14–25.

Teaching and learning mathematics in multilingual classrooms

There is a continuing debate in South African education and the public domain regarding which language should be used for teaching mathematics in multilingual classrooms in which children are still learning English. This debate is due to the fact that learners in many of these classrooms are not yet fully fluent in English which is the LoLT in their classrooms.

While the South African language in education policy (LiEP) encourages multilingualism (Department of Education, 1997) and research supports the use of the learners' home languages (Moschkovich, 2002), a recent analysis of the learners' and teachers' language choices for teaching and learning mathematics shows a preference for English (Setati, in press). In this analysis Setati argues that the desire to gain access to social goods (e.g. jobs, higher education) overrides the multilingual teachers' and learners' concern for epistemological access. The view with many teachers and learners is that English is an international language and in South Africa it is important for higher education, jobs and subsequently a better life.

Given this background, we argue that it is crucial to explore ways of drawing on the learners' home languages for teaching and learning mathematics in multilingual classrooms without denying them access to English. Previous research argues that the learners' home languages are a resource for mathematics learning (Adler, 2001; Moschkovich, 1999; Setati, 2005; Setati & Adler, 2000). The challenge, however, is that in a context such as South Africa, where the hegemony of English is so prevalent, regarding the learners' home languages as a resource tends to be seen as a threat to multilingual learners' development of fluency in English. As Sachs (1994) pointed out, in South Africa "all language rights are rights against English" (p. 1). Hence our argument that for the use of the learners' home languages in the teaching and learning of mathematics in multilingual classrooms to be successful it must ensure that the learners gain epistemological access without losing access to English. Granville et al. (1998) present a similar idea in relation to the South African language in education policy, where they argue for English without g(u)ilt. What is new about our argument is the different orientation we bring by focusing on learning and teaching rather than policy.

Debates around language and learning in South Africa tend to create a dichotomy between learning in English and learning in the home languages. They create an impression that the use of the learners' home languages for teaching and learning must necessarily exclude and be in opposition to English, and the use of English must necessarily exclude the learners' home languages. In an article entitled 'Why don't kids learn maths and science successfully?', Sarah Howie of the University of Pretoria is quoted as saying "The most significant factor in learning science and maths isn't whether the learners are rich or poor. It's whether they are fluent in English" (Science in Africa, 2003). This she said drawing on her analysis of South Africa's poor performance in the Third International Mathematics and Science Study of 1995 (see also Howie 2003, 2004). In the same article Howie makes an impassioned call on South Africa to choose only one language for teaching and learning mathematics in multilingual classrooms:

Let's stop sitting on the fence and make a hard decision. We must either shore up the mother tongue teaching of maths and sciences, or switch completely to English if we want to succeed.

Our argument in this paper is that in a multilingual country such as South Africa the choices are not as simplistic as Howie suggests. Our argument is informed by a holistic view of multilingual learners (different from Howie's monolingual view), in our view multilingual learners have a unique and specific language configuration and therefore they should not be considered as the sum of two or more complete or incomplete monolinguals. To explain this different but complete language system in multilinguals, Grosjean (1985) uses an analogy from the domain of athletics:

The high hurdler blends two types of competencies: that of high jumping and that of sprinting. When compared individually with the sprinter or the high jumper, the hurdler meets neither level of competence, and yet when taken as a whole, the hurdler is an athlete in his or her own right. No expert in track and field would ever compare a high hurdler to a sprinter or to a high jumper, even though the former blends certain characteristics of the latter two. In many ways the bilingual is like the high hurdler. (p. 471)

The use of the learners' home languages as a transparent resource that we are exploring in this paper is informed by this holistic view of multilingual learners. We accept that the idea of drawing on the learners' home languages during teaching is not necessarily new. The use of code-switching as a learning and teaching resource in bilingual and multilingual mathematics classrooms has been the focus of research in the recent past (e.g. Adendorff, 1993; Adler, 1998, 2001; Arthur, 1994; Khisty, 1995; Merritt, Cleghorn, Abagi, & Bunyi, 1992; Moschkovich, 1996, 1999; Ncedo, Peires, & Morar, 2002; Setati, 1998; Setati, 2005; Setati & Adler, 2000). These studies have argued for the use of the learners' home languages in teaching and learning mathematics, as a support needed while learners continue to develop proficiency in the LoLT at the same time as learning mathematics. All of these studies seem to be in agreement that to facilitate multilingual learners' participation and success in mathematics teachers should recognise their home languages as legitimate languages of mathematical communication. The practical manifestation of the use of the learners' home languages in these studies is through code-switching, mainly to provide explanation to learners in their home languages. In all of these studies code-switching is presented as spontaneous and reactive, the learner's home languages are only used in oral communication and never in written texts. What we are advocating in this paper is the deliberate, strategic and proactive use of the learners' home languages. This strategy recognizes the fact that learners want access to English and thus while we draw on the learners' home languages and foreground the quality of the mathematics tasks used during teaching, we also ensure that English is still available to learners and they can continue to develop fluency in it.

Research into the complex relationship between bilingualism/multilingualism and mathematics teaching and learning argues that bilingualism/multilingualism *per se* does not impede mathematics learning (Clarkson, 1991; Dawe, 1983; Stephens, Waywood, Clarke, & Izard, 1993; Zepp, 1989). This field of research has been criticised because of its cognitive orientation and its inevitable deficit model of the bilingual learner (Martin-Jones & Romaine, 1986; Frederickson & Cline, 1990; both in Baker, 1993). Learner performance (and by implication, mathematical achievement) is determined by a complex set of inter-related factors. Poor performance by multilingual learners thus cannot be attributed to the learners' limited language proficiencies in isolation from the wider social, cultural and political factors that infuse schooling.

In our view, this past research informed by a cognitive perspective presents an implicit argument in support of the maintenance of learners' home languages, and of the potential benefits of learners using their home language(s) as a resource in their mathematics learning. Multilingualism is the norm in many South African classrooms, rather than the exception. Hence the need for South African mathematics education not only to treat the multilingual learner as the norm but also to view his or her facility across languages as a resource rather than a problem (Baker, 1993). Through our work in this study we have come to recognise that separating cognitive matters from the socio-political issues relating to language and power when exploring the use of language(s) for teaching and learning mathematics in multilingual classrooms is not productive. While we accept that cognitively oriented research does not deal with the socio-political issues relating to the context in which teaching and learning takes place, we acknowledge that it is useful in helping us attend to issues relating to the quality of the mathematics and its teaching and learning in multilingual classrooms. In this study we are thus moving against dichotomies, not only of language choices but also of theoretical perspectives.

Theoretical underpinnings

This study is broadly informed by an understanding of language as "a transparent resource" (Lave & Wenger, 1991). While the notion of transparency as used by Lave and Wenger is not usually applied to language as a resource nor to learning in school, it is illuminating of language use in multilingual classrooms (see also Adler, 2001). Lave and Wenger (1991) argue that access to a practice relates to the dual visibility and invisibility of its resources:

> Invisibility is in the form of unproblematic interpretation and integration into activity, and visibility is in the form of extended access to information. This is not a dichotomous distinction, since these two crucial characteristics are in a complex interplay, their relation being one of both conflict and synergy. (p. 103)

For language in the classroom to be useful it must be both visible and invisible: visible so that it is clearly seen and understood by all; and invisible in that when interacting with written texts and discussing mathematics, this use of language should not distract the learners' attention from the mathematical task under discussion but facilitate their mathematics learning. This idea is similar to the use of technology in mathematics learning. The technology needs to be visible so that the learners can notice and use it. However it also needs to be simultaneously invisible so that the learners' attention is focussed on the mathematics problem that they are trying to solve. Like technology, language needs to be a transparent resource. As Lave and Wenger argue the idea of the visibility and invisibility of a resource is not a dichotomous distinction, it is not about whether to focus on language or mathematics, it is about recognising that the two are intertwined and are constantly in complex interplay.

We found Lave and Wenger's concept of transparency useful in conceptualising a strategy for using language in teaching and learning mathematics in multilingual classrooms. Multilingual classrooms are characterised by complex multiple teaching demands: the learners' limited proficiency in the language of learning and teaching (English); the challenge to develop the learners' mathematical proficiency as well as the presence of multiple languages. The strategy we are exploring is guided by two main principles, which are informed by the theoretical assumptions elaborated in the discussion above. First, it is the *deliberate* use of the learners' home languages. We emphasise the word deliberate because with this strategy the use of the learners' home languages is deliberate, proactive and strategic and not spontaneous and reactive as it happens with code-switching. Second, is that through the selection of real world interesting and challenging mathematical tasks, learners would develop a different orientation towards mathematics than they had and would be more motivated to study and use it (Gutstein, 2003). Many learners in multilingual classrooms in South Africa have what Gutstein (2003) describes as "the typical and well documented disposition with which most mathematics teachers are familiar – mathematics as a rote-learned, decontextualised series of rules and procedures to memorise, regurgitate and not understand" (p. 46). In this study we selected high cognitive demand tasks (Stein, Smith, Henningsen, & Silver, 2000), that present real world problems that the learners can find interesting and useful to engage with.

The study

The study presented in this paper was undertaken in a Grade 11 class taught by Terence, the second author. The data we are presenting here was collected in Terence's classroom in a multilingual high school in Soweto, Johannesburg. There were 36 learners in his class and they had the following home languages: Setswana, Xitsonga, IsiZulu and Tshivenda. Each of the learners was able to communicate in at least four languages and they were learning English as a subject at second language level as well as their respective home languages as subjects at first language level. Terence is multilingual and fluent in eight languages[1], which includes all the home languages of his learners as well as English. His home language is Setswana. At the time of the study Terence had been teaching mathematics at secondary school level for 15 years.

Data was collected through lesson observations and individual learner interviews. Lessons were observed and video recorded for four consecutive days. At the end of the four days four learners from different home language groups were interviewed by Mampho, the third author and a former teacher in the school who was not present during lesson observations. The interview focused on their reflections and views about the lessons.

On the next page is the task that the learners were working on during the lessons observed and analysed in this paper. This task was translated into the four home languages of the learners in the class.

During the lessons learners were organised into seven home language groups: two Setswana groups, two Tshivenda groups, two IsiZulu groups and one Xitsonga group. Six of the groups had five learners and one group had six learners and they were given tasks in two language versions (English and their home language). Learners were explicitly made aware of the two language versions of the task and encouraged to communicate in any language including their home languages at any stage during the lessons.

All the lessons and learner interviews were video recorded and then transcribed. To analyse the video-recording and the transcribed data we looked

---

[1] This kind of multilingualism is not unusual. Given the integration of different ethnic groups, a majorty of African teachers (indeed African people in general) in the Gauteng province are multilingual and can communicate in at least four languages.

COST OF ELECTRICITY

The Brahm Park electricity department charges R40,00 monthly service fees then an additional 20c per kilowatt-hour (kWh). A kilowatt-hour is the amount of electricity used in one hour at a constant power of one kilowatt.

1. The estimated monthly electricity consumption of a family home is 560 kWh. Predict what the monthly account would be for electricity.

2. Three people live in a townhouse. Their monthly electricity account is approximately R180,00. How many kilowatt-hours per month do they usually use?

3. In winter the average electricity consumption increases by 20%, what would the monthly bills be for the family home in (1) above and for the townhouse?

4. In your opinion, what may be the reason for the increase in the average electricity consumption in (3) above?

5. Determine a formula to assist the electricity department to calculate the monthly electricity bill for any household. State clearly what your variables represent and the units used.

6. a) Complete the following table showing the cost of electricity in Rand for differing amounts of electricity used:

Consumption (kWh)	0	100	200	300	400	500	600	700	800	900
Cost (in Rand)										

b) Draw a graph on the set of axes below to illustrate the cost of different units of electricity at the rate charged by the Brahm Park electricity department.

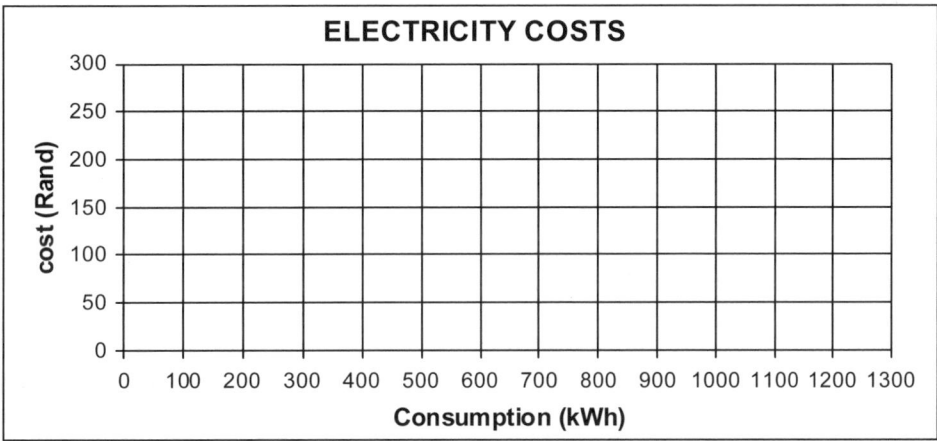

After careful consideration, the electricity department decided to alter their costing structure. They decide that there will no longer be a monthly service fee of R40,00 but now each kilowatt-hour will cost 25c.

7. What would be the new monthly electricity accounts for the family home and the townhouse?

8. a) Complete the following table showing the cost of electricity in Rand for differing amounts of electricity used using the new costing structure:

Consumption (kWh)	0	100	200	300	400	500	600	700	800	900
Cost (in Rand)										

b) Draw a graph on the same set of axes in question 6 b) to illustrate the cost of electricity for different units of electricity using the new costing structure.

9. Do both the family home and the townhouse benefit from this new costing structure? Explain.

10. If people using the electricity had the option of choosing either of the two costing structures, which would you recommend? Clearly explain your answer using tables you have completed and graphs drawn in questions 6 a) and 6 b) and 8 a) and 8 b) above.

Malati Materials (30-04-2005)

for presences (what was visible) and absences (what was invisible) in what the learners were talking about. In the lesson observation data, what was most visible was learners' attempts to find the solutions to the questions in the task without much focus to the language. There was only one incident during the lessons observed, which we discuss in more detail below, where language became visible but not simultaneously invisible in one of the groups.

All the lessons and learner interviews were video recorded and then transcribed. To analyse the video-recording and the transcribed data we looked for presences (what was visible) and absences (what was invisible) in what the learners were talking about. In the lesson observation data, what was most visible was learners' attempts to find the solutions to the questions in the task without much focus to the language. There was only one incident during the lessons observed, which we discuss in more detail below, where language became visible but not simultaneously invisible in one of the groups. Language was, however, constantly visible for Terence, the teacher. For instance when asking learners to read he would specify which language they should read in. In the section that follows we discuss, with evidence from lesson observation data, incidents during the lessons when language was transparent (visible and invisible) and also when it was visible but not simultaneously invisible (i.e. not transparent).

When language was visible and invisible

In our analysis we found that when language was transparent learners' interactions were conceptual – learners' interactions were focused not only on what the solution is but also why it is correct. The two extracts we are analysing in this section happened in the Tshivenda group and are typical of how language functioned as a transparent resource during interactions between Terence and the learners and also between the learners themselves. Both extracts are taken from the first lesson at the time when the learners were beginning their work on the task and they needed to understand the following statement in the problem:

> *The Brahm Park electricity department charges R40,00 monthly service fees then an additional 20c per kilowatt-hour (kWh). A kilowatt-hour is the amount of electricity used in one hour at a constant power of one kilowatt*

The extract below shows the interaction between Terence and the learners in the group. Here Terence is working with them on the two charges mentioned in the problem, the R40 monthly service fee and the additional 20c per kilowatt-hour.

1. Terence: *Forty rhanda heyi, vhoibadhala when* [When is the forty rand paid?]
2. Sipho: In a month.
3. Terence: *Twenty cents yone* [What about the twenty cents?]
4. Given: *Twenty cents yo ediwa.* [Twenty cents is added.]
5. Terence: *Why i ediwa* [Why is it added?]
6. Learners: (*Silent*).
7. Terence: *Vhoi edela mini? Twenty cents vhoi edela mini* [Why is it added? Why is twenty cents added?]
8. Learners: (*Inaudible*).
9. Terence: *Okay, if you use electricity ukho bhadala forty rand?* [Okay, if you use electricity will you pay forty rand?]
10. Learners: Yes *meneer* [sir]
11. Terence: *If unga shumisanga electricity ukho bhadala forty rand* [If you did not consume electricity, will you pay forty rand?]
12. Sipho: No, no no …
13. Given: *Haena, whether ushumisile ore haushumisanga, ukhobhadala forty rhanda* [No, whether you have consumed electricity or not, you pay the forty rand.]
14. Terence: *Whether ushumisile ore haushumisanga?* [Whether consumed or not?]
15. Sipho: *Eya*, yes, it is a must.
16. Terence: It is a must?

One very noticeable thing about the extract above is the fact that it is in a mixture of English and the learners' home language (Tshivenda). This as indicated earlier was typical of interactions during the lessons observed in this class. The unproblematic move between Tshivenda and English without explicit negotiation between interactants is an indication that language is functioning as a transparent resource. Whilst it is visible it is also invisible enough to be used without distracting attention from the task. This invisibility of language as a mediating tool allows focus on and thus supports visibility of the mathematics the learners are discussing (Lave & Wenger, 1991). At the same time, the visibility of language (i.e. tasks given in two languages) is necessary for allowing its unproblematic invisible use.

In the extract above, the learners are struggling to understand the phrase "an additional 20c per kilowatt-hour (kWh)". While they understand that everyone who has electricity is supposed to pay the R40 monthly cost and also that 20c is added, they seem to be having difficulty in understanding why the 20c is added. In utterances 5 to 7 Terence asks them why 20c was added. Seeing that they are not able to answer his question, he moves back to asking them about the R40 in utterance 9. By doing this Terence is separating the R40 from the 20c so that the learners can see that while everyone who has electricity is required to pay R40, as to how much they pay thereafter depends on the amount of kilowatt hours they used and 1 kWh costs 20c. The extract above ends with Terence in utterance 16 having established with the learners that the R40 payment is a mandatory service fee for everyone who has electricity. In the extract below the learners carry on with the discussion (on their own) about when and why 20c is added.

17. Given: *Hei, nayo ... ar ... (giggles) ... So forty rhanda hi monthly cost ne, then ba yieda nga twenty cents kha kilowatt for one hour. Then after that, angado shumisa ..., baibidza mini? Heyi ... ndoshumisa one kilowatt nga twenty cents kha one hour* [Hey, this question ... ar ...(giggles) ... So forty rand is the monthly cost, then they add twenty cents per kilowatt-hour. ..., they use..., what do they call it? Hey ... they use one kilowatt-hour for twenty cents.]
18. Sipho: *Eya* [Yes]
19. Given: *Boyieda, maybe boshumisa twenty cents nga one hour* [They add it, maybe they use twenty cents per hour.]
20. Sipho: *Eya, yantha* [Yes, one hour]
21. Given: *Iba ...* [It becomes...]
22. Given and Sipho (*together*): Forty rand twenty cents.
23. Sipho: *Yes, vhoibadela monthly, ngangwedzi ya hona. Yo fhelela, yes. Sesiyaqubheka.* [Yes, they pay it monthly, each month. It is complete, yes. We continue.]

The transparent use of language continues in the above extract. The learners do not focus on what language is used for what; they are focused on communicating their understanding. This transparency of language enables conceptual interactions between the learners and the teacher and also among learners themselves. Using their home language, Tshivenda, as a legitimate language of interaction together with English made it possible for them to understand that in this case 1 kWh costs 20c. The learners are not concerned about the correctness of their grammar in Tshivenda and in English, they are more focused on gaining an understanding of the problem and having both language versions serves as a resource that they can draw as and when they need to.

In the next section we focus on the incident during lesson one, when language was not a transparent resource.

When language was visible but not simultaneously invisible

As indicated earlier when a resource is too visible it distracts attention from the subject matter. The extract below indicates learners in the Tshivenda group interacting with each other on the answer to questions 3 and 4 in the task sheet, which stated as follows:

3. In winter the average electricity consumption increases by 20%, what would the monthly bills be for the family home in (1) above and for the townhouse?
4. In your opinion, what may be the reason for the increase in the average electricity consumption in (3) above?

In the extract they are trying to attend to what the questions are asking them to do but they are also dealing with the issue of language - what language to give the answer in. The extract suggests that the struggle with questions in this instance is worsened by their struggle first with understanding what the problem is asking them to do and second with finding the Tshivenda words to use.

1. Given: *Di ya benefita. Hapfa neh, kha summer, kha winter vha badhela seven hundred neh, kha townhouse. Then kha...kha mudi kha winter, kha botshibadhela one-fifty two, kha botshibadhela vhugai, one-forty. So vha benefitha ngavhugai? The amount...* [They benefit. Here, in summer... in winter they pay seven hundred for the townhouse. Then the household in winter they pay one-fifty two, and in summer how much do they pay, one-forty. So, how much do they benefit? The amount...]
2. Patience: (*Interrupts Given*) *Eya, mara I think vha budzisa huri vha kho inkhriza ngavhugai. Apfa vhobadhela one-fifty two and apfa summer vha kho vhadela one-forty.* [Yes but I think what is being asked is by how much will it increase. Here they pay one-fifty two and here in summer they pay one-forty.]
3. Given: *Ndikhongwala nga Tshivenda zwino.* [I am now going to write in Tshivenda.]
4. Sipho: *Eya, ngwala nga Tshivenda ngwananga* [Yes, write in Tshivenda baby!]
5. Given: (*Writes in tasksheet*) *Ndikhoneta nga English. Ritshi... kana ndimini u...u...u...* [I am tired of English. We say... by the way what is to...to...to...]
6. Sipho: *Khezwo! ngwala nga English* [Aha! write in English.]
7. Given: *Hae, kana ndimini u...* [No, by the way what is to...]
8. Patience: *Kwitani* [To do what?]
9. Sipho: *Inkhriza* [To increase?]

10. Given: *kutanga... kutanganisa, tshitangadzisa mudi kha...kha summer na mudi kha winter, ritshi mini? Mudi... Hu tanganisa ndimini ngaTshivenda* [To add... to add, we add the household in...in summer and in winter, what do we say? Household... What is to 'tanganisa' (add) in Tshivenda?]

11. Patience: *Ndingutanganisa* [Is to 'tanganisa' (add).]

During this interaction, language was made visible by Given in utterance 3 when she indicated that she is going to write the answer in Tshivenda. Until then language had been a transparent resource in the group as their focus was on providing answers to the questions in the task. By making language visible Given moved the learners' attention from the mathematical content of the task to the language. As we can see, from utterance 5, Given indicates that she is now going to write the answer in Tshivenda because she is "tired of English". This was responded to by Sipho in a seemingly sarcastic manner in utterance 4 by saying "yes, write in Tshivenda baby!" It is the inclusion of the word "baby" which suggests sarcasm in Sipho's tone. In utterance 6 Sipho responds to Given's struggle for a Tshivenda word for *increase* by saying "Aha! write in English". The inclusion of the word "aha!" in this case suggests that Sipho knew that Given would not be able to write the answer in Tshivenda. It is important to remember that these learners study their home languages as subjects at first language level. However, doing mathematics, which includes reading-writing-speaking-listening mathematically is something that they have never done in their home languages but only in English despite their limited fluency in it. So the challenge that Given is experiencing here is about doing mathematics in Tshivenda only, a constraint that she imposed on herself as Terence did not force them to give their solutions in their home languages only. According to Terence's instructions their answer would have still been accepted if it were given in English or in a mixture of English and Tshivenda.

In utterance 9 Sipho gives the transliterated word for increase, which is inkhriza, however Given does not accept it as he is looking for a pure Tshivenda term for increase. In utterance 11 Patience offers "tanganisa", this means 'add' and not 'increase'. The word for 'increase' in Tshivenda is 'huengedza', which literally means 'to make more'. It is evident from these learners' interactions in the above extract that even if learners know the answer, if they do not have the

words to express it then their answer may never be known. It needs to be noted that while these learners wanted to respond to the question in Tshivenda only, it was neither necessary nor required. With the invisibility of language, it is possible that these learners may not have felt the need to respond in one language only. They could have used both Tshivenda and English to write their answer. By insisting on using only Tshivenda, language was no longer an invisible resource but an obstacle they needed to overcome by for example, finding the pure Tshivenda term for increase to use in their response to the question.

It is important to note what precipitated this visibility of language in the extract above. In utterance 5 Given indicated that she is "tired of English". But why is Given tired of English? Why does she feel empowered to express her feelings against English now when they are supposed to be solving a problem? We raise these questions here not to answer them but to signal our view that the use of language as a transparent resource is not necessarily achieved overnight especially in multilingual classrooms such as in South Africa where the political role of language both historically and in the current practices cannot be ignored. In the section that follows we explore language as a transparent resource in the learners' reflections on the lessons.

Language as a transparent resource in the learners' reflections on the lessons

As indicated earlier, reflective individual interviews were conducted with learners after the first week of lessons. Terence selected four learners from different home language groups for interviewing so as to get their reflections on the lessons observed. The interviews were semi-structured and were conducted by Mampho Langa (the third author) who used to be the head of the mathematics department in the school. We decided that Mampho conduct the interviews because we thought that the learners would be free to talk frankly about their experiences of the lessons. In our view Mampho was the best person to conduct the interview also because she was not part of the team that collected the lesson observation data. Furthermore the learners were not only familiar with her but also with her position as head of the mathematics department and thus comfortable about talking with her about their mathematics learning. Learners selected languages they wanted to be interviewed in. The interviews were video recorded and then transcribed.

In analysing the learner interviews we again looked at what was visible, i.e. presences (what the learners were talking about) and what was invisible, i.e. absences (what they were not talking about). We expected that the two main changes that we introduced (language and nature of the task) would be most visible in what the learners talk about during the interviews. In the extracts below from interviews with individual learners the interviewer asked them the same open question about what was happening in their class:

Mampho: I understand this week you had visitors in your class, what was happening?

Sindiswa: Er…, we were learning a lesson in which we can calculate electricity er …. amount … er … the way in which the electricity department can calculate the amount of electricity unit per household.

Nhlanhla: *We were learning about how to calculate …er…er… kilowatts of the electricity, how do we … like … how can we calculate them and when … at …, Besifunda mem ukuthi ugesi udleka kakhulu nini.*
[We were learning about when there is high electricity consumption.]

Colbert: Er …we were just solving for electricity, kilowatt per hour, for comparing if they are using card or the meter, which is both, I think are the same.

Sipho: Er, the visitors they were doing research. *Gošho gore ba sheba gore bothata … bothata ba rona bo mo kae, ka … ka … maths, then they found out that er… ba bang ha ba understende dilanguage, like English so, then ha ba botsa karabo then they can't find the answer. So Mr Molefe then decided to … to … make it in … in English and vernacular language to … to …, for us to understand.*

Three of the learners above point to the mathematical task that they were working on during the lesson thus suggesting that is what they found as central to what was happening during the lesson. As explained earlier, the strategy we are exploring in this class centres around two principles: (1) the deliberate use of the learners' home languages, and (2) the selection of interesting and challenging mathematics tasks. Given these principles we had thought the use of the learners' home languages in the tasks given would be the most prominent thing for the learners to notice. We thought so because language is often referred to as one of the main constraints in teaching and learning mathematics in multilingual classrooms. What is emerging in the extracts above

is that for the learners the context of the task, cost of electricity, was more prominent.

Given our expectation that the learners would point to language as most prominent about what was different about the lessons, Mampho probed further as below:

Mampho: But what was so special about the lessons?

Sindiswa: It does not include those maths … maths. It is not different, but those words used in Maths didn't occur, didn't occur but we weren't using them. … Er … 'simplifying', 'finding the formulas', 'similarities', …

Nhlanhla: *Hayi, no mem, ku-different… Okokuqala mem, ilokhuza, la sidila ngama-calculations awemali, manje ku-maths asisebenzi ngemali.* [No mam, it is different. Firstly mam, we were working with money and usually in maths we do not work with money.]

Colbert: *Iya, basenzele in order to … ukuthi ibe simple and easy to us, because most of people, uyabona, aba-understendi like i … like i-card ne meter. Abanye bathi i-meter is … i-price yakhona i-much uyabona, i-card iless i-price yakhona, that's why uyabona. So, abantu abana-knowledge, uyabona, bakhuluma just for the sake of it. So, I think for us, because we have learnt something, both are the same.* [Yes, you see they made it easy for us, because most people do not understand, like card or using a meter. Some say when using the card you pay less than when using the meter, you see. So people do not have knowledge out there, they just talk for the sake of it. So I think, for us we have learnt something, both are the same.]

Sipho: *Gošho gore ba sheba gore bothata … bothata ba rona bo mo kae, ka … ka … maths, then they found out that er… ba bang ha ba understende dilanguage, like English so, then ha ba botsa karabo then they can't find the answer. So Mr Molefe then decided to … to … make it in … in English and vernacular language to … to …, for us to understand.* [They were looking at the problem… where our problem is, with… with… maths, then they found that er … some of us do not understand languages like English, so when they ask for the answer we can't find it. So Mr. Molefe then decided to… to … make it in English and vernacular language to … to… for us to understand.]

In responding to Mampho's question above, both Sindiswa and Sipho point to language. Sindiswa points to how the task differs from the textbook tasks that they are used to. Sindiswa says of the observed lesson, that the absence of many of the terms usually associated with the mathematics classroom was significant, even though the essence of the lesson and activity remained unchanged. It is evident that for Sindiswa language played a clear role in the "feel" of the lesson. This is echoed by Sipho, whose response and choice of language is very interesting. What stood out for him about the lessons was what the introduction of learners' home languages allowed learners to do. It changed the dimensions of the interaction, increased participation and intervened at the level of meaning. Noticeably, he does not say "ha ba botsa karabo then *they don't know* the answer" ["when they ask for the answer *they don't know* the answer"], he says "ha ba botsa karabo then *they can't find* the answer" ["when they ask for the answer *they can't find* the answer"]. In Sipho's analysis, the learners may *have* the answer in one language, but their inability to *find* it in another language (English) has direct effects on their participation and performance.

On the face of it, Sindiswa and Sipho address different aspects of the changed lesson. However, both highlight the manner in which the use of language (or the absence of certain kinds of language) can either enable comprehension or constrain learning. Both see the actual mathematical activity as unchanged. For Sindiswa, when "difficult" words are minimised, then learners and teachers can get on with the usual business of mathematics, focusing on the task and allowing learners to experience mathematics differently and more fluently.

Nhlanhla and Colbert point to the nature of the task. For Nhlanhla what stood out the most is the fact that in mathematics they usually do not deal with calculations involving money and so these lessons were special because they involved money calculations. This resonates with Colbert's focus on the value of the task beyond the lesson. For him it was about clarifying a real life situation that he never understood – the fact that the cost for electricity will ultimately be the same in both costing structures. What Colbert is referring to is his learning about two different costing systems for electricity as described in the problem. In his view both options end up costing the same. While Colbert's analysis of the task is mathematically incorrect, it is clear that the context of the task presented a real life problem that, as he says, people in his neighbourhood have been arguing about. Looking at the graphs below illustrating the cost of electricity for the two options, it is clear

that they intersect at the point (800, 200), which means that if electricity consumption is more than 800 kWh then the cost of electricity will be cheaper when using the first costing structure.

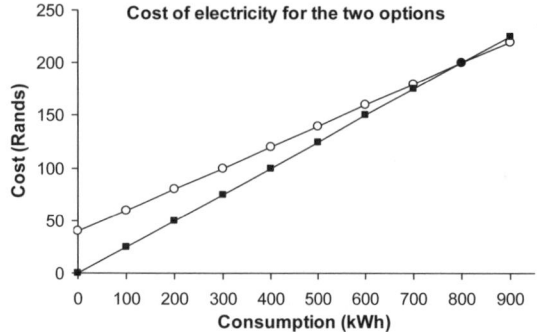

Given the learners' seeming reluctance to talk about the fact that their home languages were used, Mampho asked them a direct question about the way in which their home languages were used in the task:

Mampho: I understand that the tasks that were given were written in both your home language and English. Tell me about that.

Sindiswa: It was fine. It was just the same. It was the same as doing it in English, because I understand both languages.

Nhlanhla: *I think mem leyo kusinikeza amaphepha o i-two kuya nceda mem, ngoba, like mina, kukhona amanye ama-questions bengingawa-understandi, i-home language iyakhona ukusiza ukuthi ngiwa understande.*
[Mam I think that one of giving us tasks in two languages is very helpful mam, because like there are questions that I did not understand and my home language helped me understand.]

Colbert: *Iya, I think is a good idea, uyabona, ngoba iyenza ukuthi … iyenze izinto zibe simple, ngoba if singa-understendi ngeEnglish, sicheka ku … our languages, aba simple bese siyakhomphera.*
[Yes, I think it is a good idea, you see it makes things simple because if we do not understand in English we check in our home languages and it is simple because we can compare.]

Sipho: Iya, I did understand in English and vernac. I did benefit.

From the learners' responses above it is clear that none of them experienced the use of their home languages as a distracter or constraint. In fact Nhlanhla and Colbert explained that having their home language versions was helpful. The silences and presences in the learner interviews are interesting. We find it interesting that the interviewer had to explicitly raise the issue of language for the learners to talk about it. This to us suggests the transparency of language as a resource. While the home languages were visible in the sense that the learners were for the first time given written text during the mathematics lessons in their home languages, they are also invisible in that they are not distracting the learners' attention from the mathematics tasks they are doing. The learners are not focusing on the languages but on the mathematics of the task. As Lave and Wenger (1991) argue, for a resource to be useful it needs to be both visible and invisible. In their view the invisibility is in the form of unproblematic interpretation and integration (of the artefact – in this case the translated versions of the task) and visibility in the form of extended access to information. While the unusual use of the learners' home languages in the task can be noticed and used, when invisible, it did not distract learners from the task. The learners were at liberty to choose which language version they wanted to refer to at any time. This, we argue, contributes to the relevance of the strategy we are exploring. The learners are given an opportunity to draw on the linguistic resources they have, and at the same time the presence of English assures them of the fact that they are not loosing access to the language of power, which they so much want to gain access to.

Conclusion

> Mathematics education begins in language, it advances and stumbles because of language, and its outcomes are often assessed in language. (Durkin, 1991, p 3)

The above quotation captures the important role of language as a resource in the teaching and learning of mathematics. While it is a resource that can help advance mathematics learning, it can also be a stumbling block for successful learning. The major challenge in multilingual classrooms in South Africa is the fact that while the power of English is unavoidable, many learners do not have the level of fluency that enables them to engage in mathematical tasks set in English. In this paper we have explored a strategy for using language as a transparent resource in the teaching and learning mathematics in multilingual classrooms. This strategy is guided by two main principles – the deliberate, proactive and strategic use of the learners' main languages and the selection of real life, interesting and high cognitive demand mathematics tasks. Our analysis shows that with this strategy language becomes a visible but invisible resource in the sense that while learners can draw on different languages at any time they want, language is also invisible because it does not disturb their focus on the mathematics. We argue that our proposed strategy recognizes the political nature of English and thus while it draws on the learners' home languages, it does not present them as being in opposition to English but as working together with English to make mathematics more accessible t the learners.

Through our exploration of this strategy in this study, we became more and more aware of the challenge of translating tasks into multiple languages. Translation is never a straight-forward enterprise, it is complex. As multilingual speakers of languages from different conceptual worlds we know from experience of living in language, what monolinguals know theoretically from training, that much loss and distortion of meaning can occur in translation. Yet translation is part of how meaning is transferred, made and re-negotiated; therefore, this aspect of linguistic activity remains an important consideration. This is why the deliberate, proactive and strategic use of the learners' home languages pedagogically is so important. Much remains to be done!

Acknowledgements

The work presented in this paper was supported by a grant from the National Research Foundation (TTK2007051500040). Any ideas expressed are, however, those of the authors and therefore the National Research Foundation does not accept any liability. We are grateful to the learners in Terence's class as well as the school management for agreeing to participate in this study.

References

Adendorff, R. (1993). Code-switching amongst Zulu-speaking teachers and their pupils. *Language and Education, 7*(3), 141-162.

Adler, J. (1998). A language of teaching dilemmas: Unlocking the complex multilingual secondary mathematics classroom. *For the Learning of Mathematics, 18*, 24-33.

Adler, J. (2001). *Teaching mathematics in multilingual classrooms*. Dordrecht: Kluwer Academic Publishers.

Arthur, J. (1994). English in Botswana primary classrooms: Functions and constraints. In C. M. Rubagumya (Ed.), *Teaching and researching language in African classrooms*. Clevedon: Multilingual Matters.

Baker, C. (1993). *Foundations of bilingual education and bilingualism*. Clevedon: Multilingual Matters.

Clarkson, P. C. (1991). *Bilingualism and mathematics learning.* Geelong: Deakin University Press.

Dawe, L. (1983). Bilingualism and mathematical reasoning in English as a second language. *Educational Studies in Mathematics, 14*(1), 325-353.

Department of Education (1997). *Language-in-education policy.* Pretoria: Department of Education.

Durkin, K. (1991). Language in mathematical education: An introduction. In K. Durkin & B. Shire (Eds.), *Language in mathematical education: Research and practice* (pp. 1-3). Milton Keynes: Open University Press.

Granville, S., Janks, H., Joseph, M., Mphahlele, M., Ramani, E., & Watson, P. (1998). English without g(u)ilt: A position paper on language in education policy for South Africa. *Language in Education, 12*(4), 254-272.

Grosjean, F. (1985). The bilingual as a competent but specific speaker-hearer. *Journal of Multilingual and Multicultural Development, 6*(6), 467-477.

Gutstein, E. (2003). Teaching and learning mathematics for social justice in an urban, Latino school. *Journal for Research in Mathematics Education, 34*, 37-73.

Howie, S. J. (2003). Language and other background factors affecting secondary pupils' performance in mathematics in South Africa. *African Journal of Research in Mathematics, Science and Technology Education, 7, 1-20.*

Howie, S. J. (2004). A national assessment in mathematics within international comparative assessment. *Perspectives in Education, 22*(2), 149-162.

Khisty, L. L. (1995). Making inequality: Issues of language and meaning in mathematics teaching with Hispanic students. In W. G. Secada, E. Fennema, & L. B. Abajian (Eds.), *New directions for equity in mathematics education.* Cambridge: Cambridge University Press.

Lave, J., & Wenger, E. (1991). *Situated learning: Legitimate peripheral participation.* Cambridge: Cambridge University Press.

Martin-Jones, M. (1995). Code-switching in the classroom: Two decades of research. In L. Milroy & P. Muysken (Eds.), *One speaker, two languages* (pp. 90-112). Cambridge: Cambridge University Press.

Merritt, M., Cleghorn, A., Abagi, J. O., & Bunyi, G. (1992). Socialising multilingualism: Determinants of code-switching in Kenyan primary classrooms. *Journal of Multilingual and Multicultural Development, 13*(1 & 2), 10-121.

Moschkovich, J. (1996). Learning in two languages. In L. Puig & A. Gutierezz (Eds.), *Proceedings of the 20th Conference of the International Group for the Psychology of Mathematics Education* (Vol. 4, pp. 27-34). Valencia: Universitat De Valencia.

Moschkovich, J. (1999). Supporting the participation of English language learners in mathematical discussions. *For the Learning of Mathematics, 19*(1), 11-19.

Moschkovich, J. (2002). A situated and sociocultural perspective on bilingual mathematics learners. *Mathematical Thinking and Learning, 4*, 189-212.

Ncedo, N., Peires, M., & Morar, T. (2002). *Code-switching revisited: The use of languages in primary school science and mathematics classrooms.* Paper presented at the Tenth Annual Conference of the Southern African Association for Research in Mathematics, Science and Technology Education. Durban.

Rakgokong L (1994) *Language and the construction of meaning associated with division in primary mathematics.* Paper presented at the 2nd Annual Meeting of the Southern African Association for Research in Mathematics and Science Education.

Sachs, A. (1994). *Language rights in the new constitution.* Cape Town: South African Constitutional Studies Centre, University of the Western Cape.

Secada, W. (1992). Race, ethnicity, social class, language and achievement in mathematics. In D. A. Grouws (Ed.), *Handbook of research on mathematics teaching and learning* (pp. 623-660). New York: NCTM.

Setati, M. (1998). Code-switching in a senior primary class of second language learners. *For the Learning of Mathematics, 18*(2), 114-160.

Setati, M., & Adler, J. (2000). Between languages and discourses: Language practices in primary multilingual mathematics classrooms in South Africa. *Educational Studies in Mathematics, 43*, 243-269.

Setati, M. (2005). Teaching mathematics in a primary multilingual classroom. *Journal for Research in Mathematics Education*, 36(5), 447-466.

Setati (in press). Power and access in multilingual mathematics classrooms in South Africa. *South African Journal of Education.*

Stein, M., Smith, M., Henningsen, M., & Silver, E. (2000). *Implementing standards-based mathematics instruction: A casebook for professional development.* Reston: NCTM.

Stephens, M., Waywood, A., Clarke, D., & Izard, J. (Eds.) (1993). *Communicating mathematics: Perspectives from classroom practice and current research.* Victoria: Australian Council for Educational Research.

Science in Africa (2003). *Why don't kids learn maths and science successfully?* Retrieved 10 April 2007 from http://www.scienceinafrica.co.za/2003/june/maths.htm

Zepp, R. (1989). *Language and mathematics education.* Hong Kong: API Press.

Chapter 5

Building Young Children's Mathematics

Douglas H. Clements and Julie Sarama
University of Denver

Peter could count beyond 120 and state the number word before or after any given number word, including those in the hundreds. He could use counting strategies to solve a wide range of addition and subtraction tasks.

Tom could not count. The best he could do is say "two" for a pair of objects. Asked for the number after "six," he said "horse." Asked for the number after one, he said "bike."

Both Peter and Tom were beginning their kindergarten year. (adapted from Wright, 1991, pp. 11–12)

A large and damaging gap lies between young children growing up in higher-resource (middle to high socioeconomic status, or SES) and in lower-resource (low SES) communities. Both the income gap and the achievement gap have been increasing for decades (Reardon, 2011). For children from lower-SES communities especially, the long-term success of their learning and development requires high-quality school experience, especially in mathematics. What they know when they enter kindergarten predicts their mathematics achievement for years to come—*throughout their school career* (National Mathematics Advisory Panel, 2008). Moreover, what they know in math predicts their reading achievement later, even better than early literacy skills do (Duncan & Magnuson, 2011). Mathematics is a core component of cognition.

To help these children and their teachers, we developed a curriculum for early mathematics and a model for supporting teachers in implementing it. We conducted a series of research-and-development projects to evaluate and improve our efforts. In this chapter, we briefly describe the gaps in children's learning, the curriculum, the model, and the research studies. Each of these is a tool teachers can use to help children like Tom start school successfully.

Equity and Early Mathematics

Children from low-income families, as well as those from certain racial/ethnic groups and those who are English language learners (ELL), show specific difficulties in early mathematics (for more details, see Clements & Sarama, 2014; Sarama & Clements, 2009). As one example, the ECLS-B (Early Childhood Longitudinal Study) found that 87 percent of children in higher-SES families demonstrated proficiency in numbers and shapes, but only 40 percent of children in lower-SES families did so (Chernoff, Flanagan, McPhee, & Park, 2007).

The competencies these children lack and do not learn in school are not so much lower-level skills as more sophisticated concepts and problem-solving competencies. For example, they do not understand the relative magnitudes of numbers and how they relate to the

---

This chapter is adapted from two sources: D. H. Clements & J. Sarama (2007), Effects of a preschool mathematics curriculum: Summative research on the Building Blocks Project, *Journal for Research in Mathematics Education, 38*, 138–163, and J. Sarama & D. H. Clements (2003), Building blocks of early childhood mathematics, *Teaching Children Mathematics, 9*, 480–484. The *TCM* article is available at nctm.org/more4u.

counting sequence (Griffin, Case, & Siegler, 1994). They have more difficulty solving addition and subtraction problems (Jordan, Kaplan, Locuniak, & Ramineni, 2006). In one study, 75 percent of children in an upper-middle-class kindergarten were capable of judging the relative magnitude of two different numbers and performing simple mental additions, compared to only 7 percent of lower-income children from the same community (Griffin et al., 1994). As another example, about 72 percent of high-SES groups, 69 percent of middle-SES groups, and 14 percent of low-SES groups can answer this orally presented problem: "If you had 4 chocolate candies and someone gave you 3 more, how many chocolates would you have altogether?" Low-income children often guess or use other maladaptive strategies, such as simple counting (e.g., 3 + 4 = 5) (Siegler, 1993). In geometry, there were no significant differences between income groups on the simple tasks involving shape and comparison of shapes, but there were significant differences on *representing* shapes, *composing* shapes, and *patterning.*

The good news is that given more experience, lower-income children can learn to use concepts and strategies with the same accuracy, speed, and adaptive reasoning as middle-income children (Siegler, 1993). High-quality preschool mathematics activities can help children who start at a disadvantage catch up with their peers, but *most programs in use today do not help close this gap* (for a review, see Clements & Sarama, 2014).

The Building Blocks Curriculum	*Building Blocks* (Clements & Sarama, 2007a, 2013) is a National Science Foundation–funded mathematics project that designs curriculum using a comprehensive Curriculum Research Framework (Clements, 2007) to build numeric/quantitative and geometric/spatial ideas and skills. Woven throughout are mathematical subthemes, such as sorting and sequencing, as well as mathematical processes. General processes include communicating, reasoning, representing, problem solving, and the overarching mathematizing. Specific mathematical processes include number and shape composition and patterning. These processes were determined to be critical mathematical building blocks (see *Curriculum Focal Points* [National Council of Teachers of Mathematics, 2006]). The instructional approach of *Building Blocks* is to *find the mathematics in, and develop mathematics from, children's everyday activities.* Children are guided to extend and *mathematize* their everyday activities, from block building to art to songs to puzzles, through sequenced, explicit activities (whole group, small group, centers, including a computer center, and "throughout the day"). Thus, off-computer and on-computer activities are designed based on children's experiences and interests, with an emphasis on supporting the development of mathematical activity.

Perhaps most important, *Building Blocks* achieves these goals through the use of *learning trajectories.* Learning trajectories have three parts: a specific mathematical goal, a path along which children develop to reach that goal, and a set of instructional activities that help children move along that path. Teachers who understand learning trajectories understand the math, the way children think and learn about math, and how to help children learn it better. Learning trajectories connect research and practice. They connect children to math. They connect teachers to children. They help teachers understand the level of knowledge and thinking of their classes and the individuals in their classes as key in serving the needs of all children.

The article "Building Blocks of Early Childhood Mathematics" (Sarama & Clements, 2003; available at nctm.org/more4u) describes the program, and how it uses learning trajectories, in more detail (in particular, see a learning trajectory for geometry, p. 482.) Also, the reader can find a learning trajectory for early addition in our 2007 *JRME* article (Clements & Sarama, 2007b, p. 141; another view of the shape composition learning trajectory is on pp. 144–145). Finally, all of the learning trajectories are available in a recent book (Clements & Sarama, 2014).

The user reads the description that appears on the right. If she chooses "More info" the screen "slides over" to reveal the expanded view shown below.

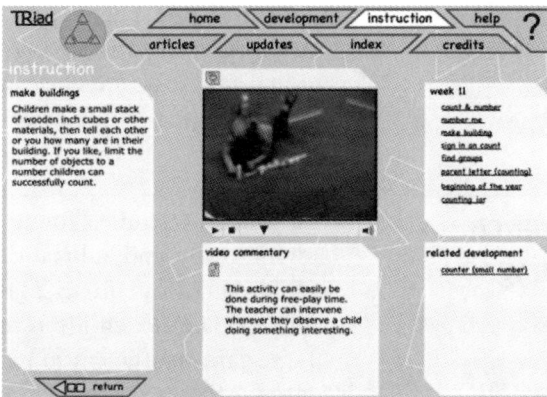

Here the user can see multiple video examples, with commentary. Clicking on the related developmental level (child's level of thinking) yields the view on the next page.

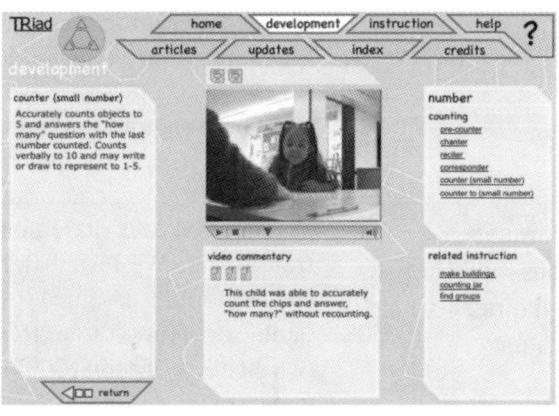

This developmental view also provides a description, video, and commentary on the developmental level. The video here is of a clinical interview task in which a child displays that *level of thinking*.

Fig. 5.1. Sample screens from the *Building Blocks Learning Trajectories (BBLT) web application*

Note: We are redesigning and reprogramming BBLT to be available to all teachers on most modern devices (e.g., iPhones and iPads), funded by the Heising-Simons Foundation. Additional information on the learning trajectories can be found in Clements and Sarama (2014).

Support for Teachers: The TRIAD Model

Although designed to support teachers' learning and implementation, *Building Blocks* was not intended to be used in isolation from professional development. Our TRIAD (Technology-enhanced, Research-based, Instruction, [formative] Assessment, and [professional] Development) model has ten research-based guidelines (for a full description, see Sarama & Clements, 2013); here we focus just on professional development. We provided thirteen days of professional development across two years focused on the learning trajectories for each mathematical topic. To understand the goals, teachers learned core mathematics concepts and procedures for each topic. To understand the developmental progressions of levels of thinking, teachers studied multiple video segments illustrating each level and discussed the mental "actions on objects" that constitute the defining cognitive components of each level. To understand the instructional tasks, teachers studied the tasks, and they viewed, analyzed, and discussed video of the enactments of these tasks in classrooms. Each of these professional development contexts used the software application Building Blocks Learning Trajectories (BBLT), which presents and connects all components of the innovation (see fig. 5.1 and www.TRIADScaleUp.org).

The professional development sessions were sequenced following the *Building Blocks* curriculum. Throughout this study, teachers learned how to use the learning trajectories as a basis for formative assessment, which is a key to high-quality teaching (see National Mathematics Advisory Panel, 2008). Formative assessment has been shown to be particularly difficult for teachers to enact without substantial support. In this case, study participants discussed and practiced how to interpret children's thinking and select appropriate instructional tasks for the class and for individuals. In addition, coaches observed and provided in-class support to teachers and completed implementation fidelity evaluations.

Research Findings

A series of studies found that the *Building Blocks* curriculum and the TRIAD model significantly and substantially increase the mathematics knowledge of low-SES preschool children. In our first, small-scale study (Clements & Sarama, 2007b) in four classrooms, *Building Blocks* children increased significantly more than the comparison children, with large gains (sufficient to raise a child at the 50th percentile to the 84th to 98th percentile) as large as those with individual tutoring. The curriculum was especially effective in supporting children's learning of often-neglected competencies such as subitizing, sequencing, composing shapes, and using more sophisticated numerical strategies and spatial imagery. The message for teachers is that even a small number of experiences (four to six sessions) focusing on these often-ignored topics can have large benefits for children.

Why was it successful? How can the results help teachers?

We believed that these positive effects resulted from *Building Blocks'* core use of learning trajectories. We tested that assumption in a larger study, in which thirty-six classrooms were randomly assigned to the *Building Blocks,* to another preschool curriculum that used a "unit" structure (rather than learning trajectories), or to "business-as-usual" control groups (Clements & Sarama, 2008). The *Building Blocks* group increased significantly more than the comparison group score (a rise from the 50th to about the 68th percentile) and the control group (a rise to about the 85th percentile), supporting the contention that learning about all three components of learning trajectories—the mathematics (goal), children's learning and thinking (the developmental progression), and teaching (the instructional tasks and strategies)—helped teachers become more effective.

A final study confirmed that the *Building Blocks* curriculum—along with the TRIAD model—could be successfully scaled up (Clements, Sarama, Spitler, Lange, & Wolfe, 2011). Preschoolers from 106 classrooms in two states again made larger gains than children in control classrooms (a rise from the 50th to about the 77th percentile). Again, *Building Blocks* children were relatively more successful on challenging mathematical topics and competencies.

This study had two other important findings. First, African American children in the control averaged *lower* gains than other control children; in *Building Blocks*/TRIAD classrooms, African American children averaged higher gains than other children. Curricula built on learning trajectories may be particularly effective in increasing the expectations for

African American children, as teachers see their untapped potential for learning mathematics (National Mathematics Advisory Panel, 2008).

Second, other analyses showed that three specific characteristics of the *Building Blocks/*TRIAD classrooms accounted for increased learning.

1. The more teachers used the *Building Blocks* software, the more their children learned.

2. Similarly, children learned more in classrooms with a positive classroom culture, which includes signs of mathematical activity and teachers who are knowledgeable and enthusiastic about mathematics and who frequently interact with and respond to children.

3. The more separate mathematics activities that a teacher conducted (how long they were mattered less), the more children learned.

Does "more math" mean "less literacy and language learning"?

The preschool teachers loved their children's math learning, but some worried that their coordinators and principals cared more about literacy scores. Would the focus on mathematics take away from language and literacy learning? To check, we assessed children in these areas (Sarama, Lange, Clements, & Wolfe, 2012). To our relief, literacy scores did not differ between the *Building Blocks* and control groups. To our delight, language scores were higher in the *Building Blocks* group. In an assessment of their expressive language—retelling a story—they scored significantly higher on the number of key vocabulary words they used, their use of grammatically complex sentences, their willingness to reproduce narratives independently, and their inferential reasoning. One feature of *Building Blocks* mathematics is talking about mathematics, including explaining your thinking. For example, when a child identifies a square out of a set of shapes, a *Building Blocks* teacher would ask, "How do you know it's a square?" At the beginning of the school year, young children often answer such a question with simplistic responses, such as, "Because it looks like a square" or "I thought it in my head." As the curriculum progresses and the children experience repeated invitation to explain their thought processes, even young children are supported in their ability to give accurate, reasoned responses, such as, "Because it has four sides and all square corners."

Do effects persist? The importance of follow-through in the primary grades

Some studies show that mathematics programs for young children make a meaningful and positive difference initially, but that the effects appear to "fade" over time. Skeptics have suggested that the effort of implementing these programs is not worthwhile if the effects fade. Although we need more research to illuminate this issue, we believe this negative point of view ignores the existing evidence. Programs that are continued into elementary school and that offer substantial exposure to early interventions have the most persistent long-term effects (Brooks-Gunn, 2003). Without follow-through, short-term early interventions cannot realistically be expected to last indefinitely. This is especially so because most children at risk attend the lowest-quality schools. It would be surprising if these children did not gain less than their more advantaged peers year by year. Our TRIAD projects support research, already reviewed, showing that most teachers of young children are not responsive to the needs of those with high mathematical knowledge. Thus, the mathematics-experienced preschoolers who go to kindergarten are often given tasks that do not challenge—or teach—them. Their development is stalled because no new mathematics is offered to them.

Therefore, we randomly assigned schools to three groups (Clements, Sarama, Wolfe, & Spitler, 2013; Sarama & Clements, 2013; Sarama, Clements, Wolfe, & Spitler, 2012). In pre-kindergarten, the two experimental interventions were identical (so we combined them in the results reported in the previous section), but only one of the two included follow-through in the kindergarten and first grade years, including knowledge of the pre-K intervention and ways to build upon that knowledge using learning trajectories. Moving into kindergarten and first grade is where the two experimental groups differed. Only the "TRIAD Follow-Through" teachers learned about learning trajectories and how to use them to help their students build on the pre-K gains in mathematics (see Clements & Sarama, 2014). The result was that, although students in both experimental groups scored significantly higher than control students, the gains were less for the TRIAD group without the follow-through. Being in the TRIAD non-follow-through group raised the average child from the 50th to about the 61st percentile and being in the Follow-Through group raised the child to the 69th percentile. The difference between TRIAD Follow-Through and non-follow-through students was statistically significant. Further, TRIAD Follow-Through students scored significantly higher than non-follow-through students (to about the 60th percentile).

Support from other studies

Separate researchers confirmed the effectiveness of the *Building Blocks* curriculum. One found the largest positive effects on pre-K mathematics taken to scale (Weiland & Yoshikawa, 2012). Given this study was independent of the developers, it confirms that teachers throughout a district can successfully develop children's mathematical learning.

Final Words

Early mathematics is surprisingly important, especially to children from low-resource communities, from underrepresented racial/ethnic groups, and who are English language learners (ELL) —children such as Tom, described at the beginning of this chapter. Research such as that described here can help teachers build classroom environments and activities that can help all young children begin a joyful and successful trajectory in learning mathematics, with benefits in language as well.

References

Brooks-Gunn, J. (2003). Do you believe in magic? What we can expect from early childhood intervention programs. *Social Policy Report, 17*(1), 1, 3–14.

Chernoff, J. J., Flanagan, K. D., McPhee, C., & Park, J. (2007). Preschool: First findings from the third follow-up of the early childhood longitudinal study, birth cohort (ECLS-B) (NCES 2008-025). Washington, DC: National Center for Education Statistics.

Clements, D. H. (2007). Curriculum research: Toward a framework for "research-based curricula." *Journal for Research in Mathematics Education, 38*, 35–70.

Clements, D. H., & Sarama, J. (2007a). *Building Blocks—SRA Real Math teacher's edition, grade preK*. Columbus, OH: SRA/McGraw-Hill.

Clements, D. H., & Sarama, J. (2007b). Effects of a preschool mathematics curriculum: Summative research on the *Building Blocks* project. *Journal for Research in Mathematics Education, 38*, 136–163.

Clements, D. H., & Sarama, J. (2008). Experimental evaluation of the effects of a research-based preschool mathematics curriculum. *American Educational Research Journal, 45*, 443–494.

Clements, D. H., & Sarama, J. (2013). *Building Blocks, Volumes 1 and 2*. Columbus, OH: McGraw-Hill Education.

Clements, D. H., & Sarama, J. (2014). *Learning and teaching early math: The learning trajectories approach* (2nd ed.). New York, NY: Routledge.

Clements, D. H., Sarama, J., Spitler, M. E., Lange, A. A., & Wolfe, C. B. (2011). Mathematics learned by young children in an intervention based on learning trajectories: A large-scale cluster randomized trial. *Journal for Research in Mathematics Education, 42*, 127–166.

Clements, D. H., Sarama, J., Wolfe, C. B., & Spitler, M. E. (2013). Longitudinal evaluation of a scale-up model for teaching mathematics with trajectories and technologies: Persistence of effects in the third year. *American Educational Research Journal, 50*, 812–850. doi:10.3102/0002831212469270

Duncan, G. J., & Magnuson, K. (2011). The nature and impact of early achievement skills, attention skills, and behavior problems. In G. J. Duncan & R. Murnane (Eds.), *Whither opportunity? Rising inequality and the uncertain life chances of low-income children* (pp. 47–70). New York, NY: Russell Sage Press.

Griffin, S., Case, R., & Siegler, R. S. (1994). Rightstart: Providing the central conceptual prerequisites for first formal learning of arithmetic to students at risk for school failure. In K. McGilly (Ed.), *Classroom lessons: Integrating cognitive theory and classroom practice* (pp. 25–49). Cambridge, MA: MIT Press.

Jordan, N. C., Kaplan, D., Locuniak, M. N., & Ramineni, C. (2006). Predicting first-grade math achievement from developmental number sense trajectories. *Learning Disabilities Research and Practice, 22*(1), 36–46.

National Council of Teachers of Mathematics. (2006). *Curriculum focal points for prekindergarten through grade 8 mathematics: A quest for coherence.* Reston, VA: Author.

National Mathematics Advisory Panel. (2008). *Foundations for success: The final report of the National Mathematics Advisory Panel.* Washington DC: U.S. Department of Education, Office of Planning, Evaluation and Policy Development.

Reardon, S. F. (2011). The widening academic achievement gap between the rich and the poor: New evidence and possible explanations. In G. J. Duncan & R. Murnane (Eds.), *Whither opportunity? Rising inequality, schools, and children's life chances* (pp. 91–116). New York, NY: Russel Sage Press.

Sarama, J., & Clements, D. H. (2003). Building blocks of early childhood mathematics. *Teaching Children Mathematics, 9,* 480–484.

Sarama, J., & Clements, D. H. (2009). *Early childhood mathematics education research: Learning trajectories for young children.* New York, NY: Routledge.

Sarama, J., & Clements, D. H. (2013). Lessons learned in the implementation of the TRIAD scale-up model: Teaching early mathematics with trajectories and technologies. In T. G. Halle, A. J. Metz, & I. Martinez-Beck (Eds.), *Applying implementation science in early childhood programs and systems* (pp. 173–191). Baltimore, MD: Brookes.

Sarama, J., Clements, D. H., Wolfe, C. B., & Spitler, M. E. (2012). Longitudinal evaluation of a scale-up model for teaching mathematics with trajectories and technologies. *Journal of Research on Educational Effectiveness, 5*(2), 105–135.

Sarama, J., Lange, A., Clements, D. H., & Wolfe, C. B. (2012). The impacts of an early mathematics curriculum on emerging literacy and language. *Early Childhood Research Quarterly, 27,* 489–502. doi:10.1016/j.ecresq.2011.12.002

Siegler, R. S. (1993). Adaptive and non-adaptive characteristics of low income children's strategy use. In L. A. Penner, G. M. Batsche, H. M. Knoff, & D. L. Nelson (Eds.), *Contributions of psychology to science and mathematics education* (pp. 341–366). Washington, DC: American Psychological Association.

Weiland, C., & Yoshikawa, H. (2012). *Impacts of BPS K1 on children's early numeracy, language, literacy, executive functioning, and emotional development.* Paper presented at the School Committee, Boston Public Schools, Boston, MA.

Wright, B. (1991). What number knowledge is possessed by children beginning the kindergarten year of school? *Mathematics Education Research Journal, 3*(1), 1–16.

Chapter 6

A Letter to Those Who Dare Teach Mathematics for Social Justice

Eric "Rico" Gutstein
University of Illinois—Chicago

With every single thing about math that I learned came something else. Sometimes I learned more of other things instead of math. I learned to think of fairness, injustices and so forth everywhere I see numbers distorted in the world. Now my mind is opened to so many new things. I'm more independent and aware. I have learned to be strong in every way you can think of. (Lupe, Chicago, grade 8)

Dear Teachers of Mathematics,

I'm writing this as a letter to you who work in the classroom every day with your own self-contained room, or your 150 high school students, or whatever are your conditions. My hat is off to you for doing this work because teachers are part of our collective heroes (and in my town, Chicago, their union is a hero too). I've been teaching one way or another for the past forty-plus years, and my mother was a teacher in my neighborhood junior high school (which my father attended!) in inner-city New York (the west edge of Harlem), where I grew up.

It's been twelve years since I wrote "Teaching and Learning Mathematics for Social Justice in an Urban, Latino School" (Gutstein, 2003), so I ask you to reflect back with me. In terms of equity and justice, some things have improved, while others have worsened. For example, way too much testing, with too-high stakes, is still the norm, but now, districts and principals are not only assessing your students with them, they're even using tests to evaluate you, their teachers. Merit pay turns teachers against each other in a competitive race to somewhere over someone, and you're judged on your students' test scores. For those of us who are teacher educators, we're soon to be "graded" on the test scores of the students of our students—talk about a dubious chain of causality. Across the United States, the Common Core is breathing down everyone's neck, especially yours, despite many questions and much resistance, and without genuine support for teachers and real money for education, which constrains you and hurts children. Education privatization is a big issue, though it's concentrated in larger cities through charter school expansion. Around the United States, especially in places where mayors appoint school boards (such as Philadelphia, Chicago, Detroit, and New York), districts close huge numbers of neighborhood public schools in low-income communities of color, predominantly African American ones. Chicago broke the U.S. record in 2013, closing fifty schools—more than a thousand teachers lost their jobs—while Philadelphia closed fewer schools but an even greater percentage of them. This is linked to the destabilization of working-class communities, because neighborhood schools are neighborhood anchors, and closings contribute to gentrification and massive displacement of low-income people of color from major cities. Teacher bashing is rising, pensions are being chopped (without hers, my mother would be broke), unions are being broken (Michigan, Wisconsin), and children are being treated as commodities by corporate oligarchs and hedge fund

This chapter is adapted from E. Gutstein (2003), Teaching and learning mathematics for social justice in an urban, Latino school, *Journal for Research in Mathematics Education, 34*, 37–73.

investors as charter schools enter global financial markets. You're under attack as a professional, and you're under attack as a worker. Both.

These are some of the conditions we live under today, and we can't just close our classroom doors and ignore them. This doesn't even begin to touch on the context outside education—poverty, racism, lack of affordable health care, economic stagnation, which all seriously impede learning—and, most pressing of all, the very survival of our planet, given today's ecological crises. And yet, we still talk about teaching (math) for social justice, however we interpret that. In fact, I'm arguing that given the conditions of work and life, we have no choice. On many levels, things are beyond serious. There are many dangers ahead, but at the same time, opportunities—people are looking for answers and there is an openness that we haven't seen for decades. Teaching mathematics for social justice can—and should—play its part. I like historian Howard Zinn's (1994) words, "you can't be neutral on a moving train," to capture our political responsibilities as educators in the current moment.

Teaching Math for Social Justice

When I wrote the original article, I proposed that "teaching math for social justice" meant that we teach students to learn and use mathematics to develop a sociopolitical consciousness of the roots of injustice in their lives and broader society, so they can eventually act to change the things they believe are wrong. Paulo Freire (1994) called this "reading [understanding] and writing [changing] the world," and we can do this with math, even in K–12 public schools. I wanted students to study their social reality with math and ask questions—their questions—about why things were the way they were. Students can make meaning of and begin to answer their questions, in part, through using mathematics. When students want to know how racial profiling works, why farm workers' wages have stagnated for thirty years, or whether their families and neighbors will be able to stay in their community—and they discover that mathematics can help them comprehend things they care about—then they take a different perspective on why they should learn math. We learn it to read our worlds, to make sense of what's happening around us, and to begin to see ourselves as "subjects" of history who shape the present and future (and even act, when and where appropriate for young people).

In my article, I discussed my seventh- and eighth-grade class at "Rivera," a school in a low-income/working-class Mexican immigrant community in Chicago. I was a full-time university professor, so I taught only one class. I had those students for almost two years, and most of the time (about 75 to 80 percent), I used the *Mathematics in Context* curriculum (MiC) (NCRMSE & FI, 1997–98), aligned with the NCTM Standards. For the rest of the time, I wove in "real-world projects" in which students used math to investigate issues they cared about—immigration, gentrification, working conditions for tomato pickers, wealth inequality, racism in house pricing, disproportionate world map projections, and much more. I, not students (see below), chose all these contexts, even though the students related to the projects because of their own experiences dealing with injustice. I wrote about the need to go beyond mathematics, even in math class, and to co-create, with students, a space that allowed us to really explore and discuss things that were sometimes painful but that we could begin to understand through mathematical analyses.

After two years in that class, students did begin to read and write the world with mathematics—which included learning math. But it wasn't an easy or linear process, or (obviously) the same for everyone. I learned much along the way, fell on my face a lot, figured out some of what not to do, and eventually wrote a book about my four years of teaching at Rivera (Gutstein, 2006a). From there, I went on to help start the Social Justice High School (aka Sojo) in Lawndale (a Black and Latina/o community) on Chicago's West side, a neighborhood (non-selective-enrollment) school. (Both Rivera's and Sojo's communities are economically battered but culturally and spiritually strong and resilient with deep wells of resistance.) Sojo's math teachers and I co-developed and co-taught several social justice math projects, and I worked closely with students in their classes. In 2008–2009, I taught a twelfth-grade "math for social justice" class at Sojo, and learned a lot more (see Gutstein, 2012, 2013b).

In this chapter, I summarize my original article, discuss some of what I learned, and make some suggestions as to how you could begin this work—or, if you are already a veteran, perhaps propose some new ideas. Rather than prescribing, I do this in the spirit of sharing. We cannot just import others' experiences into our unique contexts, but we can learn with each other and draw lessons. Freire (Freire & Macedo, 1987) referred to this as "reinventing" in our situations what others do in theirs. That's a major purpose of this volume, as I understand it. There is much to do in the area of social justice mathematics teaching, and we have only scratched the service. And while there's a good deal of theorizing and articles written about (and by) teacher educators trying to develop these ideas with preservice teachers (Foote, 2010; Gau Bartell, 2013; Spielman & Mistele, 2012), there are few articles written by or about teachers and students actually doing this work. Many of these are in *Rethinking Mathematics* (Gutstein & Peterson, 2013) and elsewhere (e.g., Gregson, 2013; Turner, 2003), but we have work to do.

An Example from My 2003 Article

I'd like to provide a seventh-grade example from my 2003 article, then move on to more recent work. This is from the article:

> For example, the following was part of the *Racism in Housing Data?* project. I gave students the data for the highest median house price in the area at the time (in 1997, the suburb of McFadden [a pseudonym], $752,250) and asked them these questions:
>
> How could you use mathematics to help answer whether racism has anything to do with the house prices in McFadden. Be detailed and specific!! Describe:
>
> 1. What mathematics would you use to answer that question.
>
> 2. How would you use the mathematics.
>
> 3. If you would collect any data to answer the question, explain *what* data you would collect and *why* you would collect that data.
>
> 4. Give examples of data that would cause you to believe that racism *is* involved in the McFadden housing price, and explain why you reached that conclusion based on the analysis of the data.
>
> 5. Give examples of data that would cause you to believe that racism *is not* involved in the McFadden housing price, and explain why you reached that conclusion based on the analysis of the data.
>
> One week later, I assigned the second part of the project, in which I reproduced the students' various responses in summary form. I asked students to pick two of the responses, explain in writing why those were good data with which to analyze if racism was involved, and answer the question, "How can you use the data you picked out to know whether or not racism is involved?" (Gutstein, 2003, p. 47)

I use this example for several reasons. One, it's clear that students were the ones to decide whether or not racism was a factor. Two, they had to back up their social and political arguments with mathematics. It was insufficient to make a claim; they also had to substantiate it. Three, it was open-ended, with no "right answer." Four, it put students in the position of analyzing reality, which is where they belong. Five, the project built on students' lives, even though I chose the topic and designed the project. I began a class with a brief story about what I thought was "everyday racism." That sparked a complex whole-class discussion about the interconnections of housing inequality, access to education, income, immigrant status, race, and racism, and that led me to develop the project, which I gave students the next day. And six, racism was an issue close to my students' hearts, and so they wanted to better comprehend it. Since mathematics was a way to understand racism, learning and using mathematics made sense to them. They came up with many different types of data to collect, developed multiple rationales and arguments, and saw the complex and even contradictory explanations that they themselves generated. As I wrote in the article, students argued:

> If the same houses were higher priced in neighborhoods of color, then racism existed—that is, prices were unfairly raised to make people of color pay more. However, other students claimed just the opposite! They stated that if house prices for the same houses were higher in white communities, this was racism because realtors knew that people of color could not afford as much and were keeping neighborhoods white by raising house prices beyond what families of color could afford. (p. 52)

It became apparent that the question of whether and how racism was a factor was difficult to definitively answer—and they reached this conclusion through a mathematics project.

Developing, Then Teaching, Math for Social Justice

But while developing curriculum like the above is essential, it is only a part, and maybe the lesser part at that—teaching it is key. For me, the starting point was that I was quite explicit about trying to teach so that my students better understood their lives and learned mathematics. Then we had to collaboratively co-create the classroom space. I could seed it—but students had to take it up. Anyone who has taught seventh graders knows that if they want, they can shred the best plans of the best teachers. You cannot impose a social justice classroom on students. For me, this involved several interrelated pedagogical commitments, including "normalizing politically taboo topics," which I allude to above and describe below, and creating a "pedagogy of questioning" (Freire & Faundez, 1992), which I describe more in Gutstein (2006b). It also meant developing (for lack of a better term) "political relationships" with students, which essentially meant going beyond the role of caring and listening adults, to

> include taking active political stands in solidarity with students and their communities about issues that matter. Political relationships also entail teachers sharing political analyses with students as much as possible. Finally, they include talking with students about social movements, involving students themselves in studying injustice, and providing opportunities for them to join in struggles to change the unjust conditions. (Gutstein, 2006a, pp. 132–133)

You can think of this work as hearing students about what matters to them, explaining how you are teaching and why (i.e., your political relationships with students and their communities), and designing mathematics curriculum built upon their concerns and situations, even if you choose the contexts—all this so they learn more about reality and math at the same time. This demands—and helps develop—space for honesty, openness, listening, sharing, learning, mutual respect, and deep questioning, along with sharp critique, analysis, and transformation. I'm not saying my class was totally there, but I know what direction we were headed. Virtually all students I've taught this way say they learned both mathematics and about the world, and the data supports their claims. If you want your students to learn mathematics to read and write the world, then on one level, it's straightforward—you have to have them do it. In my 2003 article, I wrote, "If teachers want students to develop [using math] a deeper understanding of society with all its complexities, they need to engage them in doing so" (p. 63). And we could expect that as students act to shape the norms and build a classroom where they learn in this way, their identities and ways of participating change as the class itself evolves (Lave & Wenger, 1991).

So How to Start? Or Go Further?

One thing I learned through reflection, which I tried to capture in the article, was that we made conversations about issues of (in)justice a normal part of math class. I call this "normalizing politically taboo topics." That meant, for example, telling students about what I thought was "everyday racism" while also being open to them doing the same—or to disagree. I realize that if students feel resistant, they can steer conversations to non-math topics, but my experience is that creating space for students to initiate these conversations is key to establishing a classroom where students seriously investigate issues they feel are unjust. In my classrooms, discussions about racism, immigration raids, dead-end jobs, gentrification, and more became as normal as talk about weekends or field trips, and they did not detract from instruction or take up an inordinate amount of time. In fact, as I describe above, they often led to mathematical investigations of things important to students.

As another example, in a later seventh-grade class, I was teaching during the one-year commemoration of September 11 (Gutstein, 2006a). Everything stopped at noon for three minutes of silence, and I then asked my somber students if they had any questions about September 11. I wrote their questions on the overhead projector for all to see, without comment or discussion. Questions came slowly at first, but eventually more quickly. This activity led into a conversation later that period between myself and a student whose brother was in the navy because, she said, he had no money for college. And that led to a project on the cost of the B-2 bomber compared to the cost of college (Gutstein 2006a). So you can see that, like the project on racism in housing, ideas emerged from classroom discussions. I only had to turn them into math lessons—not a trivial challenge, but doable.

There are other ways to get started, or to go further. You can take an idea, provided that it relates to your students, and create a small project and link it to the math you need to teach. Some years ago I did a workshop where I showed how to teach a basic idea, at different grades, that affects us all—gender pay inequality. At the primary level, the mathematics of 75 cents or three "quarters," compared to 100 cents or four quarters (or one dollar), leads to possibilities for place value, addition and subtraction (how much more would a girl making 75 cents need to earn to make the same as a boy?), multiplication (how much is three 25s?), comparisons, simple fractions (quarters), and more. In intermediate and middle grades, problems emerge with more complicated fractions, percentages, decimals, rates, and algebra, including graphing, examining slope, and solving equations. In high school, students can explore lifetime earnings between women and men, including unequal pay raises and inflation (exponentiation), to analyze the cumulative impact of gender pay inequality, and can create mathematical models and analyze real data. A simple idea can play out across different grades, and teachers can figure out how to connect social issues with the mathematics. In the newly revised *Rethinking Mathematics*, I wrote a chapter, with similar ideas, about using math across grades to investigate minimum wage and CEO pay (Gutstein, 2013a).

Of course, things don't always happen smoothly. Because of the challenges of teaching about complex social issues, we need to be "patiently impatient" (Freire, 1998) and accept change as gradual—both for us and for students. A critical lesson for me, and obvious in hindsight, is to start small and go slow. In my 2003 article, I described how I used a simulation of world wealth by continent (Peterson, 1995). We used cookies as "wealth" and randomly assigned students to the various continents by population proportion. For example, because Asia was 60 percent of the world's population, 60 percent of the students went to the "Asia" location. And so on. Then we doled out the wealth (cookies) by the percentage of wealth each continent had, and students saw graphically the unequal distribution of average wealth (i.e., North Americans had fourteen cookies each, Africans each had one-quarter of a cookie, etc.). This led to a rich mathematical exploration of proportionality, ratio, percentage, and measures of central tendency, and to a fascinating discussion. But one of the first times I tried this was in a class I hardly knew, and it bombed. Students mainly just wanted the cookies. In retrospect, that makes sense, because we had not developed relationships, political or otherwise, and the project came out of the blue. I have heard many similar stories from teachers who tried comparable activities without the necessary preparation (e.g., normalizing taboo topics, creating a pedagogy of questioning, developing political relationships, or whatever may be needed). Sometimes it works, but often students feel that the social justice project is just stuck in. So starting small and going slow makes sense.

Final Thoughts

I can't end this chapter without talking about race—mine and that of my students. When I've taught social justice mathematics, it has been to "other people's children" (Delpit, 1988), because I am a white male professional and my students have all been working-class/low-income students of color (who make up almost all students in Chicago neighborhood public schools). All experiences are racialized in the United States, it seems to me, and while my students "drive while Black or Brown" (e.g., are racially profiled when driving, shopping, etc.), I am racially profiled because I "drive while white"—I am never stopped when I drive my good-condition, late-model car, in my white skin. We live in a society marked by white privilege and racism, and it affects us all in myriad, complex ways. Being aware of it, naming and owning the relative privilege that we may have and, most important, using it to disrupt the unequal status quo is central to this work. We have, in my view, the responsibility to make social justice math teaching an explicitly antiracist pedagogy that is against all forms of discrimination and oppression.

When I left Rivera and moved onto Sojo, there were two major differences beyond that they were in distinct, though similar, communities and at different grade levels. First, my Rivera students did social justice projects less than a quarter of the time and studied MiC the rest of the time. That is, we did not have a full-year-long social justice mathematics curriculum. Second, though the projects related to students because of the sense of justice that they carried with them into the classroom, and I listened closely to students, I—not the students—chose the contexts. Both of these things always concerned me. But at Sojo, everything we did when I taught in 2008–2009 was for students to learn and use mathematics to study their social reality and become "subjects" of history—and students themselves chose, or agreed to, all five units we studied and made decisions about the timing, order, and altering of curriculum. I do not have room to explain this, but this was a big, intentional transition for me, from which I learned a tremendous amount (Gutstein, 2012, 2013b). My brief summation is that at Rivera, I learned that it was difficult, messy, complicated, but possible for students to begin to read and write the world with mathematics, and to learn mathematics, using a reform curriculum as the basis, teacher-selected real-world projects, and a co-created classroom culture supporting social justice pedagogy. At Sojo, I saw that it was just as complex, but with added challenges because I developed a whole-year social justice mathematics curriculum based on issues that students chose to study. Though there were differences in each setting, I think what I wrote in my book—which should not be read to minimize the difficulties of learning to teach this way—captures what I learned:

Finally, it is significant that students from immigrant, bilingual, working-class Latino/a [and, at Sojo, nonimmigrant Black] families developed aspects of mathematical power, achieved some academic success, viewed mathematics as a way to understand unjust social arrangements, *and* began to read and write the world using mathematics. That these students are in a better position to challenge a racist and sexist, class-based system that excludes them and others like them is a contribution to the causes of social justice and liberation. (Gutstein, 2006a, p. 127)

I end this with a short writing from a student in my Sojo class, who responded to my impromptu request: "What is reading and writing the world with mathematics and why do we do it?" His written response captures how many of my students came to understand this:

Reading and writing the world with mathematics means a lot. It means that you look at any issue happening anywhere in the world. When you read the world, you are getting background information and seeing why whatever problem you see is occurring. You then find a way to resolve it. This then brings in writing the world with mathematics. When writing the world, you are ready to use mathematics to prove your point. Also, every point you have will not be a solution. It will sometimes just be a way for you to bring light to a situation that no one knows about. So to me this is what reading and writing the world with mathematics means.

We do this for a reason. There are big corporations trying to take advantage of people. There are also plain old injustices that happen everyday. We do this to educate ourselves on global or local problems that can be solved with mathematics. We also do this to learn more advanced mathematics. Lastly, we do this so that we can take our knowledge back to our friends and family to educate them. Once we educate the ones that are closest to us, we then go out and educate our community on how to prevent things from happening to them and how to catch things before they are taken advantage of.

—George Carr, 2009 (After writing this, he read it aloud in a session at the 2009 annual meeting of the American Education Research Association, San Diego.)

References

Delpit, L. (1988). The silenced dialogue: Power and pedagogy in educating other people's children. *Harvard Educational Review, 58,* 280–298.

Freire, P. (1994). *Pedagogy of hope: Reliving* Pedagogy of the oppressed (R. R. Barr, Trans.). New York, NY: Continuum.

Freire, P. (1998). *Pedagogy of freedom: Ethics, democracy, and civic courage* (P. Clarke, Trans.). Lanham, MD: Rowman & Littlefield.

Freire, P., & Faundez, A. (1992). *Learning to question: A pedagogy of liberation* (T. Coates, Trans.). New York, NY: Continuum.

Freire, P., & Macedo, D. (1987). *Literacy: Reading the word and the world.* Westport, CT: Bergin & Garvey.

Foote, M. Q. (Ed.). (2010). *Mathematics, teaching, and learning in K–12: Equity and professional development.* New York, NY: Palgrave-MacMillan.

Gau Bartell, T. (2013). Learning to teach mathematics for social justice: Negotiating social justice and mathematical goals. *Journal for Research in Mathematics Education, 1,* 129–163.

Gregson, S. A. (2013). Negotiating social justice teaching: One full-time teacher's practice viewed from the trenches. *Journal for Research in Mathematics Education, 1,* 164–198.

Gutstein, E. (2003). Teaching and learning mathematics for social justice in an urban, Latino school. *Journal for Research in Mathematics Education, 34*, 37–73.

Gutstein, E. (2006a). *Reading and writing the world with mathematics: Toward a pedagogy for social justice.* New York, NY: Routledge.

Gutstein, E. (2006b). "So one question leads to another": Using mathematics to develop a pedagogy of questioning. In N. S. Nasir & P. Cobb (Eds.), *Increasing access to mathematics: Diversity and equity in the classroom* (pp. 51–68). New York, NY: Teachers College Press.

Gutstein, E. (2012). Mathematics as a weapon in the struggle. In B. Greer & O. Skovsmose (Eds.), *Opening the cage: Critique and politics of mathematics education* (pp. 23–48). New York, NY: Sense Publishers.

Gutstein, E. (2013a). "I can't survive on $8.25": Using math to investigate minimum wage, CEO pay, and more. In E. Gutstein & B. Peterson (Eds.), *Rethinking mathematics: Teaching social justice by the numbers* (2nd ed.) (pp. 75-77). Milwaukee, WI: Rethinking Schools.

Gutstein, E. (2013b). Whose community is this? Mathematics of neighborhood displacement. *Rethinking Schools, 27*(3), 11–17.

Gutstein, E., & Peterson, B. (Eds.). (2013). *Rethinking mathematics: Teaching social justice by the numbers* (2nd ed.). Milwaukee, WI: Rethinking Schools.

Lave, J., & Wenger, E. (1991). *Situated learning: Legitimate peripheral participation.* New York, NY: Cambridge University Press.

NCRMSE & FI (National Center for Research in Mathematical Sciences Education & Freudenthal Institute). (1997–1998). *Mathematics in context: A connected curriculum for grades 5–8.* Chicago, IL: Encyclopedia Britannica Educational Corporation.

Peterson, B. (1995). Teaching math across the curriculum: A 5th grade teacher battles "number numb-ness." *Rethinking Schools, 10*(1), 1, 4–5.

Spielman, L., & Mistele, J. (Eds.). (2012). *Mathematics teacher education in the public interest.* Charlotte, NC: Information Age.

Turner, E. (2003). *Critical mathematical agency: Urban middle school students engage in mathematics to investigate, critique, and act upon their world* (Unpublished doctoral dissertation). University of Texas, Austin.

Zinn, H. (1994). *You can't be neutral on a moving train: A personal history of our times.* Boston MA: Beacon Press.

Teaching and Learning Arithmetic, Algebraic Reasoning, and Calculus

Introduction

This section is devoted to research related to topics in arithmetic, algebra, and calculus. As editors, we believe that this is an appropriate content clustering because in mathematics there is a "fundamental theorem" for each of the three areas. *Principles and Standards for School Mathematics* from the National Council of Teachers of Mathematics (NCTM, 2000) focuses on arithmetic and algebra in the K–12 Standards and shows a shift in emphasis from the lower to the higher grade levels. For example, in the early grades, algebraic reasoning is based on beginning concepts of patterns, relations, and functions; by grades 9–12 the focus is on extending algebraic concepts to include abstraction and structure. The progression involving arithmetic and algebra across grade levels is similar in the Common Core State Standards for Mathematics (National Governors Association Center for Best Practices & Council of Chief State School Officers [NGA Center & CCSSO], 2010). For example, the kindergarten standards mention counting, cardinality, base-ten numbers, and operations; by high school there is a shift to the real and complex number systems. Although neither the *Principles and Standards* nor the Common Core standards directly refer to calculus, facility with arithmetic and algebra concepts has an important place in calculus learning.

The placement of the nine chapters within this section was our decision as editors. This introduction contains a brief summary of each chapter and our impressions of lessons learned from the research. Although the chapters are introduced individually and we think that each has individual merit, we encourage readers to consider them analytically and synthetically as a means to spark useful and usable knowledge to inform mathematics teaching and learning.

Chapter 7. "Class Learning Zone and Class Learning Paths: Responsive Teaching in First-Grade Mathematics," by Aki Murata and Karen Fuson

The first chapter in the arithmetic section is by Aki Murata and Karen Fuson. The authors use a case study of children in a first-grade classroom at a full-day Japanese school in the United States; the children were learning how to add one-digit numbers (e.g., 9 + 4) and using a variety of strategies to do so. The chapter illustrates responsive teaching: how facilitation of different levels of understanding and fluency can support all students to learn mathematics and how teachers can coordinate different instructional supports to ensure such classroom learning possible.

Murata and Fuson present their conceptualization of *Class Learning Zone with a Class Learning Path*, based in large part on Vygotsky's zone of proximal development (ZPD)—the distance between a child's actual development and his or her potential development under the guidance of a more experienced partner (e.g., a teacher). The Class Learning Zone involves a teacher orienting students to a new instructional topic and then eliciting their methods for solving such problems or for thinking about such contexts. Then, using appropriate teaching materials, the teacher begins to move along a Class Learning Path that provides assistance to move students forward to a general solution.

The authors continue the chapter with examples of how a teacher (Mr. Otani) supported student learning through four phases of the Class Learning Zone with a Class Learning Path model. Examples of student dialogues with the teacher appear throughout. Results from this case study allow readers to see how one teacher assisted student learning by valuing informal knowledge and approaches, allowing students time and opportunities to explore different ideas, helping bridge the distance between existing knowledge and the new method, and giving students time to practice and gain fluency. The authors encourage teachers to orchestrate and coordinate unique student learning paths by connecting and relating different ideas so that students can learn from one another. By shifting the focus from "telling" math content to students to coordinating and extending different student ideas, teaching will hold new meanings for students and teachers alike.

Lessons learned

Through examples from their case study, Murata and Fuson show that it *is* possible to operationalize ideas from a theory-based research (i.e., Vygotsky's ZPD) and implement them in the mathematics classroom. The information here is a lesson in itself about the benefits of responsive teaching as a way to support the learning of all students.

Chapter 8. "Giving Change When Payment Is Made with a Dime: The Difficulty of Tens and Ones," by Cynthia C. Chandler and Constance Kamii

Cynthia Chandler and Constance Kamii's chapter continues our look at mathematics in the early grades and how a theory-based research model can add to our understanding of how to teach important mathematics to children. Their focus is on examining children's thinking about our base-ten system: specifically, thinking about how to give change when a dime or a dime and a few pennies are offered as payment of a purchase up to 9 cents. Chandler and Kamii draw upon the work of Piaget to explain the difficulties that children have with coins in mathematics problems. Their focus is on Piaget's three sources of knowledge (physical, social-conventional, and logico-mathematical), and their study reflects how children learn and use them.

The authors draw upon findings from their study of nearly 100 children in kindergarten through fourth grade. Each child was interviewed about a situation in which he or she was a storekeeper and the interviewer the customer; it was the storekeeper's job to figure out how much change to give the customer for particular purchases. The authors categorized the children's responses to seven tasks according to five levels, ranging from Level 0 (giving no change or random amounts) to Level 4 (giving the correct change in all seven tasks). Perhaps not surprisingly, they found that the majority of younger children were at Level 0, and most of the older children tended to be at Level 4.

In the next section of their chapter, Chandler and Kamii focus on a task that required giving change when a dime was offered as payment for a 6-cent purchase. Through an interview with a student (Joslyn), the authors illustrate how she had learned bits of physical and social-conventional knowledge (e.g., a dime is worth ten pennies), but not the logical-mathematical knowledge about tens and ones.

The authors conclude the chapter with suggestions about how to encourage students to think of tens and ones at the same time. The suggestions are in the form of games based on number combinations that make 10.

Lessons learned

Again, we as readers can see how a theory-based research model—this time based on the work of Piaget—can assist us in creating classroom activities and then analyzing what and how students learn from them. The importance of making connections between mathematics and real-world knowledge is also emphasized.

Chapter 9. "Multiplication Methods in the Context of the Common Core State Standards," by Bruce Sherin and Karen Fuson

Completing the section's focus on arithmetic concepts is a chapter by Bruce Sherin and Karen Fuson in which they talk about the strategies that children use to multiply single-digit numbers (e.g., 3×5, 6×9). The authors contend that learning the multiplication tables through rote memorization is not consistent with teaching mathematics with understanding. Thus, it is important to think about how students might arrive at the solution. What must they learn to be fully proficient at this type of task?

The chapter contains a summary of their research on how children invented methods used in single-digit multiplication. In particular, Sherin and Fuson focus on six computational strategies that children used when learning multiplication: *count-all* (counting from 1 to the product); *additive calculation* (based on prior learning experiences with addition); count-by (multiplication as repeated groups); *pattern-based* (e.g., those in multiplication by 0, 1, and 10); *learned product* (based on multiplication triads); and *hybrids* (particularly, combining count-by and learned product). Student work for these strategies provides concrete examples for the reader. These strategies illustrate that it is not about just rote memorization; instead, learning multiplication involves what the authors call an "integrated generative web."

The authors conclude the chapter with suggestions for instruction. For example, students should be encouraged to see and relate pattern and structure in numbers. Sherin and Fuson also relate specific parts of the Common Core standards in the Operations and Algebraic Thinking progression to multiplication (and also division) learning. They contend that such learning needs to begin intensively and early in third grade and that teachers should capitalize on the strong relationship between multiplication and division strategies as they plan lessons. Finally, student practice should be focused on the individual learning zone of the students, with time better spent on the next most difficult problems and not on a page of 100 problems, most of which the student has already mastered.

Lessons learned

Those of us who are of a certain age might recall multiplication flash cards, worksheets of 100 single-digit multiplication problems, or timed tests of multiplication facts. The lesson that Sherin and Fuson share with us is that there are more natural strategies that children bring to the task of learning multiplication with understanding instead of learning it by rote.

Chapter 10. "Arithmetic and Algebra in Early Mathematics Education," by David W. Carraher, Analúcia D. Schliemann, Bárbara M. Brizuela, and Darrell Earnest

The chapter by David Carraher, Analúcia Schliemann, Bárbara Brizuela, and Darrell Earnest represents a transition in this section from the topics in arithmetic to those in algebra. The authors begin by positing that the extreme separation of arithmetic from algebra makes less and less sense, especially in light of investigations into additive and multiplicative structures that underlie the algebraic nature of arithmetic. In the chapter, the authors present examples from their own research to illustrate how young students become engaged in algebraic reasoning and in adopting new representations for algebraic ideas.

The first example involves arithmetic operations as functions. The authors believe that one of the most underappreciated ideas in elementary mathematics is that the four basic operations of arithmetic are expressly defined as functions (e.g., $7 + 3$, $5 + 3$, $9 + 3$ are instances of the univariate "plus 3" function $x + 3$ over the set of integers). This way of thinking leads to a recognition of the important role of variables and the need for children to implicitly or overtly deal and reason with them.

Continuing the focus on the concept of variable, the authors present two examples of classroom lessons that deal with variables. The first involves introducing literals as variables and placeholders for variable quantity—that is, an amount or measure that can be assigned any element from a domain of possible variables. The lesson described was based on a problem comparing the number of candies two people can have, the candies being in two sealed boxes with three more candies in one box than in the other. In subsequent lessons the number line was introduced, both with numbers (e.g., from 0 to a particular number) and an "N-number line" with values expressed as displacements from some indeterminate value N. In all of the lessons described, the students learned to meaningfully represent variables, and they used and operated on algebraic expressions to represent functions.

The authors conclude that there are good reasons for considering arithmetic and algebra as interwoven in the early grades, and there are important opportunities for teachers and their students to examine fundamental topics of elementary school mathematics that illuminate and emphasize the algebraic character of arithmetic.

Lessons learned

Carraher, Schliemann, Brizuela, and Earnest offer important lessons about the benefit of introducing algebraic reasoning early in children's mathematics experiences with arithmetic. The examples show us that it is possible for younger students to make a relatively quick shift to using variables and engaging in algebraic reasoning.

Chapter 11. "Children's Algebraic Reasoning and Classroom Practices That Support It," by Maria Blanton and James J. Kaput

Continuing the focus on algebraic reasoning, Maria Blanton and James Kaput offer an elaboration on what algebraic reasoning means, what it looks like in the elementary mathematics classroom, and how to support classroom teachers who work to introduce such reasoning to their students. The authors use a case study of a third-grade teacher (June) as she participated in a long-term professional development project. June initially insisted she was not a "math person" but was willing to try new ideas in her classroom. The goal of the study was to explore in what ways and to what extent June was able to build a classroom climate that supported the development of students' algebraic reasoning skills.

After a brief summary of the study and its methodology, Blanton and Kaput showcase the thirteen types of algebraic reasoning that they identified from the case study and elaborate on a set of them. For example, there was evidence that June encouraged students to reason algebraically with arithmetic relationships (e.g., adding two even numbers, two odd numbers, one of each). Also, she and her students explored the use of the equal sign as expressing a relationship between equivalent quantities, solved "missing number sentences," symbolized quantities, found functional relationships, used generalizations to solve algebraic tasks, and tested conjectures.

From their work with June, the authors identified these characteristics of her practice that supported the development of students' algebraic reasoning: (1) seamless and spontaneous integration of algebraic conversations in the classroom; (2) spiraling of algebraic themes over significant periods of time; (3) integration of multiple and independently valid algebraic processes; and (4) activity engineering (i.e., adapting or developing tasks to include algebraic reasoning). The article closes with the thought that elementary teachers are critical to teaching and learning mathematics and that even those who might see themselves as math phobic have a tremendous capacity to build rich and engaging classroom environments devoted to algebraic reasoning.

Lessons learned

Through their careful collaboration with June over the course of a year, Blanton and Kaput offer important lessons about how elementary teachers and other practitioners can foster a climate of algebraic reasoning in the classroom. The characteristics of seamless integration, spiraling of themes, integration of algebraic processes, and activity engineering offer a way to begin to create this kind of classroom climate.

Chapter 12. "The Equal Sign Does Matter," by Eric Knuth and Ana Stephens

In chapter 11, Blanton and Kaput referred to a classroom activity involving an exploration of what the equal sign means. In this chapter, Eric Knuth and Ana Stephens share findings from research on the importance of students developing a relational understanding of the equal sign—that is, an understanding that the equal sign expresses an equivalence relation between two quantities. On the surface, it may seem that understanding the equal sign is relatively trivial, and little attention is paid to it beyond an initial introduction. Yet, in their *JRME* article, and an article published in *Mathematics Teaching in the Middle School* (the *MTMS* article is available at nctm.org/more4u), Knuth and Stephens found that many students fail to develop a relational understanding of the equal sign.

The authors showcase the difficulties students have by presenting tasks that emphasize the relational nature of the equal sign (e.g., "The following number sentence is true: $15 + 8 = 23$. Is $15 + 8 + 12 = 23 + 12$ true or false? How do you know?"). Only 5 percent of grade 4 students and 9 percent of grade 5 students gave answers that mentioned adding 12 to both sides of an equation or doing the same thing to both sides. Another task involving two equations ($2 \times n + 15 = 31$ and $2 \times n + 15 + 9 = 31 + 9$) showed that middle school students could solve the equations independent of one another, compare the answer, and conclude that they had the same solution; but very few provided a relational explanation (e.g., "subtracting 9 from both sides makes both equations the same").

The authors conclude by mentioning two Common Core standards that underscore the need for students to understand the nature of the equal sign. They offer suggestions to elementary and middle school teachers on how to promote such understanding—for example, by posing tasks like the examples in the chapter or taking advantage of opportunities to discuss the meaning of the equal sign when they naturally arise.

Lessons learned

Who would have thought that a mathematical symbol as ubiquitous as the equal sign could lead to difficulties in learning algebra? The lesson learned from Knuth and Stephens's careful research is that the equal sign provides an important opportunity for students to engage in relational thinking, and that just a little bit of instructional time can make a big difference.

Chapter 13. "Students' Metaphors for Limit Concepts in Introductory Calculus," by Michael Oehrtman

Michael Oehrtman's chapter turns our attention to topics in calculus—in particular, the concept of limits. The author states that limit concepts are notoriously difficult for students, but they are the single most significant unifying concept throughout calculus. This chapter summarizes his research into the rich conceptual structure of students' intuitive (but not always mathematically correct) ways of reasoning about limits. Oehrtman contends that at the very least, a deeper understanding of how students interpret what we present to them can inform our decisions about language, images, and activities we introduce in class.

In designing his study, the author used a theory of metaphor development that emphasizes "strong metaphors," which are rich in implications and to which the user of the metaphor is strongly committed. To identify metaphors that student have about limits, he studied 120 students enrolled in a yearlong introductory calculus sequence. His analysis resulted in five strong metaphors based on students' intuitions about limits: *collapse* metaphors (e.g., "$0.\overline{9} = 1$ because the [digits] eventually become zero at infinity"); *approximation* metaphors (e.g., for the slope of a tangent line with the approximations of the slopes of the secant lines, "the smaller you make your *h*, the better an approximation you would have"); *proximity* metaphors (e.g., closeness and clustering); *infinity as a number* metaphors (e.g., infinity can be used in calculations such as $\ln(\infty) = \infty$); and *physical limitation* metaphors (e.g. based on the smallest physical size possible, such as molecule or quark). The author also found three weak metaphors for limit: motion; language about quantities being sufficiently and arbitrarily close; and images of zooming in on a graph.

In discussing his results, the author notes that determining the strong metaphors did not include criteria for correctness; in fact, much of what students said when applying strong metaphors was mathematically incorrect. The three weak metaphors were used in class by the professor, but students did not adopt these ways of reasoning, preferring their own strong (and often incorrect) ways. Oehrtman concludes by suggesting that students' nonstandard and even incorrect ways of reasoning about limits are, at least, fertile sites for positive discussions.

Lessons learned

We can always learn from our students' beliefs and misconceptions about mathematics, and Michael Oehrtman's research on metaphors in introductory calculus shows us that this is true even in higher mathematics. Information about how students conceptualize critical ideas such as limits can help practitioners create experiences that can support or refute such ideas.

Chapter 14. "Infinitesimals in Student Reasoning," by Robert Ely

In chapter 13, Oehrtman found metaphors that had to do with closeness and proximity in calculus. In this chapter, Robert Ely focuses on a proximity-like question: Is it possible to chose two different points on the real number line that are infinitely close to one another? When he asked students this question, the majority said is it possible, giving examples like 0.9999... being close to 1 if adding 9s was an unending process. When he asked mathematicians, they said it was not possible based on infinitesimals, and most of them could not believe that students could have such a misconception. However, Ely found one student (Sarah) who appeared to be reasoning about numbers and distances that really were infinitesimal—being larger than zero but smaller than any finite number or distance. His chapter is devoted to a case study of Sarah's conceptual understanding of infinitesimals.

Ely continues with a brief history of infinitesimal calculus in the time of Newton, Leibniz, and others, and how Robinson in the 1960s added rigor about nonstandard real numbers, which included the standard real numbers and infinitesimal and infinite numbers. The nonstandard numbers marks a powerful and mathematically correct mode of thought, but it differs substantially from the standard real number system taught in today's classrooms. Finding a student like Sarah who talked about infinitesimal quantities in ways that paralleled those of mathematicians was a surprise to the author. Sarah used ideas (with examples) such as infinitely close numbers, an infinitely divisible number line with infinitely many infinitesimals, and properties of and operations with infinitesimals (e.g., ordering, squaring, reciprocals).

To conclude, Ely calls Sarah's conceptions about the real number line and infinitesimals nonstandard conceptions, not misconceptions. He goes on to elaborate about nonstandard conceptions and how they confirm how students can have access to deep mathematical ideas—like infinitesimals—that they are not explicitly taught. Rather than dismissing them as misconceptions, practitioners can use these nonstandard conceptions to gain insights into how a student thinks about mathematics.

Lessons learned

Ely admits that Sarah's reasoning about a mathematical concept that had not been part of her mathematics education experience was surprising. Yet, through her, he learned that it is possible for students to have access to deep mathematical ideas that have not been explicitly taught. We as practitioners would be well served to identify such students and learn from their insights.

Chapter 15. "Modifying and Constructing Diagrams in Calculus," by Elizabeth George Bremigan

The focus of Elizabeth Bremigan's chapter is on how students use visual representations to understand and make sense of calculus problems. As a high school calculus teacher, she noted the extent to which problems in the free-response section of the Advanced Placement (AP) Calculus examinations involved reasoning with visual representations. Her experiences led her to investigate relationships between the ways students used visual representations in calculus and their achievement as measured by the AP exam. Bremigan focused her study on two areas: (1) the frequency of diagram modification and construction in solutions to AP Calculus free-response questions and (2) the nature of the diagrams that students modified or constructed.

In the next sections, the author focuses on each research question, describing first what she did to collect and analyze the data and then what she learned. For the frequency question, her data came from a random sample of 600 students who took the BC AP Calculus examination; she selected six problems based on their problem presentation that included a diagram, which gave students the opportunity to modify the given diagram or construct a new one. She found that in three of the problems, more than half of the responses included modified or constructed diagrams and that high scorers' written solutions contained more evidence of diagram use than did solutions of low scorers. Females' solutions showed evidence of diagram use more frequently than those of males.

With respect to *how* students used diagrams, Bremigan reports detailed results from a sample of 180 students on the six problems. For example, the diagrams that students modified were more diverse (i.e., marked in different ways) than were the new diagrams students constructed, and the modified diagrams tended to contain more detailed markings. Females tended to mark their diagrams more than males; diagrams produced by males were less frequently labeled than those produced by females. In similar fashion, the author provides details on other ways students used diagrams.

Bremigan concludes with suggestions for teachers. Some examples include: looking carefully at the diagrams that accompany the verbal statement of problems in curriculum materials; considering when to provide a visual representation as a component of the problem statement and when not to; and broadening the meaning of "show your work" to include work with visual representations.

Lessons learned

From her research, Bremigan learned that the relationship between diagram use and problem-solving success is complex. Moreover, her research suggests that males and females approach diagram use differently. The list of practical suggestions at the end of the chapter serves as a valuable summary of the lessons learned from her research study.

References

National Council of Teachers of Mathematics. (2000). *Principles and standards for school mathematics*. Reston, VA: Author.

National Governors Association Center for Best Practices & Council of Chief State School Officers (NGA Center & CCSSO). (2010). *Common core state standards for mathematics*. Washington, DC: Author. Retrieved from http://www.corestandards.org.

Chapter 7

Class Learning Zone and Class Learning Paths
Responsive Teaching in First-Grade Mathematics

Aki Murata
University of California, Berkeley

Karen Fuson
Northwestern University

In the Japanese grade 1 classroom, students are busily figuring out how to add two numbers, 9 and 4. The teacher, Mr. Otani, writes the number sentence 9 + 4 on the board and puts 9 blue and 4 red magnets in a row on the board. Some students share different counting strategies (count all, count on, count by 2s, etc.). Others share how to make a ten (9 + 4 = 9 + 1 + 3 = 10 + 3 = 13). Koichi goes up to the board and rearranges the magnets to show his thinking (see fig.7.1).

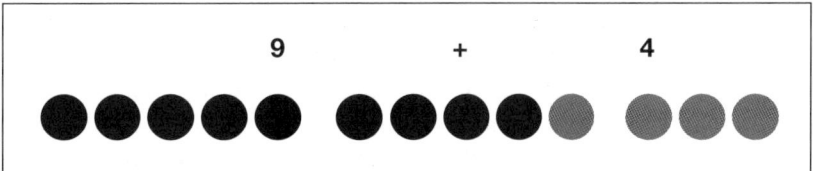

Fig. 7.1. Koichi's method of 9 + 4 = 5 + 5 + 3

Koichi: I made groups of 5 and 5.

Mr. O: I wonder what is different about this method . . .

Students: He did 5 + 5 + 3!

Mr. O: Yes. How did you think of this, Koichi?

Koichi: I thought this way: 5 and 5 is 10, so 3 more is 13. Is it OK?

Students: It is OK! *[shouting together]*

Through the case study of a Japanese classroom, this chapter will illustrate how facilitation of different levels of understanding and fluency can support all students in learning mathematics. We will discuss how different instructional supports can be coordinated for teachers to make such classroom teaching and learning possible.

National reports summarizing research describe a new view of teaching mathematics. This approach builds competence in culturally valued knowledge by relating such knowledge to what students already know and balances conceptual understanding and procedural fluency to develop mathematical proficiency. This chapter presents the ZPD Mathematical Proficiency Model to support the implementation of these views. Based on the original theory of Vygotsky (1978, 1986), Tharp and Gallimore (1988) described teaching that occurs when assistance is offered at points in the Zone of Proximal Development (ZPD) when performance

This chapter is adapted from A. Murata & K. Fuson (2006), Teaching as assisting individual constructive paths within an independent class learning zone: Japanese first graders learn to add using 10, *Journal for Research in Mathematics Education, 37*, 421–456.

requires assistance. The ZPD is defined as the distance between the child's actual developmental level and his or her potential development under the guidance of or in collaboration with a more experienced partner. Thus, teaching is a shared social activity. In our model, assistance is offered only when needed; decreasing amounts of assistance are needed as students progress through their own learning paths in any given topic, creating mutual understanding between the teacher and the learner for giving assistance adapted to the learner. This model underscores the view of learning as a constructive activity by the learner so that the internalization process does not involve rote copying of behavior.

Such teaching is well documented in many cultures and in many different activity settings around the world. However, it often occurs with one learner and one teacher. So how can such an ideal view of teaching possibly work in a classroom with one teacher and as many as twenty or even thirty-five students? We propose a perspective to illustrate how our definition of teaching could be enacted for individual students within the whole-class setting: *Class Learning Zone with a Class Learning Path*. Key to this perspective is our knowledge from research that, for many mathematics topics, a few typical errors stem from partial and incomplete understandings, and some other, more random errors arise from momentary lapses of attention or effort. Likewise, there are usually several solution methods, but these are limited in number and vary in their sophistication, generalizability, and ease of understanding. Thus, for any given mathematics topic, there are not twenty or thirty-five different learning paths or strategies for the teacher to understand and assist. Instead, there are usually three to six strategies that may have minor variations, and these can be noted in curricular materials that assist teachers in learning to assist students. Also, visual supports can be developed and shared with teachers to aid them in teaching particular topics. Of course for any mathematics topic or problem, there is always a possibility of new solutions or strategies, and not all can be anticipated, so the class Learning Path needs to be responsive to such possibilities.

When teaching with the idea of a Class Learning Zone, a teacher would orient students to the new instructional topic and then elicit from students their methods for solving such problems or for thinking about such contexts. With assistance from teaching materials, the teacher begins moving along a Class Learning Path that will provide assistance to move students forward to a good-enough and culturally valued general solution, with individual students starting from their own initial knowledge. The Class Learning Zone is the day-to-day learning zone within which the teacher organizes assistance for various students. Exceptional students (either extremely advanced or extremely delayed) may fall outside the Class Learning Zone. The former may help others but may need assistance to do so (or to want to do so). The Class Learning Path is the day-to-day sum of the learning paths of most of the students in the class (reflecting the state of growth in their methods and in their understanding each day), but this falls within manageable groups of related-enough mathematical assistance needs. A few students may not fully master a target solution method, but assistance will continue in subsequent units toward mastery or with additional help outside of class. Students may also continue to use any powerful or general-enough method of their own choice. Learning for all students includes increased understanding of how other students solve problems and increased ability to assist other students.

Figure 7.2 illustrates the ZPD Mathematical Proficiency Model. There are four stages for a learner to move through the ZPD to achieve a given performance goal: Phase 1 is assistance provided by more-capable others; Phase 2 is assistance provided by the self (as the means of assistance of others are internalized into speech-for-self); Phase 3 is internalization-automatization-fossilization; and Phase 4 is de-automatization with recursion through the stages as performance that was once mastered slips away over time. Decreasing assistance over time is part of responsive assistance.

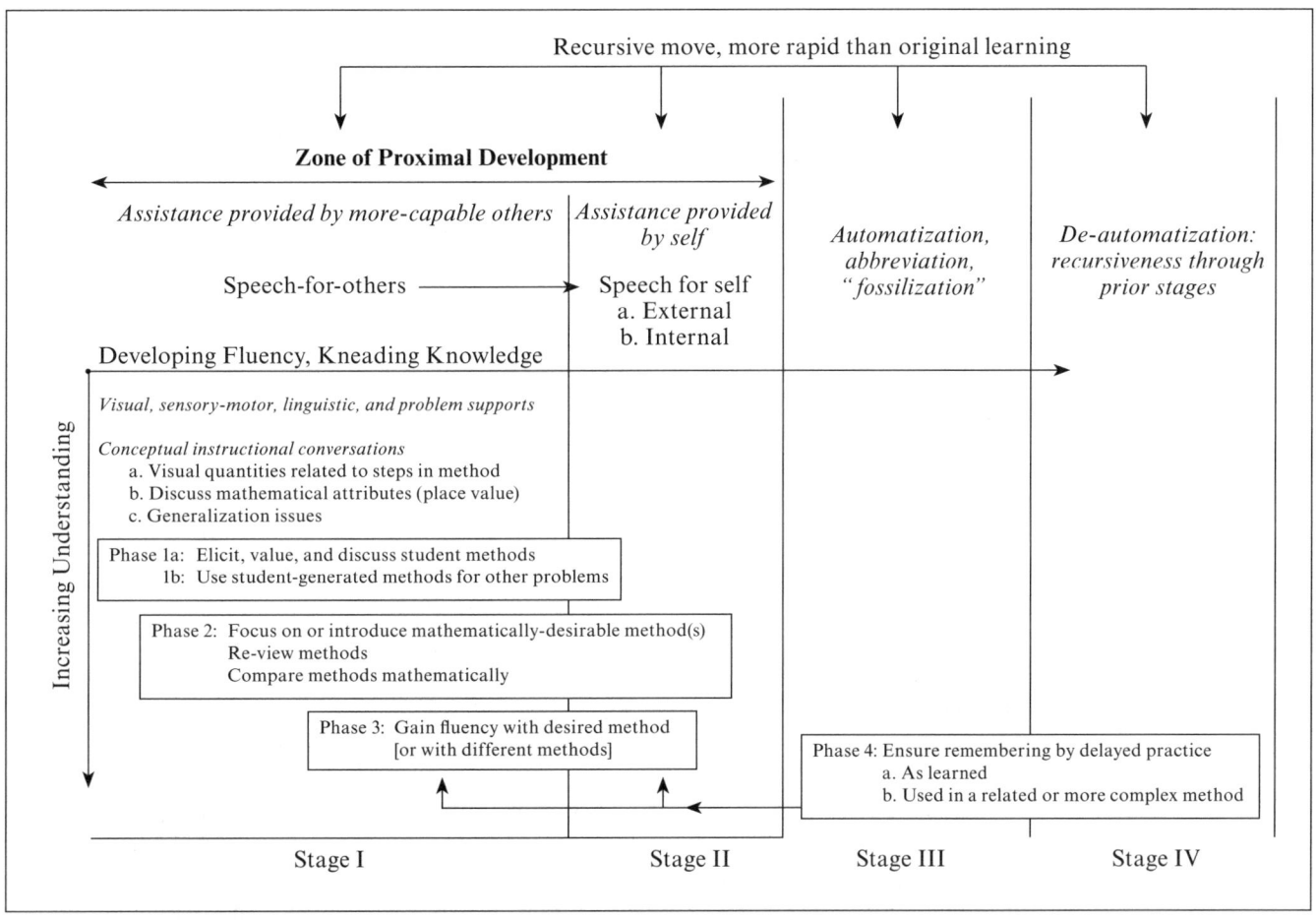

Figure 7.2. Stages of Learning and Class Learning Zone Phases in the ZPD Mathematical Proficiency Model

We identified in the ZPD model two independent but continually interacting aspects of teaching over time: developing understanding and developing fluency. Fluency moves to the right horizontally through the four stages. Understanding moves down vertically and is central in the four phases. Visual, sensory-motor, linguistic, and problem supports provide the bases for building understandings by all within conceptual instructional discussions that assist such understandings. These supports are conceptual tools that form the backdrop for all of the collective and individual functioning within the Class Learning Zone. Their use can continue at any stage or phase of learning in the ZPD. The kinds of learning supports (conceptual tools) and the mathematical points within the conceptual discussion vary with the mathematical topic. The necessity of identifying the learning supports for particular topics is part of our model.

In the following sections, we illustrate this model in action using a Japanese grade 1 classroom example. We offer an overview of the assistance that one Japanese teacher provided as students learned a culturally valued mathematics concept, how he changed the levels of support for the class and for individual students as the instructional unit progressed, and how he used teaching supports and tools. We also see how students assisted other students. This model and the teaching we see are consistent with the Common Core State Standards

(National Governors Association Center for Best Practices & Council of Chief State School Officers [NGA Center & CCSSO], 2010) and with its Standards for Mathematical Practices. The method discussed here is in a grade 1 standard (1.OA.C.6), and all eight mathematical practices are exemplified in our discussions of teaching and learning.

The goal of the unit chosen for our case study was to learn to add numbers with totals in the teens. This unit was chosen because it involves learning a complex multistep method, the Break-Apart-to-Make-Ten (BAMT) method, which is specified in the Japanese National Course of Study. The complexity of this method pushed our concept of a Class Learning Zone because this method is demanding for students. Student understanding of and fluency with the BAMT method are viewed as important for their future learning of multidigit addition and subtraction in the curriculum because it helps them make sense of and use the values of 10-ness in the number system, and it is a general addition method useful in multidigit addition, where students will be moving the new group of 10 to the next left column. This method also prepares students for related methods for subtraction.

Method

The data were collected in a grade 1 classroom at a full-day Japanese school in a suburb of a Midwestern metropolitan city in the United States. The school is operated by the Japanese Ministry of Education and closely follows the Japanese National Course of Study. Administrators and teachers are sent directly from Japan through the ministry, and the instructional language is Japanese. The school primarily serves Japanese families who are in the area for a short period of time (two to five years), and the community puts much effort into preserving Japanese culture in their lives as well as maintaining Japanese ways of teaching and learning. The grade 1 teacher, Mr. Otani, had taught in Japan prior to coming to the school. Twenty-five students were in the classroom.

Data were collected over eleven lessons in a three-week period during the fifth month of the school year. For each observation, lessons were videotaped and careful field notes were taken. The methods the students used in the classroom as they solved addition problems were also noted. Six target students (two with higher, two with medium, and two with lower performance levels, as identified by the teacher) were interviewed as they solved problems before and after the instructional unit and at the end of the school year.

Data from the observation field notes were analyzed to illustrate how Mr. Otani (1) provided assistance as students learned the steps of the BAMT method and (2) changed his support levels for individual students and also over time. Figure 7.3 shows the steps that students used to carry out this method for 9 + 4 as well as a representational drawing taken from the Japanese teachers' manual (Tokyo Publishing, 2000) and used in the classroom. Figure 7.4 shows the levels of support Mr. Otani provided for students across lessons.

Field notes were coded for the external problem-solving steps that students took in the whole-class context and individually in independent work as they learned, and for the kinds of support Mr. Otani provided for the steps. Other means of support were also identified. Videotaped data were reviewed to verify the data coded from the field notes. For individual student learning, the different methods and the steps of the BAMT method used by the six target students were also analyzed.

Teaching Phases and Teacher Support

In the following sections, we describe how Mr. Otani supported student learning through the four phases of the instructional model.

Steps of the BAMT method	Step 1 Find that 9 needs 1 more to make 10	Step 2 Separate 4 into 1 and the rest (3)	Step 3 Add 9 and 1 to make 10	Step 4 Add 10 and 3 to make 13
Counter use makes objects change from 9 + 4 to 10 + 3	Count 9 counters and 4 counters, then move 1 from 4 to make a group of 10 with 9.	See 3 left in 4.	Just see and think 9 and 1 is 10 (making of 10 already happened in step 1).	See 10 counters and see 3 counters and think ten-three or count on "ten-one, ten-two, ten-three."
Finger use makes each step visible separately	Open 9 fingers, see 1 more finger is folded to reach 10.	Open 4 fingers, fold 1, and see 3 fingers are still left.	Open 9 fingers, open 1 more, and see 10 fingers, or remember it has been done already with step 1.	Open 10 fingers, say "ten," fold them again, and open 3 more fingers and count-on as they are folded, ["ten-one, ten-two, ten-three."] or know 10 and 3, 13.
Visual representational drawings, make a numerical trace of old and new problems and of steps in the change process visible	9 + 4 / 1 The line under 4 toward the place between 9 and 4 helps students know they need to think of 9's partner to make 10.	9 + 4 / \\ 1 3 Two lines under 4 indicate how the number 4 is separated into two partners.	10 (9 + 4) / \\ 1 3 Circling of 9 and 1 shows how two numbers are combined to make 10.	10 (9 + 4 = 13) / \\ 1 3 Shows 3 is the only number that is not yet a part of 10. So 10 + 3.

Note: The first two steps are also facilitated by the linguistic support of the term "partners" for the two addends that form the totals. The final step is also facilitated by the Japanese linguistic form of 13 as "ten three."

Fig. 7.3. Teaching supports and their facilitation of the BAMT steps (example: 9 + 4)

Phase 1: Elicit, value, and discuss student methods (lessons 1 and 2)

For the initial introduction of the unit, Mr. Otani showed a group of 9 blue and 4 red magnetic counters in a row on the board. Some students immediately shouted out the answer, "13!" Mr. Otani then initiated discussion by saying, "Some of you are quick in telling the answer, but who can share with the class your thinking?" Several students raised hands to share their ideas. Sakiko went to the board when called and moved 1 red counter to add to the blues.

Sakiko: From 4, I add 1 to 9 to make 10.

Mr. O: So, the 9 became 10 and 4 became 3?

Sakiko: Then we know 10 and 3 make 13. We learned that before.

Mr. Otani summarized Sakiko's method on the board and continued to ask for other students' contributions to drive the discussion.

Mr. O: Did anyone do this differently? 9 + 4? Nobuhiko?

Nobuhiko: 9 + 4 is . . . at first, 3 and 4 is 7.

Students: What? What are you saying? We don't understand!

Mr. O: Will you say it again, Nobuhiko?

Nobuhiko: I took 3 from 9 . . . [moves 3 counters from 9]

83

Goals	Make 9 into 10 with part of the other addend	Find how many more are left to add to 10	Make 10 for new problem, 10 + 3,	In new problem. 10 + 3, find total
Steps	Step 1 Find that 9 needs 1 more to make 10	Step 2 Separate 4 into 1 and the rest (3)	Step 3 Add 9 and 1 to make 10	Step 4 Add 10 and 3 to make 13
Visual representational support in textbooks and on the board	$9 + 4$ / 1	$9 + 4$ /\ 1 3	10 $\overbrace{9 + 4}$ /\ 1) 3	10 $\overbrace{9 + 4} = 13$ /\ 1) 3
Level A Support: Steps 1, 2, 3, 4	"9 and what number make 10?" (Teacher points to 9.)	Teacher draws sticks to elicit break-apart partners for 4. "What two numbers are you separating 4 into (to make 10)?"	"What do 9 and 1 make?" (Teacher circles 9 and 1, writes 10 next to the circle.)	"What do 10 and 3 make?" (Teacher points to numbers 10 and 3, says "ten and three" to make connections to the total "ten-three.")
Level B Support: Steps 2, 3, 4		Teacher draws sticks to elicit break-apart partners for 4. "What two numbers are you separating 4 into (to make 10)?"	"9 and 1 make . . . ?" (Teacher points to 9 and 1).	"10 and 3 make . . . ?" (Teacher points to 3.)
Level C Support: Steps 2 and 4		Teacher draws sticks to elicit break-apart partners for 4. "4 is what number and what number?"		"10 and 3 make . . . ?" (Teacher points to 3.)
Level D Support: Step 2 visually		Teacher draws sticks to elicit break-apart partners for 4. No verbal guiding.		
Level E Support: Step 4 visually and with partners		Break-apart partners are filled in from level D. No verbal guiding.		(no guiding question)

Notes: Each level supports fewer steps. Levels D and E often occurred in combination to support the learning process. For level D, Mr. Otani all but once elicited only step 2 and students typically gave break-apart partners of the addends. Following this step, level E support occurred when the break-apart partners remained on the board for a visual cue while students stated answers to problems without verbal guiding.

Fig. 7.4. Steps, drawing, and levels of teacher assistance for learning the BAMT method

Mr. O:　You took 3 from 9?

Nobuhiko:　Add that 3 and 4 . . . *[puts 3 and 4 counters together]*

Mr. O:　So, 9 is . . .

Nobuhiko:　9 became 6. Separate 9 into 3 and 6 . . . Then, 7 . . . I mean . . . Because 7. 4 and 3 became 7, so 7 and 6 is 13. *[points to the counters]*

Mr. O: 13. 7 and 6 make 13. You like 7 + 6, don't you?

Students: Nobuhiko has remembered that way for a long time!

Mr. O: Is that so?

Students: Yeah, he always does that way. This is Nobuhiko's secret method!

Students worked to understand Nobuhiko's solution method. His idea, which seemed incomprehensible at the beginning ("at first 3 and 4 is 7"), was gradually clarified with guiding questions by Mr. Otani. As other students made sense of his thinking, they accepted his method as a valid mathematical approach. As Nobuhiko worked to articulate his idea to his classmates, he explained one step at a time. This careful reflection on his own thinking helped develop mutual understanding in the community. Other students actively evaluated the quality of Nobuhiko's explanation by demanding that he clearly describe the process of his thinking. In these ways, Mr. Otani and his students negotiated and established shared understanding of what a good mathematical solution and explanation of such a solution should be. The fact that this method is an atypical method (it was not listed in the Teachers' Manual) emphasizes how a Class Learning Zone can involve idiosyncratic student thinking and not just typical methods.

In this first lesson of the unit, Mr. Otani encouraged students to share their ideas based on their spontaneous thinking and prior knowledge and allowed room for diverse methods, including counting all (see our original 2006 article for all methods shared). The sharing process at the beginning of the unit provided opportunities for students to review previously learned concepts, demonstrate their competence, and set the stage for future exploration. Mr. Otani carefully directed student discussion to focus on the process of solving the problem, thus providing opportunities both for the students who already knew the answer and for those who were experiencing such a complex problem for the first time.

Phase 2: Focus on the BAMT method (lessons 2 to 4)	Figure 7.5 shows the changing levels of support provided by Mr. Otani related to changing lesson activities in the instructional unit. In the unit, the problems were introduced by the same first addend, $9 + n$, as it is easier for students to see how 9 needs 1 more to make 10 and to decompose the second addend into $1 + n$ (step 2 of the BAMT method). By the middle of the unit, students re-viewed[1] the ten partners of 9, 8, and 7 as a way to explore $6 + n$ problems. These re-views not only helped refresh students' memories but also connected their previous knowledge to the new topic. When a new first number was introduced, it was typically introduced as an extension of solving the problem with a more familiar number. For example, in introducing, $7 + n$ addition problems in lesson 5, Mr. Otani first supported students with $9 + n$ and $8 + n$ problems on the board by asking individual students to solve them. He then wrote $7 + n$ problems on a small portable board, placed it next to the $8 + n$ problems, and continued the previous questioning pattern. Placing different types of problems side-by-side highlighted the similarities between the problem sets. From day 2 to day 4, Mr. Otani shifted the conceptual emphasis of the re-views from the first step to the general pattern in the second step for all problems beginning with 9 to a short-cut way of thinking about these problems.

Mr. Otani also led the students to compare different methods shared in the discussion and had them vote on what they each considered to be the most accessible method. While the votes split at the beginning of the unit, over time the students came to see the value of the BAMT method.

---

[1] We write "re-view" with a hyphen to convey the substance of these sessions in Japanese schools. The "pre-view and re-view" learning routine is embedded in daily practices.

	Activities	Level of Support Used						
		A	B	C	D	E	No	V
1	1. Whole-class exploration of different methods for 9 + 4							
	2. Voting for the easiest method [IC]							
2	1. Whole-class re-view of methods (9 + 4) [IC]							
	2. Voting for the easiest method [IC]							
	3. Discussion of place-value and the BAMT method [IC]							
	4. Whole-class intro for 9 + #							
	a. Step 1 for the set of 6 problems (discussion of 9's partner to make 10 [IC])	As						
	b. Step 2 for the set of 6 problems	As						
	c. Step 3 for the set of 6 problems	As						
	d. Step 4 for the set of 6 problems	As						
	5. Voting for the easiest method [IC]							
3	1. Whole-class re-view of methods (9 + 4) [IC]							
	2. Whole class practice of 9 + # (3 problems) steps 1, 2, 3, 4	Ap						
	3. Individual practice, 9 + # (4 problems)							
	4. Individual-in-whole-class practice of 9 + # (problems from 3)							V
	a. Step 1 for the set of 4 problems	As						
	b. Step 2 for the set of 4 problems	As						
	c. Step 3 for the set of 4 problems	As						
	d. Step 4 for the set of 4 problems	As						
	5. Discussion of 9's partner to make 10 [IC]							
	6. Voting for the easiest method [IC]							
4	1. Whole-class re-view of the BAMT method, 9 + 3, steps 1, 2, 3, 4 [IC]	Ap						
	2. Whole-class re-view of 9 + # (6 problems) steps 2, 4 [IC]							
	3. Individuals-in-whole-class review of 9 + # (problems from 2) steps 2			Cp				
	and 4			Cp				
	4. Individuals-in-whole-class practice of 9 + # (problems from 2), BA partners written on the board (other things erased), Ss say answers, 6 problems					Ep		
	5. Individual-in-whole-class practice of 9 + # (problems from 2), BA erased, Ss say answers				Dp			
	6. Whole-class intro for 8 + # (8 + 3), steps 2, 3, 4 [IC]							
	7. Individuals-in-whole-class practice, 8 + # (7 problems)		Bp					
	a. Step 2 only, with break-apart sticks							
	b. BA written on the board, Ss say answers, T points to random problems				Ds	Ep		
5	1. Whole-class re-view of 9 + 5 and 8 + 6, steps 2, 3, 4 [IC]		Bp					
	2. Individual-in-whole-class practice of 15 mixed 9 + #, 8 + #							
	a. Step 2 only, with break-apart sticks				Ds			
	b. BA written on the board (other things erased), Ss say answers, T points to random problem					Ep		
	3. Individual-in-whole-class intro of 7 + # (6 problems) [IC]							
	a. Step 2 only, with break-apart sticks				Ds			
	b. BA written on the board (other things erased), Ss say answers					Ep		
	4. Whole-class say answers to 7 + #, with BA partners written					Ep		
	5. Individual practice of 7 + # (4 problems)					Ep		
								V

Fig. 7.5. Levels of support over eleven lessons

	Activities	Level of Support Used						
		A	B	C	D	E	No	V
6	1. Whole-class intro of 6 + 5 by discussing 10 partner for 6, re-views of 10 partners for 9, 8, 7 [IC] 2. Individual practice of 6 + # (5 problems) 3. Individual-in-whole-class report of 6 + # (problems from 2) while the rest of the class gave feedback, "It is OK!" 4. Individual practice for 16 mixed 6 + #, 7 + #, 8 + #, and 9 + #						No	V V
7	1. Individual-in-whole-class report answers on problems solved in lesson 6; T writes equation and answer as it is shared, class gives feedback, "It is OK!" 2. Whole-class intro for smaller + larger (4 + 8, equation and answer only) [IC] 3. Individual practice of 12 smaller + larger; teacher notices that many students are counting on, so shifts to 4. 4. Whole-class discussion on smaller + larger, 2 + 9, steps 1, 2, 3, 4, solved from 9 and from 2 [IC]	Ap					No	V
8	1. Individual practice of 11 smaller + larger problems 2. Individual-in-whole-class report answers on problems just solved (as in lesson 7, 1 above) 3. Individual practice of 2 word problems 4. Individual-in-whole-class report on problems just solved (disagreement on quantifiers)						No No	V V
9	1. Individual practice on 8 mixed problems 2. Individual-in-whole-class report answers on problems just solved (as in lesson 7, 1 above) 3. Individual practice on 8 mixed problems						No	V V
10	1. Like lesson 9 (no observation)							V
11	1. Whole-class report answers on 8 mixed problems solved in previous class (as in lesson 7, 1 above) 2. Individual practice on 6 mixed problems 3. Individual-in-whole-class report answers on problems just solved (as in lesson 7, 1 above)						No No	V

Fig. 7.5. *Continued*

Notes: The support always involved the drawing on the board and sometimes (especially for individuals) also involved fingers or counters. The support identified is standard support for the class. Some individuals might have received more support. Small "s" and "p" placed after the letter that shows the support level mean "steps" and "problems." For example, "As" means level A support for a step; "Ap" means level A support for solving the whole problem. "No" means no support. "V" means varied support with students (for individual practice). [IC] means instructional conversation.

Part 3: Gaining fluency with the BAMT method (lessons 5 to 11)

When the word *practice* is used in Japanese classrooms, it conveys a meaning slightly different from what it does in English. The Japanese word *practice* is written as a combination of two Chinese characters: 練習; the first character means "kneading" and the second character means "learning." Together, the characters represent the meaning of kneading different ideas and experiences together to learn. Such kneading was observed in all three of the different modes of practice identified: whole-class practice, individuals in whole-class practice, and independent practice (see fig. 7.5). Students brought different ideas, experiences, and approaches to learning the BAMT method, and the differences were "kneaded" through various practice forms to support each student's learning as well as to establish a common understanding base in the classroom.

For the whole-class practice, Mr. Otani typically stood in front of the class with a set of problems written on the board. As he asked questions to support students to take a specific step in the BAMT method, he pointed to that part of the problem on the board, and students answered the questions together aloud. Mr. Otani often pointed to the questions on the board in order (e.g., from left to right, from top to bottom), but he sometimes pointed to the questions randomly so students could not think ahead. Sometimes, his questions assisted all of the steps to solve one problem and the same questions were asked to solve the next problem. At other times, he asked questions for one particular step for all the problems on the board, then moved on to the next step for all of the problems. Step support happened in lessons 2 and 3 when students were learning the steps of the BAMT method for the first time; in lessons 4 and 5, this support assisted their learning of step 2 (the most challenging step) for the first addend 8 (lesson 4) and then addend 7 (lesson 5) by combining level D step support and level E problem support. Students were encouraged to speak loudly for all whole-class practices, and they answered with enthusiasm and energy.

Individuals in whole-class practice followed the same pattern as the whole-class practice. With a set of problems written on the board, the same questions were asked, but students took turns answering the questions individually. Students usually answered by their seating order (e.g., starting from the students who sat at the front row of the right side of the room to the students in the back row, then to the students in the next column, etc.). As with the whole-class practice, Mr. Otani changed the order of the problem at times. The most distinctive difference for this individual in whole-class practice is that after one student answered the question, the student always asked the whole class, "Is it OK?" The whole class then answered by shouting together, "It is OK!" if they agreed with the student, or "It is not OK!" if they did not. When they disagreed, Mr. Otani guided the discussion among students to identify and resolve the disagreement.

For independent practice, students worked at their seats solving problems independently. Often, they worked on assigned problems from textbook pages, but as they finished those, they worked on a packet of worksheets Mr. Otani had prepared or a set of calculation cards (small flash cards held together by a ring). When individual students had difficulties, students who sat close by spontaneously helped them.

The interactions among these different modes of practice supported student learning in different ways. The whole-class practice provided a fun and safe group-learning environment where students shouted answers together. Individuals in whole-class practice offered opportunities for individual students to show their developing fluency with the method and get whole-class feedback. Individual practice allowed students to focus on areas where they needed more work and also created a foundation before the whole-class sharing of individual in whole-class reporting answers.

Phase 4: Delayed practice	Delayed practice happened in the re-view section of the textbook where concepts that students learned previously are revisited and practiced. Here, students re-viewed independently in familiar practice contexts. The BAMT method was also used in a related or more complex method in a subsequent unit of subtraction using 10 and in multidigit addition in grade 2.
Shifts in Teacher Levels of Assistance	The steps involved in the BAMT method were not difficult when they were taken one step at a time, because each had been learned in previous units. However, many students experienced difficulty coordinating the steps into a fluent whole. Initially, Mr. Otani supported each step by questions (see level A support in fig. 7.3). He then dropped support one step at a time,

eliminating the easier steps 1 and 3 first and keeping support for the most difficult step 2 at the final level (visual only). However, he always increased the levels of support for students who needed it.

Figure 7.5 shows how this full support decreased over time through levels B through E and how it varied with the kinds of problems. On days 4 to 6, assistance decreased as the class continued to practice a given type of problem but increased when they began a new type of problem. From day 4 through day 7, the initial level of assistance at the beginning of the day decreased.

Sometimes more-advanced students spontaneously modeled for the class the BAMT method with fewer steps than the steps Mr. Otani was supporting in the Class Learning Zone. For example, when students in the whole-class practice were experiencing level C support (steps 2 and 4), Sachiko stood up to solve the problem 9 + 5:

Mr. O: *[points to the problem 9 + 5]*

Sachiko: *[starts talking before Mr. Otani can ask guiding questions]* 10 and 4 is 14.

Mr. O: OK, OK, what did you do first?

Sachiko: Separated 5 into 1 and 4, then 10 and 4 is 14. Is it OK?

Students: It is OK!

Here Sachiko followed only step 4, but Mr. Otani elicited from her steps 2 and 4.

On day 7, Mr. Otani introduced a new class of problems in which the smaller number was the first addend (e.g., 2 + 9 instead of 9 + 2). In the individual practice, many students solved these problems by counting on and did not use the BAMT method. When Mr. Otani realized this, he initiated a conversation shifting back to level A support to discuss BAMT solutions for 2 + 9 (making a 10 with the second addend) and related the solutions of 2 + 9 and 9 + 2 to each other. He drew on the board full representational drawings for 2 + 9 (where 2 was broken into 1 and 1 to make a 10 with 9) and then for 9 + 2 (where the same partners of 1 and 1 were shown under 2 but now on the right). He then guided student discussion of these solutions by using two groups of 2 and 9 counters and asking students, "Can we move counters like this and make 10 on this side [for 1 + 1 + 9 = 11]?" and then for the counters 9 and 2, "Can we move counters to make 10 to make 11 this way [for 9 + 1 + 1 = 11]?" He then wrote 2 + 9 and 9 + 2 on top of each other and led a discussion by questioning to help students analyze which of these was easier and to see the similarities between the new situation and the larger-plus-smaller-addend addition situations they had been solving using the BAMT method. Most students quickly went back to use the BAMT method, and most started with the larger number even if it was the second addend.

Mr. Otani's questions also shifted through levels to become more abstract and informal. His questions at the beginning of the unit were explicit directives (e.g., "What number do 1 and 9 make together?"). As the unit progressed, he was more likely to state the same question as a process in action (e.g., "9 and 1 is . . . ?"), or sometimes he only pointed to the numerals on the board as an implied nonverbal question (see fig. 7.4).

Data from individual student interviews and classroom work showed how each target student moved along a unique learning path, which illustrated his or her own learning trajectory. A full discussion of the target students appears in our *JRME* article (Murata & Fuson, 2006).

Table 7.1

Aspects of teaching for understanding and fluency: Examples from the Japanese Class Learning Zone Classroom

Focus on meaning supports (representational and cultural/visual tools) and on conceptual discussion	
Visual, linguistic, and sensori-motor representational support for learning steps	• Use visual representations (physical objects, drawings, and fingers along with oral explanations) to strengthen students' understanding of crucial steps: – Move objects to show 9 becoming 10 – Circle numbers to make 10 – Draw upside-down "v" to show break-apart partners – Emphasize the critical conceptual step by using a colored ten in the drawing • Help students make connections between different representations
Focus on individual mathematical thinking	
Discuss, value, and assist students' ideas and thinking	• Allow students to share ideas and different approaches • Ask questions to guide student thinking • Maintain students' ownership of ideas (call different methods with students' names, vote for different methods after discussing advantages and disadvantages)
Support students' different learning paths	• Vary questioning patterns to meet different levels of understanding of individual students and provide modeling and explanation when needed • Include less-advanced students in whole-class practice to allow them to experience the whole process rapidly, but support their individual solving as necessary with questions and modeling as needed • Consider differences among students as strengths, and create situations where they benefit from the differences
Focus on mathematics	
Support generalization and focus on the mathematics of their learning	• Support generalization of problems with the smaller number first – Introduce and practice problems by their mathematical structure (e.g., 9 + # problems, then 8 + # problems, etc.) to support initial learning – Discuss the similarities and differences of problems according to their mathematical structure (size of first addend) to support generalization • Discuss mathematical aspects of methods (e.g., the new unit of 10 related to place value) – Discuss whether to start with the smaller or larger addend
Focus on assisting all students to speech-for-self, abbreviation, and automatization	
Facilitate fluency	• Provide opportunities to practice with decreasing visual and question support • Pair students and encourage them to practice using flashcards that mix the problem types (practice with immediate feedback) • Send a worksheet packet home that explains to parents what students are learning, and asks them to time the students as they finish their homework. This helps the teacher understand the fluency level of each student.

Teaching for Understanding and Fluency

Our case study gave life to the ZPD Model of Mathematical proficiency. It illustrated how one teacher assisted student learning by valuing students' informal knowledge and approaches, allowing students time and opportunities to explore different ideas, helping bridge the distance between their existing knowledge and the new method, and giving time for students to practice and gain fluency with a newly learned method. Table 7.1 summarizes aspects of the teaching. The focus on meaning supports and on conceptual discussion of individual mathematical thinking, on aspects of the mathematics, and on a learning path helped students develop understanding of the overall method, coordinate and verbalize steps in the multistep method, and develop and move toward fluency. Mr. Otani assisted students of different fluency levels to work together, and he helped individuals to move forward within their own learning path. The visual and verbal question teaching supports were gradually internalized by students as they used them to provide self-assistance to coordinate or carry out particular steps. Mr. Otani assisted community and individual interaction for everyone's learning, including his own, as the class developed shared understanding. The use of consistent visual representational supports kept the community together, as it helped reduce the differences between individual students during whole-class and individual practice.

The students in this grade 1 classroom were always willing and eager to support and adjust their own levels to the levels of their peers whose learning paths were different from their own. The emphasis on relationship and "sameness" in the culture helps create an environment in which students understand difference as a norm but also a changeable characteristic, and thus they try to be like one another. Helping one another is a part of their identities, and that is well supported in various classroom rituals and activities. There are similar examples in U.S. classrooms where teachers work diligently to create a collaborative learning environment to help students learn mathematics. In these classrooms, teachers value collaboration among students over competition, and they create a safe environment in which students are encouraged to share their mistakes, ask questions when they do not understand, and provide explanations when they are not completely sure. Teachers can model these actions first (making mistakes, show confusion, etc.), recognize and support students when they also try to share their incomplete ideas, and explicitly discuss why these actions are important in helping everyone learn.

In Murata (2013) and Murata et al. (under review), the model was further developed and focused on responsive interactions within urban classrooms in the United States. Teachers can orchestrate and coordinate unique individual student learning paths by connecting and relating different ideas so that students push and stretch each other's ZPD. In planning a lesson, teachers can carefully consider the differences and the coordination of the differences, so students will learn from one another. We consider this interaction among different ideas to be a critical aspect of responsive teaching, when we maintain high academic expectations by taking advantage of diversity among students. In classrooms, different teachers' facilitation styles and different combinations of student ideas can drive classroom learning on somewhat different learning paths. Student diversity will always be there. By making these otherwise invisible student learning paths visible in planning and in orchestrating classroom discussion, we take advantage of the wealth of ideas students bring into classrooms. When we shift our focus from telling math content to students to coordinating and extending different student ideas, teaching will hold new meanings for teachers and for students, as everyone helps along the Class Learning Path to important grade-level understanding.

References

Murata, A. (2013). Diversity and high academic expectations without tracking: Inclusively responsive instruction. *Journal of Learning Sciences, 21*, 312–335. doi:10.1080/10508406.2012.682188

Murata, A., & Fuson, K. C. (2006). Teaching as assisting individual constructive paths within an interdependent class learning zone: Japanese first graders learning to add using 10. *Journal for Research in Mathematics Education, 37*, 421–456.

Murata, A., Siker, J., Kang, B., Kim, H-J., Baldinger, E., Scott, M., & Lanouette, K. (under review). *Classroom group learning and individual student learning: Case of two first grade math talk lessons.*

National Governors Association Center for Best Practices & Council of Chief State School Officers (NGA Center & CCSSO). (2010). *Common core state standards for mathematics.* Washington, DC: Author. Retrieved from http://www.corestandards.org.

Tharp, R. G., & Gallimore R. (1988). *Rousing minds to life: Teaching, learning, and schooling in social context.* New York, NY: Cambridge University Press.

Tokyo Publishing. (2000). *New mathematics.* Tokyo, Japan: Author.

Vygotsky, L. S. (1978). *Mind in society: The development of higher psychological processes.* Cambridge, MA: Harvard University Press.

Vygotsky, L. S. (1986). *Thought and language* (A. Kozulin, Ed. & Trans.). Cambridge, MA: MIT Press.

Chapter 8

Giving Change When Payment Is Made with a Dime

The Difficulty of Tens and Ones

Cynthia C. Chandler
Black Hills State University

Constance Kamii
University of Alabama at Birmingham

"I can't do anything with this!" This is what a second grader, Joslyn, said when we gave her a dime to pay for 6 cents worth of candy in a store game. Her statement was surprising because we had asked her what a dime was worth, and she had clearly replied that it was worth 10 cents. As it turned out, the problem was about tens and ones. In this chapter, we explain the reasons for the child's difficulty and discuss their educational implications.

First and second graders are taught to recognize coins and to add their values. This instruction is sometimes given not only to teach about money but also to reinforce "place-value concepts." For example, Hatfield, Edwards, Bitter, and Morrow (2000) stated, "Money reinforces place-value concepts because it is composed of a base-10 system. In counting numbers, exchanges of 10 individual units for a set of ten can be compared to exchanging 10 pennies for 1 dime" (p. 315). Kamii (2000, 2004), on the other hand, does not recommend the use of coins, base-ten blocks, or Unifix cubes for teaching tens and ones, reasoning that children need to abstract tens out of the ones that are in their heads, rather than from objects in the external world. Van de Walle, Karp, and Bay-Williams's (2013) position, on the other hand, is an unfortunate attempt to combine two theoretical positions that cannot be combined. In the authors' words, "Physical models for base-ten concepts play a key role in helping students develop the idea of 'a ten' as both a single entity and a set of 10 units. Remember, though, that the models do not 'show' the concept to the students; the students must mentally construct the 'ten makes one relationship' and impose it on the model" (p. 195). Educators are thus not in agreement about the value of coins, base-ten blocks, and Unifix cubes for instructional purposes.

The main purpose of our study was to examine children's thinking about tens and ones by finding out how they give change when a dime or a dime and a few pennies are offered as payment for a purchase of up to 9 cents. The work of Jean Piaget, especially his theory about logico-mathematical knowledge and the distinction he made between empirical and constructive abstraction, explains the difficulties that children like Joslyn have with coins.

Theoretical Background: Three Kinds of Knowledge

The three kinds of knowledge Piaget (1950) distinguished according to their ultimate sources are physical knowledge, logico-mathematical knowledge, and social-conventional knowledge. Physical knowledge is knowledge of objects in external reality. Our knowledge of the weight and color of a coin is an example of physical knowledge. Another example of this type of knowledge is the fact that a glass is likely to break if it is dropped on the floor. By contrast, the fact that a penny is worth 1 cent but a dime is worth 10 cents is an example of social-conventional knowledge. Languages like Spanish and English are also examples of social-conventional knowledge. The ultimate source of social-conventional knowledge is conventions, which are made by people, but the ultimate source of physical knowledge is objects.

This chapter is adapted from C. C. Chandler & C. Kamii (2009), Giving change when payment is made with a dime: The difficulty of tens and ones, *Journal for Research in Mathematics Education, 40,* 97–118.

Physical and social knowledge thus have sources in the external world. The ultimate source of logico-mathematical knowledge, however, is in each individual's head. Logico-mathematical knowledge is made up of mental relationships that each person creates. For example, if we are presented with a penny and a dime, we can say that the two objects are different. But if we decide to ignore their size and color, we can also say that the two coins are similar (because they are both round and made of metal). The reason it is just as true to say that these objects are different as it is to say that they are similar is that "different" and "similar" are mental relationships that each person makes by thinking about the two coins.

A third mental relationship an individual can create between the two coins is the numerical relationship "two." Each coin is empirically observable (physical knowledge), but "one" is not. Only when we think about a coin as "one" does it become "one." "Two" is likewise a mental relationship we can create when we think about two coins. And if "two" is constructed by each individual, all the other numbers (like "eight," "ten," "one ten," and "two tens") must also be constructed by each individual.

| Abstraction and representation | Piaget made a distinction between empirical (or simple) abstraction and constructive (or reflective or reflecting) abstraction. In empirical abstraction, we focus on one or more properties of objects and ignore the others. For example, when we categorize objects by color, we focus on color (physical knowledge) and ignore all the other properties. In constructive abstraction, by contrast, we create mental relationships (logico-mathematical knowledge) that do not exist in objects. For example, when we are presented with a dime and a penny and say that they are "different," we are creating a mental relationship that does not exist in the objects in the external world. This relationship is created by our thinking. The numbers "two," "eight," "ten," or any others are likewise created by each individual through constructive abstraction. |

The verb *to represent* is often used carelessly, as when we hear that base-ten blocks represent the base-ten system. Base-ten blocks do not represent anything by themselves. Representing is what an individual does. When adults are presented with 4 ten-blocks and 6 one-blocks, we can represent 46 to ourselves because we have already abstracted a system of tens out of our system of ones. When young children see the same blocks, however, they cannot represent 46 to themselves unless they have constructed a system of tens through constructive abstraction. They can give to these objects only the meaning that their level of abstraction permits.

When children first construct number, they construct a system of ones. Therefore, for most first graders, 23 is 23 ones, as can be seen in figure 8.1a. When these children go on to second grade, most of them begin to construct higher-order units of tens out of the system of ones that they constructed before (see fig. 8.1b).

Figure 8.1c is an example of what happens when children are taught the social knowledge of saying "10, 20, 30 . . ." before they have constructed higher-order units of tens. These children have learned to recite or memorize the words (social knowledge) without representing their own logico-mathematical knowledge. Because they cannot yet think about tens and ones simultaneously, they cannot shift to ones after saying "20." When adults say "10, 20, 21, 22, 23," as shown in figure 8.1b, they can shift to "ones" because they have been thinking about ones while saying "10" and "20." If they could not think about tens and ones simultaneously, they would not be able to shift to ones after saying "20."

We were surprised when Joslyn insisted that she could not do anything with a dime, because we knew she could count by tens and add a few cents to give the value of a dime and a few pennies. Joslyn's insistence resulted from the fact that she had not yet constructed tens out of her own system of ones. For her, a dime was one thing, and ten pennies were something else, in spite of the words she had learned.

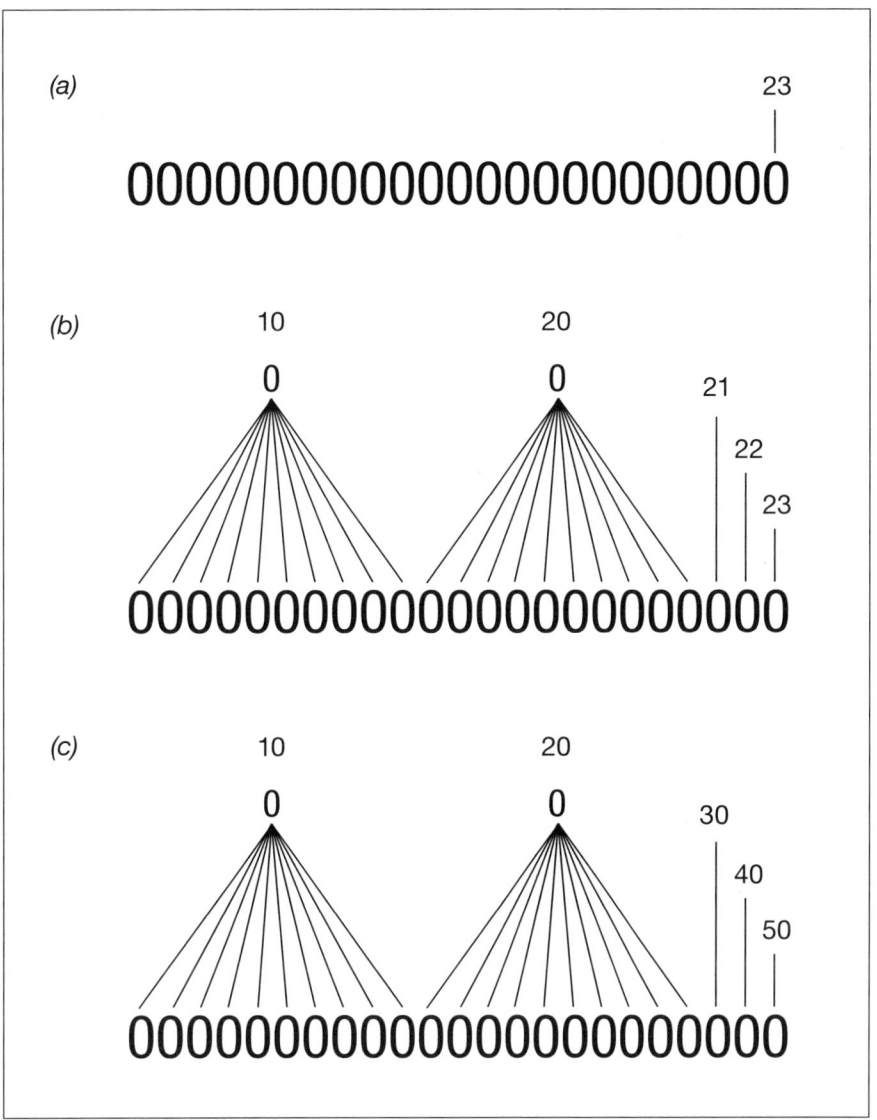

Fig. 8.1. Counting twenty-three objects (*a*) with a system of ones, (*b*) with well-coordinated systems of tens and ones, and (*c*) with only words implying a system of tens

Our Study

Method

Ninety-eight children in kindergarten through fourth grade were interviewed in one urban elementary magnet school in the South (about twenty students at each grade level). No child was excluded because of race, gender, or ability. The sample consisted of 66 percent African American, 31 percent Caucasian, 2 percent Asian, and 1 percent Hispanic participants. Sixty-five percent of the participants were receiving free or reduced-price lunches.

The children were individually interviewed with a store game adapted from Berti and Bombi (1981/1988). The child was in the role of "storekeeper," and the interviewer was the "customer." Each interview began with the child being asked to state the name and value of the coins in a box (the "cash register") that contained 2 dimes and 20 pennies. The interviewer also had a box, but it contained 4 dimes and 20 pennies. After each transaction, the child was asked how he or she figured out the cost and change (if any). All the interviews were videotaped, and anecdotal notes were written on a form.

As can be seen in table 8.1, the first purchase was for one small piece of candy (S). The child was asked to state the cost (2 cents) and was paid the correct amount. This amount was offered to see whether the child might try to give "change" for the correct payment. The second purchase was for a large piece of candy (L). As in the previous transaction, the child was asked to state the cost (3 cents) but was purposely paid 4 cents. The third purchase was for two small candies (SS), and 8 pennies were offered to pay for the 4-cent purchase. The second and third purchases were made to find out if and how children gave the correct change when payment was made only with pennies. In the fourth and sixth tasks (LL and LS), a dime was offered for a 6-cent purchase and a 5-cent purchase, respectively. In the fifth and seventh purchases (LLL and LLS), 1 dime and 2 or 4 pennies were offered, respectively, for a 9-cent and an 8-cent purchase.

Table 8.1
Sequence of transactions in the store game

Candy purchased	Cost	Coins used as a payment by interviewer
1. One small (S)	2 cents	2 pennies
2. One large (L)	3 cents	4 pennies
3. Two small (SS)	4 cents	8 pennies
4. Two large (LL)	6 cents	1 dime
5. Three large (LLL)	9 cents	1 dime and 2 pennies
6. One large and one small (LS)	5 cents	1 dime
7. Two large and one small (LLS)	8 cents	1 dime and 4 pennies

Data analysis

The following four levels were conceptualized for categorization of the children according to their responses to the task:

> **Level 0.** Level 0 children were of three types: (a) those who did not give any change, (b) those who returned random amounts as change, and (c) those who returned exactly the same amount as what was paid.

> **Level 1.** These children gave the correct change only when payment was made exclusively with pennies (tasks 2 and 3).

> **Level 2.** These children gave the correct change when payment was made with pennies only (tasks 2 and 3) and with a dime only (tasks 4 and 6), but not when it was made with a dime and a few pennies (tasks 5 and 7).

> **Level 3.** These were children who gave the correct change in all seven of our tasks.

After all the children were categorized, a retired professor was asked to review 50 percent of the tapes independently (ten selected randomly from each grade level) in order to check the categorizations. The coefficient of inter-rater reliability was found to be .90, thus confirming the correctness of the categorizations.

Results

As can be seen in table 8.2, the majority of the kindergartners (58 percent) were found to be at Level 0. In first grade, 38 percent remained at Level 0, but 48 percent went up to Level 1 (where they could give the correct change when payment was made exclusively with pennies). The second graders were distributed at all four of the levels, and the highest level was reached by 48 percent of the third graders and 63 percent of the fourth graders.

Table 8.2
Number and percentage of children categorized at each level by grade

Level	K	1	2	3	4	Total
0	11 (58%)	8 (38%)	5 (25%)	1 (5%)	0 (0%)	25 (26%)
1	7 (37%)	10 (48%)	5 (25%)	4 (21%)	3 (16%)	29 (30%)
2	0 (0%)	1 (5%)	2 (10%)	5 (26%)	4 (21%)	12 (12%)
3	1 (5%)	2 (9%)	8 (40%)	9 (48%)	12 (63%)	32 (32%)
Total	19 (100%)	21 (100%)	20 (100%)	19 (100%)	19 (100%)	98 (100%)

Task 4 was of special interest because it required being able to give change when a dime was offered for a 6-cent purchase. Below is part of the interview with Joslyn, the second grader mentioned at the beginning of this chapter. (A videotape of this interaction can be viewed on www.constancekamii.org.) Because she did not meet the criteria for Level 2, Joslyn was categorized at Level 1.

The interviewer had just tendered a dime for a 6-cent purchase.

Joslyn: That's ten. I need 6 cents. I don't need 10 cents *[returning the dime.]*

Interviewer: I can't pay with 10 cents? This is 10 cents. But last time you gave me some money back.

J: Oh! *[She counts 6 pennies out of her "cash register" and puts them in one hand and the dime in the other.]*

I: You got 6 pennies because?

J: Because you need them.

I: So?

J: Here. *[She gives the dime back to the interviewer and 6 pennies to herself.]*

I: How much is that altogether?

J: Sixteen cents. . . . *[She takes a different dime out of her "cash register."]* You take the 6 cents out of this dime.

I: OK. . . . How?

J: Give it to the person and tell them to take 6 cents out of it to buy the suckers.

I: You take 6 cents out of it.

J: *[no response]*

I: How much is the dime?

J: Ten cents. . . . You count out 6 pennies and give it to the person.

I: I count out 6 pennies, or you count out 6 pennies?

J: Me! *[She counts 6 pennies out of her "cash register," gives them to the interviewer, and keeps her dime.]*

As could be seen in this account, Joslyn knew that a dime was worth 10 cents and that 10 cents was more than 6 cents. She easily added 6 cents to 10 cents, but a dime was clearly something different from 10 pennies.

What surprised us the most about children like Joslyn was the fact that none of them thought about exchanging a dime for 10 pennies. They could have met the criteria for Level 2 if they had thought about this exchange. Although a countersuggestion was specifically made in many interviews, such as "I thought you said that a dime was worth 10 pennies," none of the Level 1 children thought about making the exchange.

Educational Implications

Joslyn's problem was that she had learned bits of physical and social-conventional knowledge (e.g., a dime is worth ten pennies), but not the logico-mathematical knowledge about tens and ones. She seemed to have constructed a system of ones (illustrated in fig. 8.1a), and the teacher's objective for Joslyn should be that she construct a system of tens out of her system of ones that was in her head so that the two systems would function together (as shown in fig. 8.1b).

A teacher can encourage children to construct a logico-mathematical system of tens out of their system of ones by encouraging them to think about tens and ones at the same time. For example, if the teacher asks for "a quick and easy way to answer $8 + 6$," children can change the problem to $10 + 4$ if they know all the combinations of two numbers that make 10 (e.g., $9 + 1$, $8 + 2$, $7 + 3$, etc.). In other words, when children think about $8 + 6$, they are thinking only about ones, but when they think "$10 + 4$," they are thinking about 1 ten and 4 ones. There are many games that children can play that encourage them to think about combinations that make 10, such as the five described in the following sections: Tens with Nine Cards, Find Ten, Draw Ten, Go Ten, and Tens Concentration. Other easy games can be found in Kamii (2000).

Tens with Nine Cards

In this game, as well as in the other four games described below, playing cards up to 9 are used (ace = 1). The thirty-six cards are shuffled, and the first nine cards are arranged randomly. The object of the game is to find pairs of cards that make 10; the combinations that can be picked up are $1 + 9$, $8 + 2$, and $7 + 3$. After taking these six cards, the first player replaces them with cards from the drawing pile, and the turn passes to the next player. The person who makes the most pairs is the winner.

The number of players in these games must be limited to two or three because if there are more players, children have to wait longer to have a turn. What is important in these games is that the players think, and having to wait reduces each player's possibility of thinking.

Find Ten

In this game, players pick up any two cards that make 10. The game begins with the thirty-six cards (described in the first game above) being dealt to three players, who keep them in facedown piles. The first player turns over the top card of his or her pile. The first player must always discard his or her card in the middle of the table because there is no other card that can be picked up. If the first discarded card is a 2, for example, and the second player turns over a 6 on his or her pile, the 6 must also be discarded in the middle of the table because $6 + 2 \neq 10$. If the third player then turns over a 4, that player can pick up the 6 with the 4 because $6 + 4 = 10$. The person who collects the most cards is the winner.

Draw Ten	This game is played like "Old Maid" except that each pair must make 10. One of the thirty-six cards is removed from the deck before the other thirty-five cards are dealt to the players. Each player begins by making all possible pairs of 10 with the cards received. When all the pairs have been made and placed on the table, the first player draws one of the cards that the person to his or her right is still holding. Play continues in this way until all the pairs have been made. The person left with the last card loses the game.
Go Ten	This game is played like "Go Fish," except that there is no "pond." All the cards are dealt to the three players, who put down all the pairs of cards that make 10. The first player then asks, for example, "Suzie, do you have a 4?" If Suzie has a 4, she has to give it to Johnny. Johnny can keep asking for cards as long as he keeps getting them. This game offers many possibilities for logical deductions. For example, if Johnny asks Suzie for a 4 and Suzie answers that she does not have any, first graders do not necessarily deduce that the third player must have it.
Tens Concentration	This is a good two-player game, and the teacher can experiment to decide how many cards to use and how to arrange them spatially. One card each of cards 1 (ace) through 9 may be arranged face down in two rows at the beginning. The players alternate in turning over two cards trying to come up with a pair that makes 10 (e.g., $6 + 4$). If a player turns over such a combination, the player can keep both cards. Otherwise the two cards must be returned to their original positions (face down), and the turn passes to the second player. The person who collects the most cards is the winner.
	For children who play these games with ease we recommend advancing the games to include problems like $8 + 6$, which players change to $10 + 4$ as stated before. Another similar problems is $7 + 6$ (which children change to $10 + 3$ or to $12 + 1$). The next progression may then be problems like $10 + 14$, $14 + 14$, $14 + 16$, and $14 + 18$.
A Word of Caution about Subtraction	For traditional math educators, subtraction is only the inverse of addition, and textbooks introduce subtraction soon after addition. For constructivists, by contrast, subtraction is much harder than addition, and it becomes easy for children only when the corresponding sum has become well known (Kamii & Lewis, 2003). For example, $12 - 6$ becomes easy only after $6 + 6$ becomes easy. In first grade, therefore, we recommend concentrating on sums rather than playing games like "Salute!" that involve subtraction.
A Concluding Remark about SES Differences	As can be seen in Chandler and Kamii (2009), we replicated our store task in two schools in upper-middle-class neighborhoods. The percentages of second graders who were found to be at Level 1 were 0 percent and 8 percent, respectively. These percentages are much lower than the 50 percent of second graders in our main study who were found to be at Level 1 or below. These SES differences should be the focus of a future study.

References

Berti, A. E., & Bombi, A. S. (1988). *The child's construction of economics* (G. Duveen, Trans.). Cambridge, England: Cambridge University Press. (Original work published in 1981)

Chandler, C., & Kamii, C. (2009). Giving change when payment is made with a dime: The difficulty of tens and ones. *Journal for Research in Mathematics Education, 40,* 97–118.

Hatfield, M. M., Edwards, N. T., Bitter, G. G., & Morrow, J. (2000). *Mathematics methods for elementary and middle school teachers* (4th ed.). New York, NY: John Wiley & Sons.

Kamii, C. (2000). *Young children reinvent arithmetic* (2nd ed.). New York, NY: Teachers College Press.

Kamii, C. (2004). *Young children continue to reinvent arithmetic, 2nd grade* (2nd ed.). New York, NY: Teachers College Press.

Kamii, C., & Lewis, B. A. (2003). Single-digit subtraction with fluency. *Teaching Children Mathematics, 10*(4), 230–236.

Piaget, J. (1950). *Introduction a l'epistemologie genetique* [Introduction to genetic epistemology]. Paris: Presses Universitaires de France.

Van de Walle, J. A., Karp, K. S., & Bay-Williams, J. M. (2013). *Elementary and middle school mathematics: Teaching developmentally* (8th ed.). New York, NY: Pearson Education.

Chapter 9

Multiplication Methods in the Context of the Common Core State Standards

Bruce Sherin and Karen Fuson
Northwestern University

This chapter is about the strategies that children use to multiply single-digit numbers. Given problems that require them to figure out the products of 3×5 or 6×9, how do they go about figuring out the solution? What must they learn in order to be fully proficient at this type of task?

A big question here has to do with the role of "rote memorization." In the United States, learning all single-digit multiplications has often been called "learning the multiplication tables" or "memorizing the multiplication tables." As mathematics educators, we are committed to teaching mathematics with understanding. Given that commitment, how should we understand the crucial task of learning single-digit multiplications and their related divisions? Looking at a multiplication table can be helpful in thinking about this task (see fig. 9.1). We immediately see that there are many number-specific patterns in the multiplication table. Each column shows a vertical list of the products in numerical order for a given factor. These lists are often called "count-by" lists, because children can learn to say these patterns aloud. Some of these count-by lists have an easy pattern: 5, 10, 15, 20, 25, ... Others have a more difficult pattern: 7, 14, 21, 28, 35, ... The 9s list is particularly rich in patterns. The nature of the pattern depends on the relationship of the count-by number to ten. Identifying and discussing such patterns is a worthwhile and important mathematical endeavor. The lists can also be seen in the rows of the table, but the numerical patterns are a bit easier to see vertically.

1	2	3	4	5	6	7	8	9	10
2	4	6	8	10	12	14	16	18	20
3	6	9	12	15	18	21	24	27	30
4	8	12	16	20	24	28	32	36	40
5	10	15	20	25	30	35	40	45	50
6	12	18	24	30	36	42	48	54	60
7	14	21	28	35	42	49	56	63	70
8	16	24	32	40	48	56	64	72	80
9	18	27	36	45	54	63	72	81	90
10	20	30	40	50	60	70	80	90	100

Fig. 9.1. A multiplication table

This chapter is adapted from B. Sherin & K. Fuson (2005), Multiplication strategies and the appropriation of computational resources, *Journal for Research in Mathematics Education, 36*, 347–395.

Students can use a count-by list to find specific products, such as six fives (6×5) is 5, 10, 15, 20, 25, 30. But generating products in this slow way is not sufficient for most multiplication needs, so students must progress over time to produce specific products rapidly. This is a complex process we discuss next. We can see that there is a great deal of number-specific learning required and that considerable practice will be involved. But it is also clear that this activity need not be simply "rote memorization"; there are patterns here, and children can learn to attend to and use these patterns.

We now turn to how children approach single-digit multiplication. This chapter summarizes our *Journal for Research in Mathematics Education* (*JRME*) article published in 2005, which was based on data collected in earlier years. In that article, we devoted much of our effort to exploring children's invented methods, particularly those that make use of drawings and finger counting. However, much has changed in the intervening years. In the Common Core State Standards for Mathematics (National Governors Association Center for Best Practices & Council of Chief State School Officers [NCA Center & CCSSO], 2010), teaching and learning single-digit multiplication and division begins and comes to fluency in grade 3. This means that we must help students move rapidly from slow and possibly inaccurate methods to meaningful but more rapid methods that can culminate in fluency for multiplication and division of single-digit numbers.

We draw here on the extensive research discussed in the original article (Sherin & Fuson, 2005), which was based on a corpus of 230 interviews with third-grade students that were conducted before, during, and after instruction in multiplication. These interviewed students were in classrooms of the Children's Math Worlds Project (CMW), which combined the design of curricular materials and professional development for teachers with a range of more traditional research activities such as interviews and intensive observations of classrooms. The relevant portions of our interview data were digitized, transcribed, and coded for the methods used. Insights from experience with *Math Expressions* (Fuson, 2009/2013), the published form of the Children's Math Worlds program, are also included here.

Computational Strategies in Addition

We believe it is helpful to start by looking briefly at computational strategies that children use for addition, in part because the story for multiplication differs in some important ways. For addition, children progress through three conceptual levels. These levels capture changes in children's ability to conceptualize the relationships among quantities that are at the heart of the addition task. The list below gives these three levels (we use the language in Fuson, 1992, p. 250), and examples of each follow the list:

1. **Perceptual unit items.** Children must present addition or subtraction situations to themselves using things they can see, such as drawn quantities.

2. **Embedded integration.** All three quantities involved—the two addends and the total—can be simultaneously represented by embedding entities for the addends within the total.

3. **Ideal unit items.** The addends are not embedded within the total but can be conceptualized as outside and can be compared to the total. Numbers become units that comprise numerical triads—two known addends and a known total. This permits recomposition of the addends so that a problem can be transformed into an easier total of different addends.

As children move through the levels, they develop computational strategies that are characteristic of the level. At each level there are general strategies that work for all numbers that a child might be given to add or subtract. Children at the first level use a *count-all* strategy: They count out items for each of the addends, then they count all of the items, starting at

1 and proceeding to the total. In contrast, students at level 2 are capable of using a *count-on* procedure. They begin with the first addend and count on from there, stopping when they have counted on the number of the second addend. For example, 8 + 6 would be solved as 8, 9, 10, 11, 12, 13, 14. Finally, at level 3, students can use a recomposing procedure: One addend is broken apart to make a related addition problem whose total is known. For example, 8 + 6 would be solved as 8 + (2 + 4) = (8 + 2) + 4 = 10 + 4 = 14.

The important point here is that changes in the addition strategies are produced by fundamental changes in conceptual understanding, and not only practice with specific number problems. Children come to understand numbers and addition differently, and this drives the changes in strategies.

Computational Strategies in Multiplication

As we outlined above, the learning of addition strategies is driven primarily by fundamental conceptual developments in how children conceptualize relationships among quantities. In contrast, in our research we found that the learning of multiplication strategies is driven primarily by the learning of specific knowledge about specific numbers. In our 2005 paper, we called this new knowledge *number-specific computational resources*. Such knowledge was important for all but the most time-consuming and earliest of our six main types of strategies. We also found that as children approached fluency for specific numbers, the main strategies merged so that they were not easily identifiable. Children seemed to have developed a web of related knowledge on which they drew to give answers. The six strategies we identified are summarized below.

Count-all

When children first start to learn multiplication, they can make use of knowledge that they have acquired during their time learning addition. In the first type of strategy, count-all, a student can be seen counting from 1 to the product as he performs the computation. Example 9.1 describes an incident in which a student, Danny, was presented with the task of finding the total number of children, given that 4 children are seated at each of 3 tables. He solved this problem by first drawing a picture, and then counting all of the children he had drawn.

Example 9.1. Danny, pre-interview

Task: There are 3 tables in the classroom and 4 children are seated at each table. How many children are there altogether?

Description: Initially Danny was unsure how to proceed. Following the suggestion of the interviewer, he drew the situation. When the interviewer asked, "So, how many children are there altogether?" he counted quietly without pointing, but his head moved and he nodded a bit, as if in the direction of each drawn child.

An important feature of count-all strategies is that they are time-consuming and difficult to enact correctly, especially when the factors are large. To multiply using count-all, a child must keep track of three simultaneous counts. Consider the task of multiplying 3 × 4. One way to do this is to count to 4 three times, and then count the total. This approach requires that we enact and coordinate the three counting sequences shown in figure 9.2: (1) We need to count from 1 to 3 to keep track of the number of groups; (2) we need to count from 1 to 4 three times, to keep track of where we are within each group; and (3) we need to count from 1 to 12, thus keeping track of the running total.

1				2				3				Count of the number of groups
1	2	3	4	1	2	3	4	1	2	3	4	Count of entities in a group
1	2	3	4	5	6	7	8	9	10	11	12	Count of total

Fig. 9.2. The three coordinated counting sequences for multiplying 3 × 4

Children employ different techniques to keep track of the three separate counts. For that reason, children's use of count-all strategies can look different in different circumstances. In example 9.1, Danny used a drawing to help him keep track. Children also use their fingers to track counts. Or they make use of *rhythmic counting* and emphasize each value that is associated with the completion of a group. So a student multiplying 3 × 4 might say: "One two three <u>four</u>, five six seven <u>eight</u>, nine ten eleven <u>twelve</u>."

Additive calculation

Students also have prior learning experiences using addition. This knowledge can provide the basis of strategies that are less time-consuming and easier to enact than count-all strategies. We call these strategies that are based on addition-related techniques *additive calculations*. Example 9.2 shows this strategy. In this example, Ellen multiplies 3 × 4 by first adding 4 + 4 to get 8, and then 8 + 4 to get 12. This is clearly different than a count-all calculation. Ellen had to keep track of the three 4s she added, but she did not have to count from 1 to the total.

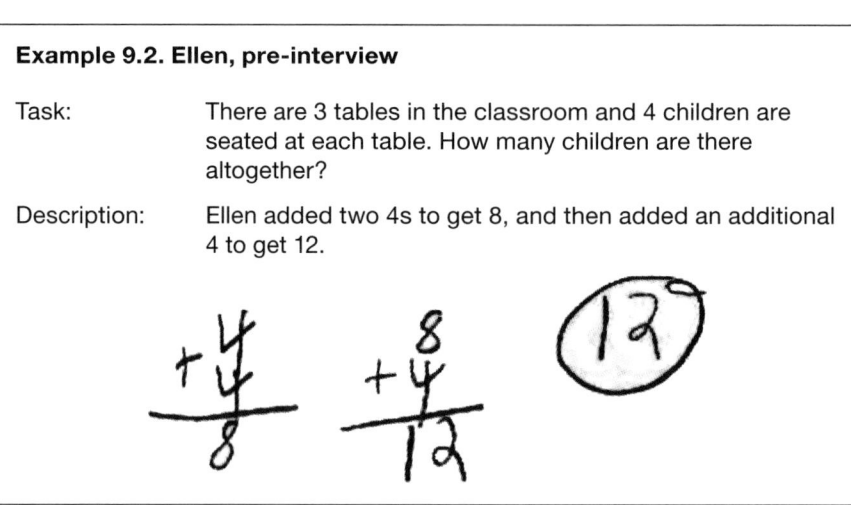

Example 9.2. Ellen, pre-interview

Task: There are 3 tables in the classroom and 4 children are seated at each table. How many children are there altogether?

Description: Ellen added two 4s to get 8, and then added an additional 4 to get 12.

Count-by

The first two strategies make use of knowledge about numbers that students have before they receive instruction in multiplication. Such instruction emphasizes the meaning of multiplication as repeated groups. Then students begin the extended task of learning the various number-specific computational resources that can support more efficient and accurate strategies. One such resource is the collection of *count-by sequences*; students learn to say sequences such as "5, 10, 15, 20, . . ." and later "6, 12, 18, 24, . . ." and "9, 18, 27, 36 . . . ." Knowing these sequences makes it possible for students to use count-by strategies to multiply single-digit numbers. In example 9.3, we describe an episode in which a student used a count-by strategy to multiply 8×4: She counts by 4s to 32, putting up a finger on each hand to keep track of the number of groups.

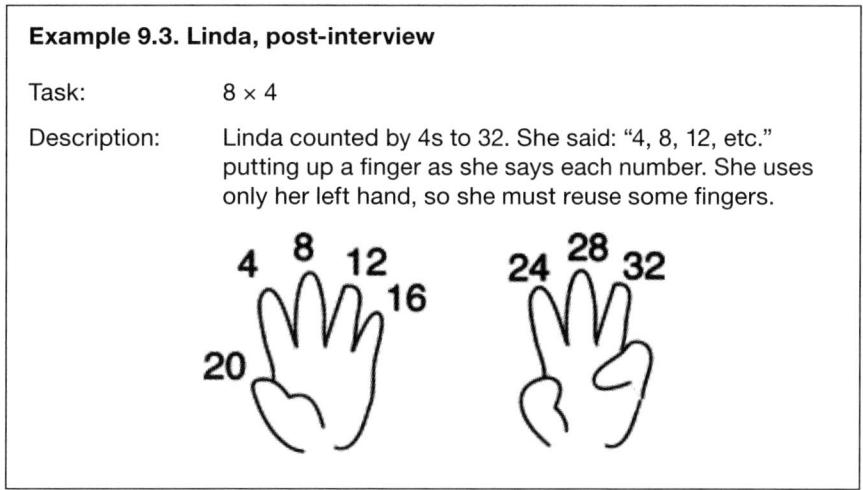

Example 9.3. Linda, post-interview

Task: 8×4

Description: Linda counted by 4s to 32. She said: "4, 8, 12, etc." putting up a finger as she says each number. She uses only her left hand, so she must reuse some fingers.

As in count-all, count-by strategies require children to keep track of multiple counting sequences. However, in the case of count-by, there are only two sequences, a reduction that greatly lessens the difficulty of accurately enacting count-by strategies. The tradeoff is that a count-by sequence must be learned for each number. Figure 9.3 depicts a count-by sequence for the case of 8×4.

4	8	12	16	20	24	28	32	Count of total
1	2	3	4	5	6	7	8	Number-of-groups count

Fig. 9.3. Two sequences to be coordinated for multiplying 8×4

Pattern-based

Many single-digit multiplication problems become easy once children learn to recognize certain simple patterns. These *pattern-based* strategies are another type of strategy children use. There are clear patterns associated with multiplication by 0, 1, and 10. These three patterns allow students to produce certain results rapidly and without visible work. For example, when students are asked to multiply 7×1, they may very quickly respond by saying "seven."

Beyond the 0s, 1s, and 10s patterns, students may learn other patterns that may support them in multiplication computations. These patterns are visible in the multiplication table and can be the focus of continued discussion by students. The 9s products are particularly rich with useful patterns, and students' recognition of these patterns can reduce the difficulty of multiplication tasks involving 9. In CMW, students first considered 9s patterns based on 9 as $10 - 1$ (e.g., $6 \times 9 = 6(10 - 1) = 60 - 6 = 54$). After working through all of the related patterns and discussing them using tens and ones, students summarize these using a finger shortcut that captures the patterns. The pattern shortcut works in this way: If a student wants to multiply $9 \times N$, the student holds up both hands and puts down the Nth finger, counting from the left. The tens digit of the result is then given by the number of fingers to the left of the finger that was put down (because it is always 1 less than the number of tens in $10 \times N$), and the ones digit is given by the number of fingers to the right (because those fingers show how many ones remain after N ones are taken from $N \times 10$).

Learned product	The learned product strategy requires number-specific resources that are multiplication triads, such as $9 \times 6 = 54$ and $4 \times 7 = 28$. These triads can be remembered as results from any other strategies. The count-by sequences especially are rich sources for learning multiplication triads as students link the product and the multiplier used in the keeping-track process to the multiplied factor (e.g., 7, 14, 21, 28, four 7s are 28). The multiplication-triad resources are acquired bit by bit, with some triads being learned earlier than others, especially those for smaller or easier factors. Learning all such multiplication triads takes time and practice because the triads are number specific.
Hybrids	Because the two main strategies, count-by and learned product, are learned gradually, we sometimes saw hybrid strategies that combine the use of a count-by or learned product strategy with count-all or additive calculation. For example, for 7×6 one student said, "6, 12, 18, 24, 32, 36, 37, 38, 39, 40, 41, 42," and another student said, "6 times 6 is 36, and 4 more is 40 and 2 more is 42."
Relationships among the strategies	Our discussion above presented the individual, number-specific computational resources—such as count-by sequences and multiplication triads—as if they are clear and distinct elements of knowledge. For example, we implied that each count-by sequence is learned separately, and each number triad is a separate multiplication "fact" to be learned. However, relationships exist among all of the strategies. When students solve problems with additive calculations or with rhythmic counting, these strategies can support the learning of specific count-bys. Analysis of patterns can underlie any of the strategies. Learned products can emerge from the use of any of the other strategies. So students are always building a web of integrated and related knowledge rather than separate, discrete bits of knowledge. This web of knowledge rests on the meanings for multiplication and division that students are developing from the Common Core Operations and Algebraic Thinking (OA) standards 1 through 7; these meanings (e.g., equal groups, arrays, area) help students relate strategies.
Is It All about Rote Memorization?	We now return to the question with which we began this chapter: *Is the learning of single-digit multiplications and divisions just about rote memorization?* No. Instead, we have seen that it is about acquiring knowledge that is linked to specific numbers—number-specific computational resources, such as count-by sequences and multiplication triads—and that these resources are related to form an integrated generative conceptual web. A student may respond that 7×5 equals 35 by drawing on many visual, oral, and reasoning experiences from different parts of this web: 35 is one more 5 past 6 fives are 30; the product ends in 5 because 7 is odd and 35 seems about far enough along; five 5s is 25 and 30, 35. These thoughts may occur very quickly and may not functionally differ much from a learned-product response. Similar points have been made elsewhere in the research literature. For example, Heege (1985) states that students can become so skilled "that the border between 'figure out' and 'know by heart' seems to blur" (p. 386).

The OA learning progression, written by members of the Common Core Standards Writing Team (2011), came to a similar conclusion:

> All of the understandings of multiplication and division situations, of the levels of representation and solving, and of patterns need to culminate by the end of Grade 3 in fluent multiplying and dividing of all single-digit numbers and 10. Such fluency may be reached by becoming fluent for each number (e.g., the 2s,

the 5s, etc.) and then extending the fluency to several, then all numbers mixed together. Organizing practice so that it focuses most heavily on understood but not yet fluent products and unknown factors can speed learning. To achieve this by the end of Grade 3, students must begin working toward fluency for the easy numbers as early as possible. Because an unknown factor (a division) can be found from the related multiplication, the emphasis at the end of the year is on knowing from memory all products of two one-digit numbers. As should be clear from the foregoing, this isn't a matter of instilling facts divorced from their meanings, but rather the outcome of a carefully designed learning process that heavily involves the interplay of practice and reasoning. All of the work on how different numbers fit with the base-ten numbers culminates in these "just know" products and is necessary for learning products. Fluent dividing for all single-digit numbers, which will combine just knows, knowing from a multiplication, patterns, and best strategy, is also part of this vital standard 3.OA.7. (p. 27)

Instruction

What does all this mean for instruction? It is important for us to emphasize, once again, a point made in the introduction. In this chapter, we have presented a variety of strategies, including students' invented strategies. But given the need to move from understanding to fluency in grade 3, it does not seem to us to be possible or wise to spend instructional time on a leisurely exploration of student strategies. Third grade instruction must support students quickly to meaningful, accurate, and efficient strategies. We hope we have made clear, however, that though number-specific knowledge must be practiced in order to be learned, this knowledge does not need to be acquired through rote memorization. The goal can be met by helping students see and relate pattern and structure in numbers.

Before concluding, we want to introduce a few additional thoughts about instruction that arise from the second author's extensive experience with the *Math Expressions* program and with the Common Core standards. Additional discussion of multiplication/division instruction and the Common Core standards can be found in the OA progression (Common Core Standards Writing Team, 2011) and in a twenty-minute webcast developed by the second author: "Math Expressions and Operations and Algebraic Thinking (OA) in the Common Core State Standards Part 3: The Grade 3 Learning Path for OA x/÷" (available at http://www.brainshark.com/hmhsupp/vu?pi=134411335).

First, multiplication and division learning needs to begin intensively and early in grade 3. Group and individual practice needs to continue throughout the year. Initially the class can move through each number, from the easier numbers (e.g., 2, 5, 10, 9) to medium (3, 4) to difficult (6, 7, 8), with the general patterns for the 0s and 1s folded in somewhere along this path. Practice needs to occur for each number separately for count-by sequences, pattern analysis and discussion, and known products. Then, students need to practice known products mixed across numbers—for example, mixed across multiples 2, 5, 10, and 9. Practice on new larger numbers separately and then mixed numbers continues throughout the year. This learning path requires a complex and sustained social organization of support and motivation for students to maintain their focus throughout the year. Students learn at different rates. Students who learn more slowly must be given support in learning the easier numbers so that they do not fall completely behind the class. This support needs to be given outside of and in addition to class so that students can participate in the discussions of problem solving and reasoning about new numbers that occurs in class.

Second, division strategies are closely related to multiplication strategies. Counting-by to divide is the same process as counting-by to multiply, but the student monitors the count-by sequence and stops when she or he hears the known product. For example, $32 \div 4$ looks and sounds like Linda's method in example 9.3, but one would listen for and stop at 32 and then look at the 8 fingers raised to find the unknown factor 8. This is actually easier than multiplying because the student can just look at the product while saying the count-bys to help keep track of when to stop. The strong relationship between these multiplication and division count-by methods means that students can practice multiplications and divisions involving the same count-by in a related fashion, and these can strengthen each other.

Finally, it is vital that student practice be focused on the individual learning zone of the student: on the count-by sequences or individual known products that the student does not yet know firmly. Practice time is often wasted by using resources such as a page of 100 multiplication problems or a general computer game. Both of these resources often have many products known to a given student and only a few problems that the given student needs to learn next. Student time is better spent on their next most difficult problems, whether those involve the 9s or the 3s or the 7s. Individualized piles of not too many flash cards can provide individualized practice on what a given student needs to practice. Supports such as count-by sequences written on the back of a flash card can provide the grounding a student needs to advance more quickly.

References

Common Core Standards Writing Team. (2011). *The operations and algebraic thinking (OA) progression for the Common Core State Standards in Mathematics*. Retrieved from http://ime.math.arizona.edu/progressions/.

Fuson, K. C. (1992). Research on whole number addition and subtraction. In D. A. Grouws (Ed.), *Handbook of research on mathematics teaching and learning* (pp. 243–275). New York, NY: Macmillan.

Fuson, K. C. (2009/2013). *Math Expressions K to grade 6*. Boston, MA: Houghton Mifflin Harcourt.

Heege, H. T. (1985). The acquisition of basic multiplication skills. *Educational Studies in Mathematics, 16*, 375–388.

National Governors Association Center for Best Practices & Council of Chief State School Officers (NGA Center & CCSSO). (2010). *Common core state standards for mathematics*. Retrieved from http://www.corestandards.org/.

Sherin, B., & Fuson, K. C. (2005). Multiplication strategies and the appropriation of computational resources. *Journal for Research in Mathematics Education, 36*, 347–395.

Chapter 10

Arithmetic and Algebra in Early Mathematics Education

David W. Carraher
TERC

Analúcia D. Schliemann
Tufts University

Bárbara M. Brizuela
Tufts University

Darrell Earnest
University of Massachusetts at Amherst

Several decades ago mathematics educators widely agreed that algebra should follow arithmetic. Considerable attention was devoted to "prealgebra" instruction aimed at easing the difficult transition from arithmetic to algebra that students were expected to face (see reviews by Carraher & Schliemann, 2007, and by Schliemann, Carraher, & Brizuela, 2007). Over the years, however, the extreme separation, indeed, isolation, of arithmetic from algebra appeared to make less and less sense. Investigations into "additive and multiplicative structures" increasingly drew attention to the underlying algebraic nature of arithmetic. Researchers identified a variety of factors and features of word problems that appeared to play critical roles in students' approaches to and success with problems. In doing so, they increasingly made use of concepts and representations regularly identified with algebra. Among these concepts were functions, variables, relations, ratio and proportion, and a set of field axioms including, among others, the commutative, associative, and distributive properties. Representations on the line and in the plane, function tables, and algebraic notation served to clarify the nature of arithmetic. As a result, mathematics educators, policymakers, and researchers came to realize that algebra belongs in the elementary education curriculum (e.g., Ball, 2003; Cai & Knuth, 2011; Carraher & Schliemann, 2007, 2016; Carraher, Schliemann, & Brizuela, 2000a, 2000b, 2005; Kaput, Carraher, & Blanton, 2008; National Council of Teacher of Mathematics [NCTM], 2000; Schliemann, Carraher, & Brizuela, 2007; Schoenfeld, 1995). This view has been backed by studies showing that young students can grasp concepts such as variables and functions that were thought to be beyond their reach (e.g., Blanton, 2008; Carpenter & Franke; 2001; Carraher, Schliemann, & Brizuela, 2000a, 2000b, 2005; Carraher, Schliemann, & Schwartz, 2008; Davydov, 1991; Dougherty, 2008; Moss & Beatty, 2006; Schliemann, Carraher, & Brizuela, 2007, 2012).

In this chapter, we look at examples from our own research to illustrate how young students, in classroom activities, become engaged in algebraic reasoning. Along the way, we highlight how multistep word problems are used to support discussions about functions and variables, to promote deeper and stronger learning of arithmetic, and to prepare students for more advanced mathematics. We argue that there are important opportunities for teachers and students to examine fundamental topics of elementary school mathematics that illuminate and emphasize the algebraic character of arithmetic. Our approach is guided by the view that generalizing lies at the heart of algebraic reasoning, that arithmetical operations are inherently functions (Carraher, Schliemann, & Brizuela, 2000a, 2000b, 2005; Carraher, Schliemann, & Schwartz, 2008; Schliemann, Carraher, & Brizuela, 2007), and that abstract mathematical objects (in particular, functions) should be introduced through multiple

This chapter is adapted from D. W. Carraher, A. D. Schliemann, B. M. Brizuela, & D. Earnest (2006), Arithmetic and algebra in early mathematics education, *Journal for Research in Mathematics Education, 37,* 87–115. Materials and publications from the research reported in that article and from other related publications are available at http://ase.tufts.edu/education/earlyalgebra/default.asp.

representations (verbal statements, tables, number lines, graphs, and algebraic notation) (e.g., Brizuela & Earnest, 2008; Earnest, 2014; Schliemann, Carraher, & Brizuela, 2007). We treat algebra as a *generalized arithmetic of both numbers and quantities* in which functions and other relations assume a major role (see Carraher & Schliemann, 2007; Kaput, 2008).

The examples come from a longitudinal study of sixty-nine students, in grades 2 to 4, in a multiethnic community (roughly 75 percent Latino) in the Greater Boston area. The lessons were designed and implemented to deepen students' understanding of mathematics content appropriate to their grade levels. Although the research was undertaken as part of a National Science Foundation project, from the students' perspective the researchers were mathematics teachers.

Each semester, from the start of the second semester in second grade to the end of fourth grade, we implemented six to eight classroom activities of about ninety minutes each. Students worked with variables, functions, algebraic notation, function tables, graphs, and equations as they related to addition, subtraction, multiplication, division, integers, fractions, ratio, and proportion. Here we will describe how third graders were introduced to variables and variable notation (in lesson 2) and to number lines (in lessons 3–6), drawing from videotaped discussions and students' written work. We will then examine how, in one third-grade classroom, students used number lines and algebraic notation to solve a multi-step word problem (lesson 7) and to represent the *differences* in heights among three children, without revealing their actual heights (lesson 8). Our aim is to illustrate how, with certain forms of support, young elementary students not only can solve problems with an underlying algebraic structure but also can come to explicitly make use of variables.

Arithmetic Operations as Functions

One of the most underappreciated ideas in elementary mathematics education is that the four basic operations of arithmetic—addition, subtraction, multiplication, and division—are defined in mathematics as functions. To be sure, this idea has been alluded to in various works about "additive and multiplicative structures" in recent decades. However, the idea merits being explicitly acknowledged and examined.

A function is, first and foremost, a relation among the elements of two sets, referred to as a domain and a codomain (or range). What distinguishes a function from other relations is the fact that each and every element of the domain is assigned a single element from the codomain. For example, each instance of an operation, e.g., $5 + 3$ or 7×4, is associated with a unique output (e.g., 8 and 28, respectively). One may think of the domain as containing inputs and the codomain as outputs. Hereafter, we will suppose, as we did with our students, that each arithmetic operation has as input a single number (e.g., 7). In this manner, $7 + 3$, $5 + 3$, and $9 + 3$ are interpreted as instances of the univariate "plus 3" function, $x + 3$ (over the integers).

Acknowledging that the operations of arithmetic are functions implies that they involve variables. It is of key importance that we determine whether young students, as they discuss the problem, merely handle variables implicitly or overtly express them in language or notation. As a corollary question we may ask, "Can young students reason not only about particular instances but also about sets of logical possibilities (all of the inputs and outputs of a given function)?"

Introducing Variables and Number Lines in Third Grade

Lesson 2 was designed to expressly introduce literals as variables, that is, placeholders that can be assigned any element from a domain (a set of possible values). The instructor initiated the lesson by introducing a problem of comparing the number of candies that John and Mary had. She displayed two small sealed boxes and assured the students that each box contained the same amounts. However, John and Mary did not have the same total amounts, as the wording of the problem reveals:

> John and Mary each have a box containing the same number of candies. Mary has three additional candies. What can you say about how many candies they each have?

The problem asserts a relation between John's total and Mary's total. (It can also be construed as entailing two functions of the number of candies in a box.) By refraining from assigning a value to the number of candies in a box, we initially left open the issue of whether the number of candies in a box, hence John's and Mary's amounts, should be regarded as unique.

Over ten to fifteen minutes, the teacher and students took part in a discussion about the possible values for each amount. The students proposed many possible pairs of values for John's and Mary's totals. It is fair to say that the students were attempting to guess how many candies each child had. They seemed to assume that the issue was empirical and that the definitive answer would be ascertained by opening the boxes and counting the candies. As the lesson proceeded, and with the teacher's guidance, the issue began to take on another form.

At one point the instructor listed in a table, with columns for John's and for Mary's possible amounts, the students' answers. The students realized that multiple pairs of values were consistent with the wording of the problem, whereas others (e.g., John has 9 and Mary has 11 candies) were not. Some expressed this as "whatever amount John has, Mary has 3 more" or as "the difference is always 3," thereby focusing on the relation rather than a single ordered pair of the relation. Through these discussions, the instructor identified and supported opportunities for students to incorporate a letter to represent any possible amount in the boxes. Most students easily accepted the convention of using the letter N to represent the amount in each box. In the discussion that followed, a student proposed to use algebraic notation ($N + 3$) to represent Mary's amount and a few others agreed with him. Later, most of them came to use this algebraic representation.

In an introductory lesson regarding number lines, we started with a clothesline spanning the front of a classroom, displaying large, equally spaced integers, ordered from -20 at the left to +30 at the right, where it met a wall. A third-grade student stood at the number 9 to represent his current age. At the teacher's request, he moved to positions on the line to represent his age a year later, at his birth, 10 years later, and 3 years before he was born. When the teacher asked, "How many numbers are there on the number line?" another student claimed there were fifty numbers, explaining, "They [the numbers] only go as far as the wall!" The teacher then asked, "But what if we want our number line to hold all the numbers that exist?" Various students, with increasing enthusiasm, claimed that it would need to go "beyond the sun," "out of the universe," "to infinity," "without ever stopping."

In the next lesson, using diagrams on paper and projected onto a screen, we introduced arrows linking points on the number line to represent changes in values (see fig. 10.1) as displacements. When two or more arrows were joined, tail to head, on the number line, students learned that they could simplify the information by a single arrow going from the tail of the first arrow to the head of the last arrow. They also learned to express such shortcuts in notation: For example, the composition of two additive operations, "<some number> + 5 – 3," could be a single arrow and an expression of the form "<some number>" + 2. They showed the similarities and differences between the two by making trips along the number line in the class or on paper. The large-scale number line in the class allowed students and teacher to discuss mathematical operations in a forum where students could follow the reasoning of others. This helped students deal with a range of issues, including the immensely important one of distinguishing between numbers as points and numbers as intervals.

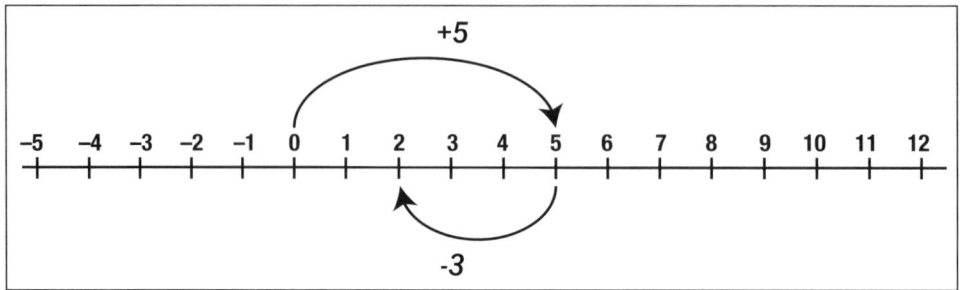

Fig. 10.1. Representing changes of values as displacements

In subsequent lessons, we introduced the third graders to a variable number line, referred to as the "N-number line," a variation on a standard number line in which values were expressed as displacements from some indeterminate value, N. The N-number line appears in figure 10.2.

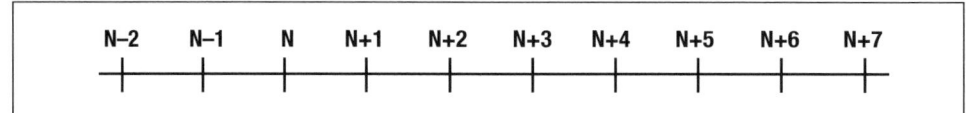

Fig. 10.2. The N-number line

Using Variables and the N-number Line to Solve Problems

In lesson 7 we presented the Piggy Bank problem (see fig. 10.3). The problem states that Mary and John start with the same, unspecified amount of money in their piggy banks, and describes changes in the amounts over the course of several days. It concludes by revealing how much Mary had in her piggy bank on the final day. The students were asked to determine how much Mary and John had at the beginning and how much they had each day. We initially displayed the whole problem, except for the sentence describing Mary's final amount. Then we covered up all days except Sunday.

Mary and John each have a piggy bank.

On *Sunday,* they both had the same amount in their piggy banks.

On *Monday,* their grandmother comes to visit them and gives 3 dollars to each of them.

On *Tuesday,* they go together to the bookstore. Mary spends $3 on Harry Potter's new book. John spends $5 on a 2001 calendar with dog pictures on it.

On *Wednesday,* John washes his neighbor's car and makes $4. Mary also made $4 babysitting. They run to put their money in their piggy banks.

On *Thursday,* Mary opens her piggy bank and finds that she has $9.

Fig. 10.3. The Piggy Bank problem

The aim of the lesson was to have the students describe relations among quantities using notation for variables and to derive new inferences from changes in the syntactical forms.

We used the "*N*-number line" to support the representation of relative, indeterminate amounts at various moments from the narrative. The *N*-number line is a variant of a number line with an origin at *N* and tick marks for . . . $N - 2, N - 1, N, N + 1, N + 2$ . . . . Unlike a standard number line, it allows one to express additive relations among indeterminate quantities. This was useful because the problem requires working with addition and subtraction on unspecified initial amounts.

Representing an unknown amount	After reading each day's statements, students worked alone or in pairs to represent on handouts the parts of the problem, from Sunday through Thursday. Members of the research team encouraged students to explain their work and to further develop their representations. Next we describe the discussions that followed and students' insights and achievements.

Sunday

After a student read aloud the information regarding Sunday, Bárbara, the researcher teaching this class, asked whether the students knew how much money each of the characters in the story had. Several students stated in unison "No" and did not appear to be bothered by that. A few uttered "*N*," and Talik stated, "*N*, it's for anything." Others called out "any number" and "anything." When Bárbara asked what they were going to show on their answer sheets, Filipe proposed, "You could make [*sic*] some money in them, but it has to be the same amount." Bárbara reminded him that the amount was unknown, to which Filipe suggested to write *N* to represent the unknown amount. As the class transitioned to work individually or in pairs, Jeffrey mentioned he was going to follow Felipe's suggestion. The first item in the handout described the situation on Sunday along with a copy of the *N*-number line. Bárbara reminded the students that they could use the *N*-number line on their paper if they so wished and she drew a copy of it on the board.

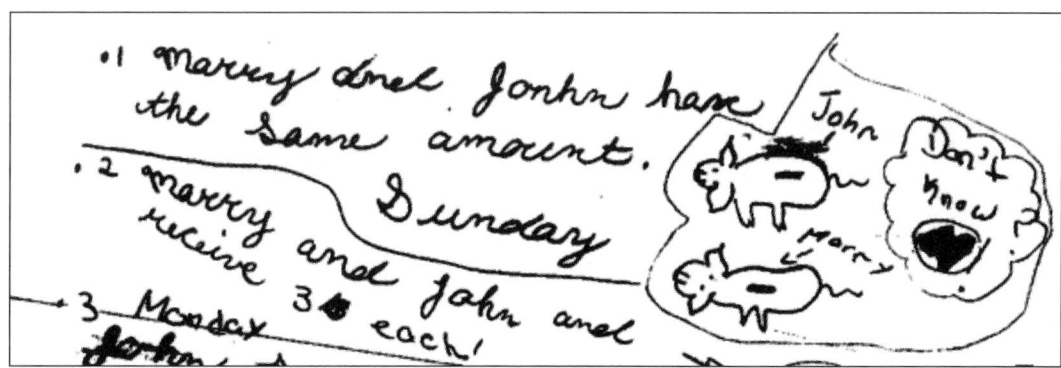

Fig. 10.4. Jennifer's initial representation for the problem

The students drew piggy banks and represented amounts in each of them. Four students attributed specific values for Mary and John on Sunday; five represented the amount as *N*, usually inside a drawing of a piggy bank; two placed a question mark inside or next to each piggy bank; and five drew piggy banks with no indication of what each would contain. Jennifer used *N* to represent the initial amount (see fig. 10.4) and drew two piggy banks, next to which she wrote a large *N* and "Don't know?" When David (present but not serving as instructor) asked Jennifer why she had done so, she answered: "Because you don't know. You don't know how much amount they have" and "*N* means any number." When further asked,

113

"Is it that they have, like, 10 dollars each?" she answered "No. . . . Because we don't know how much they have." While in prior lessons the instructor had introduced the convention of *N* to represent an unknown value, in this lesson the students made use of the convention of their own accord.

Changes in unknown amounts

Monday

From the information that John and Mary each received $3 on Monday, the students inferred that the children would continue to have the same amount of money as each other and that they both would have 3 dollars more than the day before. Talik then explained:

Talik: Because before they had the same amount of money, plus 3, they both had 3 more, so it's the same amount.

David: The same amount as before or the same amount as each other?

Talik: The same amount as each other. Before, it was the same amount.

Nathan was the first to propose that on Monday they would each have *N* plus 3, explaining, "Because we don't know how much money they had on Sunday, and they got plus . . . and they got 3 more dollars . . . on Monday." Talik proposed to draw a picture showing Grandma giving money to the children. Filipe represented the amount of money on Monday as ? + 3. Jeffrey stated that their grandmother gave them 3 more dollars and represented Mary's amount on Sunday (see fig. 10.5) by a sketch of a pig with the letter *N* drawn inside the border; underneath he wrote "*N* + 3 in monday" [*sic*]; John's bank was represented through a house-shaped figure containing the letter *N*. He also depicted the amounts on Monday through bar graphs of equal heights: a base of *N* with 3 additional units displayed on top.

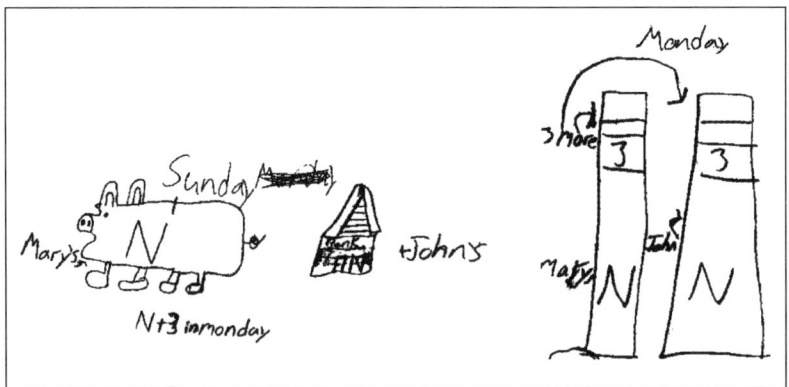

Fig. 10.5. Jeffrey's representations of the children's amounts on Sunday and on Monday

Bárbara commented to the class on Filipe's use of question marks, and he and other students acknowledged that *N* was an alternative to the question marks. She then told the class that some of the students proposed specific values for the amounts in the piggy banks on Sunday. Filipe questioned this view, adding, "nobody knows (how much they have)," and James said that these other students "are wrong." Jennifer pointed out, however, that a suggested value "*could* be correct."

By the end of the class, 11 of the 16 students had adopted *N* + 3 to represent the amounts each would have on Monday. For instance, Carolina wrote *N* + 3. Jennifer wrote *N* + 3 with the explanation "3 more for each." Carolina, Adriana, and Andy wrote *N* + 3 inside or next to each piggy bank under the heading Monday. Jeffrey wrote $N + 3$ for Monday and explained that their grandma gave them 3 dollars more. Jimmy, who first represented the amounts on Sunday as question marks, wrote $N + 3$ with connections to schematic representations of Mary's and John's piggy banks, and explained: "Because when the Grandmother came to visit them they had like, N. And then she gave Mary and John 3 dollars. That's why I say [*pointing to N + 3 on paper*] N plus 3." Only one student continued to work with specific amounts, four students produced drawings that could not be interpreted, and one wrote the equation $N + 3 = N$.

| **Operating on unknowns with multiple representations** | *Tuesday* |

Some of the students wondered whether there would be enough money in the piggy banks to cover the children's spending. A student suggested that the children in the story probably had ten dollars. Others assumed that there must be at least $5 in their piggy banks by the end of Monday (otherwise John could not have bought a $5 calendar). When asked to describe the balances at the end of Tuesday, several students concluded that Mary ended Tuesday with more money than John, and Jennifer noted that, on Tuesday, Mary would have the same amount of money that she had on Sunday "because she spends her three dollars."

Bárbara then encouraged the students to use the N-number line to represent the amounts from Sunday through Tuesday. She drew arrows from N to $N + 3$ and back to N to show the changes in Mary's amounts. She showed the same with algebraic notation, narrating the changes from Sunday to Tuesday, step by step, and getting the students's input while writing $N + 3 - 3$. She then wrote a bracket under $+ 3 - 3$ and a 0 below it, commenting that $+ 3 - 3$ is the same as 0 and extending the notation to $N + 3 - 3 = N + 0 = N$. Jennifer then explained that the 3 dollars spent on Tuesday cancel out the 3 dollars given to Mary by her grandmother. Bárbara led the students through John's transactions on the N-number line, drawing arrows from N to $N + 3$, then $N - 3$. She also used algebraic script to register the states and transformations and, with the students' input, kept track of these, eventually writing $N + 3 - 5$ to express John's amount at the end of Tuesday. Some students suggested that this amount is equal to "N minus 2," an inference that Bárbara registered as $N + 3 - 5 = N - 2$.

Bárbara asked Jennifer to show, on the number line, the difference between John's and Mary's amounts on Tuesday. Jennifer at first pointed vaguely to the interval between $N - 2$ and N. When asked to show precisely where the difference starts and ends, she correctly pointed to $N - 2$ and to N (John's and Mary's balances, respectively). When asked to show how much one would have to give John so that he would have the same amount of money as he had on Sunday, Jennifer answered that we would have to give 2 dollars to John and explained, showing on the number line that "if he is at $N - 2$ and we add 2, he goes back to N." Bárbara represented Jennifer's solution as $N - 2 + 2 = N$. Jennifer then drew brackets around the expression $-2 + 2$ and wrote 0 under it. Bárbara asked why $-2 + 2$ equals 0 and, together with Jennifer, went through the steps corresponding to $N - 2 + 2$ on the N-number line, showing how $N - 2 + 2$ ends up at N. Talik stated and explained how this works when N has the value of 150.

Nathan's drawing in figure 10.6 depicts Sunday (top), Monday (bottom left), and Tuesday (bottom right). For Tuesday, he first drew iconic representations of the calendar and the book next to the values $5 and $3, respectively, the icons and dollar values connected by an equals sign. During his discussion with an in-class interviewer, he wrote the two equations, $N + 3 - 5 = N - 2$, using the N-number line as support for his decisions.

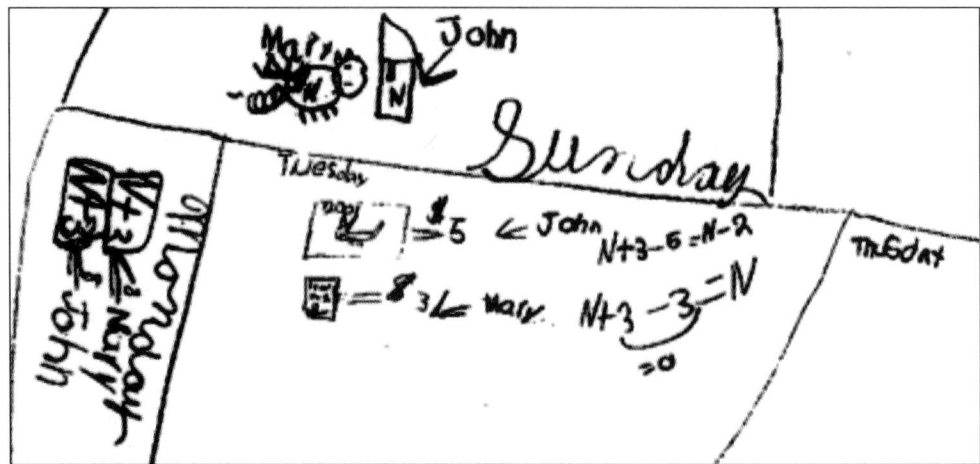

Fig. 10.6. Nathan's representation of the problem

Wednesday

Filipe read the Wednesday step of the problem. Bárbara asked whether Mary and John would end up with the same amount as on Monday. James said "No," and Adriana then explained that Mary would have $N + 4$ and John $N + 2$. Bárbara drew an N-number line and asked Adriana to tell the story using the line. Adriana represented the changes for John and for Mary, much as she would on a regular number line. Bárbara wrote $N + 4 = N + 4$ and then $N - 2 + 4 = N + 2$. Talik volunteered that "if you take 2 from the 4, it will equal 2." To clarify where the 2 came from, Bárbara represented $- 2 + 4 = 2$ on a regular number line.

Bárbara then asked if anyone could show why $N + 3 - 3 + 4$ equals $N + 4$. Talik crossed out the subexpression $+3 - 3$, saying, "We don't need that anymore." Jennifer stated that this is the same as 0. Note that we had not introduced the procedure of striking out the sum of a number and its additive inverse (although we *had* used brackets to simplify sums). Talik's *striking* instead of *bracketing* shows his understanding of the meaning held in the equation.

Next, Bárbara proposed to write out an equation conveying what happened to John's amounts throughout the week. The students helped her go through each of the steps in the story and build the equation $N + 3 - 5 + 4 = N + 2$. They did not initially agree upon the expression for the right-hand side of the equation, and Bárbara helped them visualize the operations on the N-number line. She asked Jennifer to show how the equation could be simplified. While Jennifer pondered, Bárbara pointed out that this problem was harder than the former and asked Jennifer to start out with $+ 3 - 5$; Jennifer answered "minus 2." Then, they bracketed the second part at $-2 + 4$, and Jennifer, counting on her fingers, said that it was $+ 2$. Talik then explained: "N is anything, plus 3, minus 5, is minus 2; N minus 2 plus 4, equals (while counting on his fingers) N plus 2." He also grouped the numbers differently, first adding 3 and 4 and then taking away 5. Bárbara showed that grouping $+ 3 + 4$ leads to $+ 7$ and that $+ 7 - 5$ results in $+ 2$: the answer is 2 regardless of the order of operations.

Finding a particular value and instantiating other values

Thursday

When Amir noted that Mary ended with $9.00, several students suggested that N had to be 5. Some students claimed that John (whose amount was represented by $N + 2$) has "2 *more*," apparently meaning "2 more than N." Others said that he had 7, apparently determining this by adding 5 + 2. Still others found the answer from Mary's final amount because $N + 2$ (John's amount) is 2 less than $N + 4$ (Mary's amount), and John would have to have 2 less than Mary (known to have 9). Bárbara concluded the lesson by working with the students in filling out a 2 × 4 table displaying the amounts that Mary and John had across the four days. Some students suggested expressions with unknown values and others used the actual values, inferred from the information about Thursday.

A New Context: Differences between Heights

The following week (lesson 8) students worked on the problem in figure 10.7. The problem states the *differences* in heights among three children without revealing their actual heights. The heights could be thought to vary insofar as they could take on any of a set of possible values. Addition and subtraction operations that focus on relative, as opposed to fixed, amounts are known to be challenging for many students under the age of 10 or 11 (see Vergnaud, 1982). We were particularly interested in determining the sense the *students* made of such a problem after they were somewhat familiar with working with indeterminate amounts and had certain resources (rudimentary algebraic notation and the N-number line) at their disposal.

Tom is 4 inches taller than Maria.

Maria is 6 inches shorter than Leslie.

Draw Tom's height, Maria's height, and Leslie's height.

Show what the numbers 4 and 6 refer to.

Maria   Maria's height

Fig. 10.7. The Heights problem

After discussing each statement in the problem, Bárbara encouraged the students to focus on the differences between the protagonists' heights and to represent the problem on individual worksheets. Most used vertical lines to show the three heights (see fig. 10.8). To our surprise, without any prompt, one of the students (Jennifer) chose to represent the heights on a variable number line much like the one they had been working with during previous meetings (see fig. 10.9). Bárbara then adopted Jennifer's number line as a basis for a full-class discussion of the relations among the heights. She further adopted Jennifer's postulate that Maria is located at N on the variable number line (see the middle number line in fig. 10.10).

The students concluded that if Maria's height was N, Tom's height would be N plus 4, because he is 4 inches taller, and Leslie's height would be N plus 6, because Leslie is 6 inches taller. It is remarkable that Jennifer realized that a representational tool introduced in earlier classes would help to clarify the problem at hand. It is equally impressive that the remaining students accepted this idea and easily inferred Tom's and Leslie's heights ($N + 2$, $N + 4$, respectively) from Maria's height (N). Bárbara wondered whether the students realized that calling Maria's height N was arbitrary and asked the students to assume, instead, that Leslie's height was N. The students noted in this case that Tom's height would be $N - 2$ and Maria's would be $N - 6$ (bottom line in fig. 10.9). They inferred this by making displacements on the variable number line consistent with the information given in the problem.

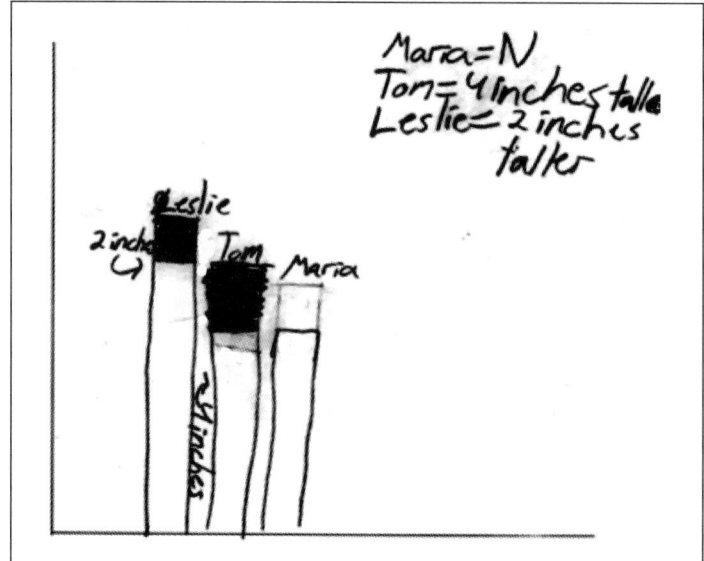

Fig. 10.8. Jeffrey's drawing and notation for the Heights problem

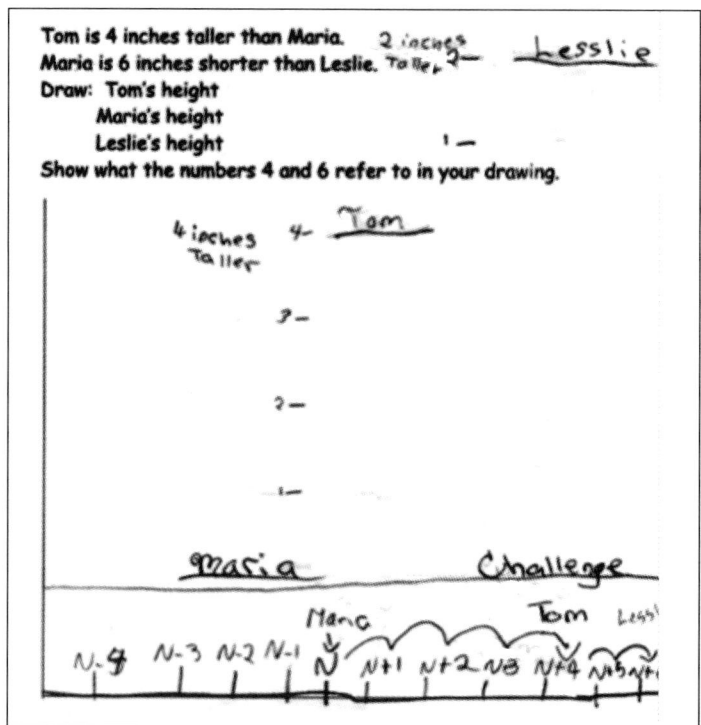

Fig. 10.9. Jennifer's drawing (notches) showing differences but no origin. She also makes use of a variable number that forms the basis of subsequent discussion.

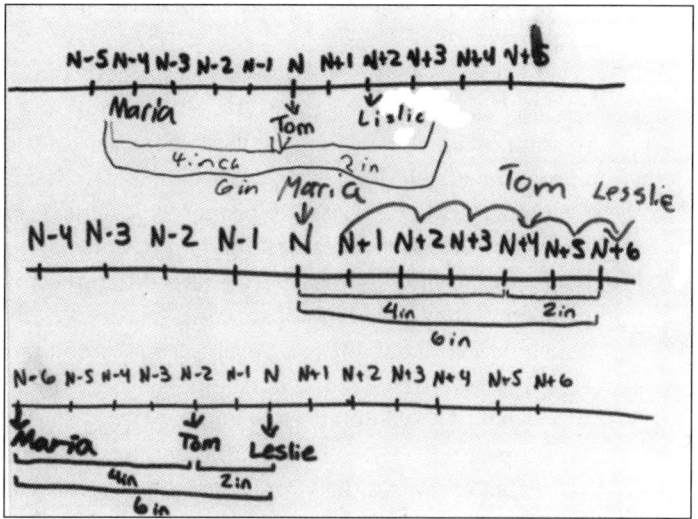

Fig. 10.10. Three variable number line representations (on overhead) used by students and teacher to discuss the cases where [middle] Maria is attributed the height of *N*; [bottom] Leslie is assigned the height of *N*; and [top] Tom is assigned a height of *N*.

Next, Bárbara asked the students to assume instead that Tom's height was *N*. Max went to the front of the class and placed Leslie at $N + 3$. He realized that the difference between Tom's and Leslie's heights is 2, but mistakenly placed her 3 units to the right of Tom (see the erasure in the top number line in fig. 10.10, under $N + 3$, where he had first incorrectly put Leslie's name). The challenge Max faced could be thought of as a "fence post" issue: Students are accustomed to the idea that a number refers to the count of elements in a set. However, the issue before them often is, "What should I count?" On a number line two sorts of elements suggest themselves. One can count the number of intervals or the number of "fence posts" or notches. In Max's case, he seemed to have counted the "fence posts" lying between *N* and $N + 3$. Other students correctly stated that Leslie should be placed under $N + 2$. When Bárbara asked Amir to show where Maria's name should be located, he placed it, without hesitation, under $N - 4$.

By the end of the lesson the students were relating the numerical differences to algebraic notation, line segments, number lines (including variable number lines), subtraction, counting, and natural language descriptions. The fluidity with which they moved across representations suggests that their understanding of additive relations was robust and flexible: The particular symbols used were inconsequential, provided that the relations between them remained invariant.

Discussion

In the lessons described, the students used letters to meaningfully represent variables and used and operated upon algebraic expressions such as $N + 3$ to represent functions. They used knowledge about the differences between quantities to formulate new algebraic expressions and to solve multistep arithmetic problems. They understood the relations between the daily amounts of each protagonist in the Piggy Bank problem and how the amounts on each day related to the starting amounts, even though these amounts were not specified. In the discussion of the Heights problem they worked with variables while maintaining an invariant relationship between them. And they generated appropriate expressions for the heights of the children, regardless of which protagonist had been assigned the initial value *N*. As such, it appears that the students were working with functions, a fundamental mathematical concept

119

(e.g., Dubinsky & Harel, 1992; Kaput, 2008; Schwartz & Yerushalmy, 1995; Yerushalmy & Schwartz, 1993).

In several senses the lessons described above are typical of the early algebra lessons we implemented each semester. At the beginning, most students attributed specific values to unknowns. They came to use algebraic notation and number line representations as a natural means of describing the events of problems they were presented. These findings have persuaded us that, given the proper experiences, even third-grade students can learn to use letters to represent variables and can operate on representations involving letters and numbers without having to assign values to the unknown. By doing so, while solving arithmetic problems, they are also considering relations between known and unknown quantities. To conclude that the sequence "$N + 3 - 5 + 4$" is equal to $N + 2$, and to be able to explain, as many students did, that N plus 2 must equal 2 more than what John started out with, *whatever that value might be*, is a noteworthy achievement. The students were able to draw inferences about these operations while fully realizing that they did not know the values of the unknown.

From lesson 2, the Candy Boxes problem, where one child was described as having 3 more candies than another, the third graders were able to treat the unknowns in additive situations as having multiple possible solutions, to accept that one child would have N candies, and to propose that the other would have $N + 3$ candies. Furthermore, they found it reasonable to view a host of ordered pairs, (3,6), (9,12), (4,7), (5,8), as *all* being valid solutions for the case at hand, even though they knew that in any given situation only one solution could be true. They even expressed the relationship between the elements in such pairs through statements such as "the number that comes out is always three larger than the number you start with." When students make statements of such a general nature, they are considering relations (among variables or sets of values) and not simply unknowns restricted to single values. In traditional arithmetic problems, such relations are routinely obscured by constraints. For example, the equation $x + 3 = 8$ will only be true if $x = 5$; x thus appears to be a missing number rather than a variable.

In the Candy Boxes lesson, the expression $x + 3$, examined by itself, invites one to treat x as a variable, that is, as a placeholder for a set of possible values. This idea can be highlighted by leaving variables "free to vary" across a large, possibly infinite, domain. This was achieved in the lesson by stating the problem in such a way that students could entertain many ordered pairs constituting the relation "3 more than." A similar point can be made about the Heights problem. In the Piggy Bank problem, it was achieved by withholding, until a later moment (Thursday), information corresponding to the assertion $N + 4 = 9$, where N represents the initial amount. In this manner, students were able to first describe phases of the problem (Sunday through Wednesday) in which the unknown was free to vary across some unspecified, nontrivial domain. This is precisely the spirit with which many students viewed the Piggy Bank problem before information was discovered, regarding Thursday, that finally allowed them to disregard the multiple possibilities and focus on the values to which the problem was now constrained.

The few students who feel the need to instantiate variables from the beginning could do so and participate in the classes from their own perspective, restricting their attention to one possible scenario from the start. This should not be a reason for concern, for we have found that such students learn from others and from class discussions; within a few weeks they comfortably welcome algebraic representations into their own personal repertoire of expressive tools.

Concluding Remarks

Our work has been guided by the view that (1) students' understanding of additive structures provides a fruitful point of departure for an "algebrafied arithmetic," (2) multiple problems and representations for handling unknowns and variables, including number lines and algebraic notation itself, can and should become part of students' repertoires early on, and (3) meaning in context and students' spontaneous notations should provide a footing for syntactical structures during initial learning, even though syntactical reasoning should become relatively autonomous over time.

Even though there are many reasons for viewing algebra as more advanced than arithmetic, we argue that there are good reasons for considering arithmetic and algebra as interwoven from the early grades. The arithmetic operations of addition, subtraction, multiplication, and division are functions amenable to description through algebraic notation. Moreover, although computational fluency is an important objective of arithmetic instruction—we would even say crucial for allowing students to reason algebraically—the teaching of arithmetical computations does not assure that students will be attentive to arithmetical properties and relations. The tools of algebra, such as algebraic notation (as well as tables, number lines, and graphs in the Cartesian space) offer a means for expressing such properties and relations clearly and succinctly. Further, if we dwell too much on the concrete nature of arithmetic, we run the risk of offering students a superficial view of mathematics, discouraging their attempts to generalize. Our data further expand our understanding of young students' capabilities to appropriate algebra notation as they represent open-ended problems. These are classroom examples of how the teaching and learning of arithmetic operations can be enhanced by a focus on variables and functions, instead of merely specific values.

By arguing that the algebraic character of arithmetic deserves a place in early mathematics education, we are not denying the developmental nature of mathematical understanding. Number concepts, the ability to use algebraic notation, interpret graphs, model situations, and so forth, develop over the course of many years. Nonetheless, most young students do not learn about variables for the simple reason that they were not given the opportunity. Were developmental constraints responsible for their attributing single values to (what we view as) variables, one would not have expected students to make the relatively quick shift to variables observed in the present study.

References

Ball, D. L. (2003). *Mathematical proficiency for all students: Toward a strategic research and development program in mathematics education.* Santa Monica, CA: RAND Corporation.

Blanton, M. (2008). *Algebra and the elementary classroom: Transforming thinking, transforming practice.* Portsmouth, NH: Heinemann.

Brizuela, B. M., & Earnest, D. (2008). Multiple notational systems and algebraic understandings: The case of the "best deal" problem. In J. Kaput, D. Carraher, & M. Blanton (Eds.), *Algebra in the early grades* (pp. 273–301). Mahwah, NJ: Erlbaum.

Cai, J., & Knuth, E. (Eds.). (2011). *Early algebraization: A global dialogue from multiple perspectives.* New York, NY: Springer.

Carpenter, T., & Franke, M. (2001). Developing algebraic reasoning in the elementary school: Generalization and proof. In H. Chick, K. Stacey, J. Vincent, & J. Vincent (Eds.), *The future of the teaching and learning of algebra. Proceedings of the 12th ICMI Study Conference* (Vol. 1, pp. 155–162). University of Melbourne, Australia.

Carraher, D. W., & Schliemann, A. D. (2007). Early algebra and algebraic reasoning. In F. Lester (Ed.), *Second handbook of research on mathematics teaching and learning* (pp. 669–705). Greenwich, CT: Information Age.

Carraher, D. W., & Schliemann, A. D. (2016). Powerful ideas in elementary mathematics education. In L. English & D. Kirshner (Eds.), *Handbook of international research in mathematics education* (3rd ed., pp. 191–218). New York, NY: Routledge, Taylor & Francis.

Carraher, D. W., Schliemann, A. D., & Brizuela, B. M. (2000a). Bringing out the algebraic character of arithmetic: Instantiating variables in addition and subtraction. In T. Nakahara & M. Koyama (Eds.), *Proceedings of the XXIV Conference of the International Group for the Psychology of Mathematics Education* (Vol. 2, pp. 145–152). Hiroshima, Japan.

Carraher, D., Schliemann, A. D., & Brizuela, B. (2000b, October). *Early algebra, early arithmetic: Treating operations as functions* [Plenary address]. XXII meeting of the Psychology of Mathematics Education, North American Chapter, Tucson, AZ.

Carraher, D., Schliemann, A. D., & Brizuela, B. (2005). Treating operations as functions. In D. W. Carraher & R. Nemirovsky (Eds.), *Media and meaning* [CD-ROM issue of *Monographs of the Journal for Research in Mathematics Education*].

Carraher, D. W., Schliemann, A. D., Brizuela, B. M., & Earnest, D. (2006). Arithmetic and algebra in early mathematics education. *Journal for Research in Mathematics Education, 37*, 87–115.

Carraher, D. W., Schliemann, A. D. & Schwartz, J. (2008). Early algebra is not the same as algebra early. In J. Kaput, D. Carraher, & M. Blanton (Eds.), *Algebra in the early grades* (pp. 235–272). Mahwah, NJ: Erlbaum.

Davydov, V. V. (1991). *Psychological abilities of primary school children in learning mathematics* (Vol. 6). Reston, VA: National Council of Teachers of Mathematics.

Dougherty, B. (2008). Measure up: A quantitative view of early algebra. In J. Kaput, D. Carraher, & M. Blanton (Eds.), *Algebra in the early grades*. Mahwah, NJ: Lawrence Erlbaum.

Dubinsky, E., & Harel, G. (1992). *The concept of functions: Aspects of epistemology and pedagogy.* Washington, DC: Mathematical Association of America.

Earnest, D. (2014). Exploring functions in elementary school: Leveraging representational contexts. In K. Karp (Ed.), *NCTM annual perspectives in mathematics (APME) 2014: Using research to improve instruction* (pp. 171–179). Reston, VA: National Council of Teachers of Mathematics.

Kaput, J. (2008). What is algebra? What is algebraic reasoning? In J. Kaput, D. W. Carraher, & M. Blanton (Eds.), *Algebra in the early grades*. Mahwah, NJ: Erlbaum.

Kaput, J., Carraher, D. W., & Blanton, M. (Eds.). (2008). *Algebra in the early grades.* Mahwah, NJ: Erlbaum.

Moss J., & Beatty, R. (2006) Knowledge building in mathematics: Supporting collaborative learning in pattern problems. *International Journal of Computer-Supported Collaborative Learning, 1*(4), 441–465.

National Council of Teachers of Mathematics. (2000). *Principles and standards for school mathematics.* Reston, VA: Author.

Schliemann, A. D., Carraher, D. W., & Brizuela, B. M. (2007). *Bringing out the algebraic character of arithmetic: From children's ideas to classroom practice.* Hillsdale, NJ: Erlbaum.

Schliemann, A. D., Carraher, D. W., & Brizuela, B. M. (2012). Algebra in elementary school. In L. Coulange & J.-P. Drouhard (Eds.), *Enseignement de l'algèbre élémentaire* [Special issue of *Recherches en Didactique des Mathématiques*], 109–124.

Schoenfeld, A. (1995). Report of working group 1. In C. B. Lacampagne (Ed.), *The algebra initiative colloquium: Vol. 2, Working group papers* (pp. 11–18). Washington, DC: U.S. Department of Education, OERI.

Schwartz, J., & Yerushalmy, M. (1995). On the need for a bridging language for mathematical modeling. *For the Learning of Mathematics, 15*(2), 29–35.

Vergnaud, G. (1982). A classification of cognitive tasks and operations of thought involved in addition and subtraction problems. In T. Carpenter, J. Moser, & T. Romberg (Eds.), *Addition and subtraction: A cognitive perspective.* New York, NY: Academic Press.

Yerushalmy, M., & Schwartz, J. (1993). Seizing the opportunity to make algebra mathematically and pedagogically interesting. In T. Romberg, E. Fennema, & T. Carpenter (Eds.), *Integrating research on the graphical representation of functions* (pp. 41–68). Hillsdale, NJ: Erlbaum.

Chapter 11

Children's Algebraic Reasoning and Classroom Practices That Support It

Maria Blanton   James J. Kaput
TERC   *Formerly at the University of Massachusetts Dartmouth*

Over the past decades, it has become widely accepted that preparing elementary students for the increasingly complex mathematics of the twenty-first century requires a different type of school experience—one that goes beyond the usual elementary grades' focus on arithmetic and computational fluency to one that cultivates algebraic forms of reasoning. The Common Core State Standards for Mathematics (National Governors Association Center for Best Practices & Council of Chief State School Officers [NGA Center & CCSSO], 2010) recently reiterated this view, following predecessors such as the *Principles and Standards for School Mathematics* (NCTM, 2000) in framing algebra learning as a longitudinal approach that should begin in kindergarten. The premise of this approach to learning algebra from the start of formal schooling is that the integration of algebraic reasoning into the elementary grades can help build a deeper conceptual understanding of more complex mathematics into students' experiences, which can result in greater student success in mathematics.

But what do we mean by "algebraic reasoning"? Algebraic reasoning can be viewed as ways of thinking that include generalizing, representing, justifying, and reasoning with mathematical relationships and structure (e.g., Blanton, in press; Blanton, Levi, Crites, & Dougherty, 2011; Kaput, 2008). Generalizing is the heart of algebraic reasoning and involves noticing mathematical structure and relationships. It can occur in a variety of mathematical contexts. For example, it might include noticing relationships that govern operations on numbers, such as the commutative property of addition, or functional relationships between two covarying quantities. Algebraic reasoning also includes expressing relationships and structure through representations such as words, variable notation, pictures, tables, or graphs. It includes building arguments that justify whether or not generalizations about the relationships children notice are true and reasoning with generalizations, accepted to be true, to build new understanding.

The increasing emphasis on algebraic reasoning places elementary teachers on the critical path to mathematics reform. However, as we suggested more than ten years ago in our *Journal for Research in Mathematics Education* (*JRME*) article (Blanton & Kaput, 2005), if we are to build classrooms that promote algebraic reasoning, we must provide appropriate forms of professional support that will effect changes in instructional and curricular practices. In part, this requires us to understand what it means for a teacher's practice to support a culture of algebraic activity in the classroom.

The focus of that *JRME* article was on a case study examining the classroom practice of one third-grade teacher, June, as she participated in a long-term professional development project. Our goal was to explore in what ways and to what extent she was able to build a classroom climate that supported the development of students' algebraic reasoning skills. In this chapter, we focus on results from the case study that illustrate the algebraic character

---

This chapter is adapted from M. L. Blanton & J. J. Kaput (2005), Characterizing a classroom practice that promotes algebraic reasoning, *Journal for Research in Mathematics Education*, *36*, 412–446.

of June's classroom, especially the ways in which she included algebraic reasoning activities in her instruction and how the students responded to them. For a more complete picture of our study and its findings (including the effects of June's practice on student performance on algebraic reasoning items from a standardized test), we refer the reader to the *JRME* article.

A Brief Background to Our Study and Its Methodology

At the time we conducted the research, June was in her second year of participating in the Generalizing to Extend Arithmetic to Algebraic Reasoning (GEAAR) project, a five-year professional development project designed to develop teachers' abilities to identify and strategically build upon students' attempts to reason algebraically and to use existing and supplemental resources to engineer instructional activities to support this (Kaput & Blanton, 1999). June initially insisted that she was "not a math person," and she described her experiences in GEAAR as her first exposure to the ideas of algebraic reasoning. However, because she was willing to try new algebraic reasoning tasks and ideas suggested by us (as the leaders of GEAAR) and by the other teacher participants, we asked to team with her to examine her teaching practice more closely in order to understand characteristics of her instruction that might support students' algebraic reasoning.

We observed June's ninety-minute mathematics class approximately twice per week for one school year, collecting information via field notes, audio recordings, June's reflections, students' written work, and classroom activities. As we analyzed the information, our goal was to characterize instances of algebraic reasoning and determine how they were embedded in instruction. We took the following as a measure of the robustness with which she integrated algebraic reasoning: (1) the *diversity* of types of algebraic reasoning that occurred; (2) their *frequency and form* of integration; and (3) *techniques of instructional practice* by which algebraic reasoning could thrive.

We first coded the data by identifying instances of spontaneous algebraic reasoning (SAR) and planned algebraic reasoning (PAR). We defined SAR as those instances that occurred without prior planning on June's part but that arose naturally in the mathematics in which the students were engaged and that June used effectively to infuse algebraic reasoning in instruction. For example, June demonstrated SAR when, during the course of reviewing homework on simple addition tasks, she spontaneously shifted the focus from computing sums to determining if the sum of two numbers would be even or odd (especially with respect to numbers that were sufficiently large so that the students could not easily do direct computation). (Details about this SAR activity appear in a later section of this chapter.)

In contrast, PAR referred to algebraic reasoning that resulted from classroom activities that June planned in advance of the class. For example, June preselected the Trapezoid problem, an activity that asks students to find a functional relationship between an arbitrary number of trapezoidal-shaped tables placed end to end and the number of people who could be seated at these tables. (Details about the Trapezoid problem appear in a later section of this chapter.)

Each episode of SAR or PAR was further categorized based on the nature of algebraic reasoning that occurred. We identified thirteen categories of what we viewed as algebraic reasoning that emerged in June's teaching; we describe them briefly, then elaborate on some of them in the next sections of this chapter. While most of these categories are organized by content areas where algebraic reasoning can occur (e.g., functional thinking), as we highlight in our discussion, they also reflect the core algebraic thinking practices of generalizing, representing, justifying, and reasoning with mathematical relationships and structure.

The Algebraic Character of June's Classroom

Frequency of PAR and SAR

From the data gathered during our year observing June's teaching, we identified 204 episodes of algebraic reasoning. Of these, 65 percent were classified as SAR; that is, these were instances in which, in response to students' thinking, June spontaneously crafted instruction that required students to reason algebraically. We thought it was significant that so many episodes occurred spontaneously in her instruction. June's flexibility in spotting opportunities for algebraic reasoning points to her growth in content knowledge and understanding of how such reasoning could be integrated into third-grade arithmetic.

Did June start with PAR and progress to SAR? Not according to our findings. Instead, her instruction integrated both planned and spontaneous episodes of algebraic reasoning throughout the year. This is perhaps because June participated in GEAAR the year prior to this study and, therefore, had some level of expertise and a growing flexibility to spontaneously integrate algebraic reasoning in instruction. In the case of PAR (planned algebraic reasoning) activities, June frequently included algebraic tasks taken from the GEAAR project and other resources.

The lesson we can take from this is to never underestimate what teachers can accomplish in their classrooms, given appropriate supports and self-confidence in their content knowledge. June evolved from a teacher with a self-proclaimed aversion to mathematics, to a skilled teacher who fluidly and frequently built rich opportunities for algebraic reasoning into her instruction (see also Soares, Blanton, & Kaput, 2006). To this end, professional development projects can offer a wealth of resources and knowledge that can be used in the classroom.

Examples of algebraic reasoning in June's classroom

Table 11.1 contains a summary of the thirteen types of algebraic reasoning we observed in the data gathered in June's classroom. Previously, for the *JRME* article, we organized these types around two core content areas where algebraic reasoning practices occurred: generalized arithmetic (Categories A–E) and functional thinking (Categories F–J). We also created a third categorization (Categories K–M) for processes central to algebraic reasoning. In preparing this chapter, findings from our *JRME* publication were organized in the same way for consistency but are discussed here in a way that more clearly highlights how the core algebraic reasoning practices (e.g., generalizing) occur within these categories. In the section that follows we showcase specific examples from selected categories, with a focus on student and teacher comments for particular tasks within the category. (Categories G and J are not discussed here.)

Categories A & B: Algebraic reasoning with arithmetic relationships

We found that in June's classroom, students reasoned algebraically about arithmetic relationships in the following ways:

- generalized about sums and products of even numbers and odd numbers;

- generalized about properties of operations on numbers, such as the result of subtracting a number from itself, expressed formally as $a - a = 0$ (the additive inverse property);

- decomposed whole numbers into possible sums and examined the structure of those sums; and

- generalized about place-value properties.

Table 11.1
Categorizing algebraic reasoning

Algebraic reasoning within generalized arithmetic	
Refers to instances where arithmetic was used as a domain for generalizing, representing, justifying, and reasoning with arithmetic relationships. This included reasoning with generalized quantities, where the end goal was not necessarily generalizing. For example, students were engaged in whole number operations on generalized forms, such as equations with unknown quantities ("missing number sentences"), or used number in a generalized, "algebraic" way.	
Category A	Exploring relationships in operations on whole numbers (e.g., sums of even numbers and odd numbers)
Category B	Exploring properties of operations on whole numbers (e.g., commutativity of addition and multiplication)
Category C	Exploring the equal sign as expressing a relationship between quantities (e.g., the algebraic role of " = ")
Category D	Algebraic treatment of number (i.e., requiring students to attend to structure and not rely on computation of specific numbers)
Category E	Solving equations (e.g., If $V + V = 4$, what is $V + V + 6$?)
Algebraic reasoning within functional thinking	
Includes instances where students were engaged in generalizing functional relationships, representing these relationships in words and symbols, justifying the symbolic rule from the problem context, and reasoning with these relationships to make predictions about data.	
Category F	Symbolizing varying quantities and operating with symbolized expressions
Category G	Representing data graphically
Category H	Finding functional relationships (e.g., exploring patterns and developing a rule)
Category I	Predicting unknown states using known data (e.g., Handshake problem for large numbers of people)
Category J	Identifying and describing numerical and geometric patterns
More with algebraic processes	
Includes instances that were not specific to the content areas of generalized arithmetic and functional thinking. Instead, they represented core algebraic thinking practices (e.g., generalizing) abstracted from particular mathematical content.	
Category K	Using generalizations to solve algebraic tasks
Category L	Justification, proof, and testing conjectures (e.g., Is 0 an even number?)
Category M	Generalizing a mathematical process

June describes a classroom conversation that involved algebraic reasoning about even numbers and odd numbers (Blanton & Kaput, 2005, p. 420; see also Blanton & Kaput, 2000):

> I asked the class what would happen if I added 2 even numbers together. Most of them said that I would get an even number. When I asked what would happen if I added 2 odd numbers together, most of them said that I would get an odd number. When asked about odd and even together, the answers were mixed. In the past I would have told them the answers by giving some examples (e.g., $5 + 5 = 10$). But . . . I wanted them to see how it really works, so that they could see that it would [generalize to all cases]. We did [an] activity combining (square) grid-paper shapes to model adding even and odd numbers. I asked the same questions again. This time they answered with more certainty. One student explained that "the sum of any two odd numbers is even" using the idea of adding square shapes: "If you have two odd numbers it makes it even because if you have leftovers the two leftovers go together."

The only confusion came when Sarah said that odd + even was odd and even + odd was even. Stephen responded that that couldn't be. He used numbers in place of odd and even and said that it (using "odd" and "even") was the same as using letters instead of numbers. Sarah explained to the class, "I thought that all the time when odd is the first one it was supposed to be odd and when even was the first it was going to be even. [But then I saw that that wasn't correct] because once you start turning them around, then it's the same thing. It doesn't make a difference."

Here, June's ability to create an instructional context that promoted algebraic reasoning is clear. Not only did she ask students to generalize and represent—in their own words—relationships about sums of evens and odds, she asked them to justify why their generalizations might be true. In this, she pushed them beyond the typical case-based arguments of testing specific examples, which children often employ, to a more generalized, representation-based argument (Schifter, 2009) that invoked visual ways—here, the use of square grid-paper shapes to model evens and odds—to reason about sums of arbitrary evens and odds. The representation helped students see that their generalization "the sum of any two odd numbers is even" was true for *any* two odd numbers because "the two leftovers go together."

Category C: Exploring the equal sign as expressing a relationship between equivalent quantities

The development of a relational understanding of the equal sign is an important objective in the Common Core State Standards (NGA Center & CCSSO, 2010), introduced as a standard in first grade. We observed June developing this concept through use of a balance scale (see Falkner, Levi, & Carpenter, 1999) and problems like $8 + 4 = \Box + 5$. Students learned to operate on both sides of the equal sign, and as a result, they began to treat equations as statements characterizing two expressions as equivalent. In the article (Blanton & Kaput, 2005), we report an episode in which June asked students to solve $(3 \times n) + 2 = 14$. She described the response from a student we called Sam (all student names used in this chapter are pseudonyms):

> Sam said that we could take the 2 away. He said that if we take the 2 from one side we have to take it from the other side. This was to make it balance. After we [took] the 2 away, he said to take the 12 tiles and put them in groups of 3. There were 4 groups, so the answer had to be 4. We tried replacing the n with 4 and it worked. (p. 421)

We can infer from Sam's explanation (as reported by the teacher) that he viewed " = " as indicating that the expressions $(3 \times n) + 2$ and 14 were equivalent, or as signaling that any action on one quantity (i.e., subtracting 2 from $(3 \times n) + 2$) required an equivalent action on the second quantity (i.e., subtracting 2 from 14). Additionally, while the operation of "dividing both sides by 3" was not an operation available to Sam (division had not been taught in this class yet), he was flexible enough to compensate by thinking multiplicatively to find the number of groups of 3 in 12. This is rather sophisticated algebraic thinking in that Sam was able to view "12" as equivalent to $3 \times n$. While the goal of this task was not to generalize per se, its significance is in Sam's reasoning with a generalized quantity (i.e., $(3 \times n) + 2$) in order to solve an equation.

Category D: Algebraic treatment of number

Numbers can actually bridge children's reasoning about known quantities to unknown quantities. Category D describes episodes in which June treated numbers in an algebraic way, that is, as a placeholder that required students to attend to structure rather than rely on computation of specific numbers. For example, in the following interchange, June challenged the use of an arithmetic strategy to deduce that 5 + 7 was even (Blanton & Kaput, 2005, p. 422):

June:	How did you get that?
Tony:	I added 5 and 7 and then I looked over there *[pointing to a visible list of even numbers and odd numbers on the wall]* and saw that it was even.
June:	What about 45678 + 85631? Odd or even?
Jenna:	Odd.
June:	Why?
Jenna:	Because 8 and 1 is even and odd, and even and odd is odd.

By using numbers large enough that students could not easily compute their sum, June required students to reason with generalized even and odd properties (e.g., as Jenna described, "even and odd is odd"). In doing so, we maintain that she was using numbers as placeholders (similar to variables) to represent *any* odd or even numbers. This illustrates how a teacher can use numbers to set the stage for the next move—representing a generalization more formally through algebraic notation.

Category E: Solving "missing number sentences"

This category included not only simple linear equations like $(3 \times n) + 2 = 14$, but also more complex equations and sets of equation. June's students sometimes spontaneously generated equations with unknowns in the course of other tasks. In our article, we included an example of Zolan's solution to a triangle puzzle. Figure 11.1 shows this puzzle, in which the regions have additive relationships between them and the goal is to "complete the triangle" by finding all missing numbers. The missing numbers are found by adding two side-by-side numbers to determine the entry above them (e.g., in the middle row, 7 plus the unknown number to its right would equal 12, the entry above the two numbers). Zolan spontaneously symbolized the problem by generating a set of equations ($7 + a = 12$; then $e + 4 = 5$ and $4 + d = 7$) from which he could solve for the missing numbers. Not only could he symbolize unknowns, he understood that different symbols were needed for different unknown quantities. He was able to correctly solve each equation for its unknown and then use that information in subsequent equations. Although June had not asked students to set up these kind of equations, that Zolan chose this process without prompting suggests that he was seeing mathematics in new, algebraic ways.

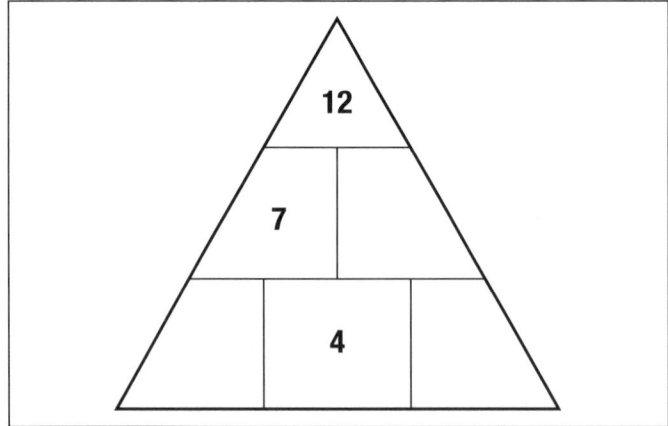

Fig. 11.1. Zolan's triangle puzzle

Category F: Symbolizing varying quantities and operating with symbolized expressions

Although connected to Category E in its emphasis on symbolized equations or expressions, Category F focuses on the use of symbols to model problems or to operate on symbolized expressions (not on solving an equation for an unknown or missing quantity). Representing unknown quantities symbolically and operating on them is central to algebraic reasoning. In Category E, for example, Zolan's activity of creating equations to represent relationships between known and unknown quantities reflects the goals not only of the mathematical practice "model with mathematics" (NGA Center & CCSSO, 2010), which states that mathematically proficient students should be able to write equations to model problem situations, but also the algebraic practices of *representing* generalized quantities through variable notation and *reasoning* with these quantities to solve equations. June's students also developed what they described as "secret messages," which were symbolic codes for unit conversions. For example, *3 ft 5 in* became 3(12) + 5. Encoding a secret message was then a process of symbolizing quantities as variable amounts. To convert feet to inches, students used the more general message $F(12) + I$, where F represented the number of feet and I represented the number of inches. Thus, this "secret message" acted as a conversion function. Students later extended these ideas by symbolizing functional relationships (see Category H).

Category H: Finding functional relationships

Category H includes instances in which students were asked to explore either recursive patterns or functional relationships and develop a rule that described a relationship. In the early part of the school year, June and her students explored simple recursive patterns (e.g., add 2 every time) in their function tables (referred to as "t-charts"). However, as June's mathematical understanding about patterns and relationships evolved, the complexity of the tasks she presented to her students evolved as well. The following is an example of how her students solved the Trapezoid problem (see fig. 11.2). Here, students have just found the number of people who can sit at twelve adjoined desks by finding a pattern in the sequence of values representing the total number of people. As June related (Blanton & Kaput, 2005, p. 425):

> The strangest thing happened! I saw another pattern. I asked the class to look at the [data] in another way. I wanted them to look at the relation between the [number of] desks and the number of people. Find the rule. No luck. Then I gave them the hint to see if there was a way to multiply and then add some numbers to have it always work. Jon suggested that we try and find a "secret message." After a few minutes, believe it or not, Anthony and Alicia started to multiply the number of desks with different numbers starting with one. The two children arrived at multiplying by 3 and then they would have to add 2. We tried many examples and it worked all of the time. We even tried some big numbers like 100. We then tried to make a "secret message." Anthony said that the 3 stays the same so use a d for [the number of] desks. This is what he came up with:
>
> $3(d) + 2$ = number of people.
>
> They realized that the 3 came from the people that could sit "on the top and bottom" and the 2 came from the two sides. This is not where I thought this was going to go.

This type of activity had become a fairly routine practice in June's classroom. The Standards for Mathematical Practice in the Common Core State Standards (NGA Center & CCSSO, 2010) describe that students should be able to make sense of quantities and the relationships between them using such tools as diagrams, tables, graphs, and formulas. This mathematically rich work can occur in function tasks like the Trapezoid problem. June's students used a systematic strategy to test what patterns would produce the desired data; conjectured a relationship, described it in everyday language ("multiplying by 3 and then . . . add 2"), then tested it on a diverse domain of numbers; symbolized the relationship by noting which quantities varied (the number of desks) and which remained constant (the number of people seated on the two sides); and described how the physical situation was represented by their model.

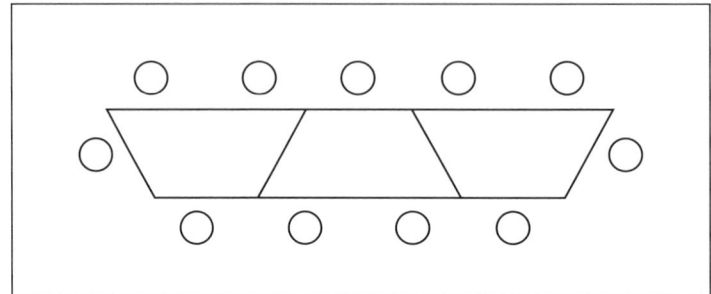

Fig. 11.2. A three-desk configuration for the Trapezoid problem

Category I: Predicting unknown states using known data

Category I describes those instances in which students made conjectures about what would happen for some unknown state, given what they knew from analyzing known data for functional relationships. For example, with the Handshake problem, June asked the class to write an equation that would give the number of handshakes in a group of twelve people, without enacting the handshakes or drawing a model from which they might count the number of handshakes. Students had already determined the number of handshakes in groups of six, seven, and eight people. An excerpt of their conversation follows (Blanton & Kaput, 2005, p. 426):

June: If there were 12 people here and they were going to shake hands, what would you do?

Ben: You could only shake 11 people's hands. *[Based on prior conversation, we infer that the student meant the first person to shake hands would shake 11 hands.]*

June: Why?

Karen: Because he can't shake his own hand *[therefore the number sentence begins with 11 as opposed to 12].*

June: So how would your number sentence change if there were 12 people?

Karen: Eleven, ten, nine, eight, seven, six, five, four, three, two, one.

Since thoughtful predictions about quantities require the analysis of relationships between numbers, not simply operations on them, June's query about an unknown state (see the first line in the excerpt above) became a point of entry into algebraic reasoning. That is, it forced children to think about relationships in the known data (the number of handshakes for groups of six, seven, and eight people) that might be applied to the unknown case of a group of twelve people. Choosing a value that was sufficiently large (twelve) enabled June to prompt children to think about the problem algebraically, or structurally, rather than use a more tedious approach with limited usefulness, such as drawing a model to represent all possible handshakes and computing the result.

Category K: Using generalizations to solve algebraic tasks

This category defines those instances in which June's students used generalizations to build other generalizations. This algebraic practice of reasoning with a generalization is a rather sophisticated level of algebraic thinking because it requires the student to see the generalization as an object that can be acted upon. The Standards for Mathematical Practice (NGA Center & CCSSO, 2010) encourage this type of thinking as reasoning "abstractly and quantitatively." One particularly compelling episode concerned generalizing about sums of even numbers and odd numbers. June had asked students to determine "if we add odd plus odd plus odd, what would the sum be?" One student argued, "The sum would be odd because two odds make an even and when you add odd plus even, you get odd." In this, the student invoked previously established generalizations ("odd plus odd is even"; "odd plus even is odd") to justify the claim that the sum of three odds would be odd. That is, students were able to reason without the use of specific odd numbers, thus achieving a level of abstraction in which their reasoning was based, not on particular cases, but on generalized quantities and their relationships.

Category L: Justification, proof, and testing conjectures

Category L addresses the practice of justifying generalizations. Broadly speaking, constructing viable arguments and critiquing the reasoning of others—one of the eight Standards for Mathematical Practice (NGA Center & CCSSO, 2010)—is an essential characteristic of any classroom culture in which deep mathematical thinking occurs. It is not necessarily unique to algebraic reasoning. There are times when children need to justify forms of reasoning, such as arithmetic or geometric, that might not necessarily entail reasoning about a generalized claim. Algebraic reasoning, however, provides a context in which students can engage their peers in thoughtful debate as a generalization is established or found to be invalid.

We observed a strong expectation of justification in June's classroom; students routinely described or justified their thinking or tested their generalizations. For example, students developed representation-based arguments to justify generalizations about sums of evens and odds (see Categories A and B). In another episode, June spontaneously shared with students a conversation she had with her colleagues in which she had argued that zero is an even number, while some of the other teachers had argued that it was simply a "special number," but not even. She invited students to share their thinking. After a protracted conversation with a variety of perspectives (too lengthy to include here), the view emerged that zero is even because it belonged to a sequence of numbers, all of which were even, and it belonged to the sequence because it could be reached by skip counting by two ("Zero is an even number because it goes 0, 2, 4, 6, 8, 10, 12, 14, 16, 18, 20 . . . "). It is worth contrasting a teaching practice in which June might simply have defined zero for children, rather than prompting a debate in which they shared and refined their thinking until they reached a justification that was accepted by their classroom community.

Characteristics of Classroom Practice That Support Algebraic Reasoning

From our work with June, we identified characteristics of her practice that supported the development of students' algebraic reasoning. Although these techniques do not form an exhaustive list, we see them as part of an emerging profile of the type of practice that teachers skilled in algebraic reasoning might exhibit in instruction.

The first technique involves the *seamless and spontaneous integration of algebraic conversations in the classroom*. We observed that over time June was able to create spontaneous "algebraic conversations" by which students engaged in some form of algebraic reasoning— whether generalizing, representing, justifying, or reasoning with mathematical relationships.

She did this spontaneously by instinctively transforming what seemed to be a routine arithmetic task into one that required algebraic reasoning. Her conversation with Jenna (see Category D in this chapter) about whether "45678 + 85631" was odd or even illustrates this characteristic. In other words, we take instances of spontaneous algebraic reasoning (SAR) as one indicator that a teacher's practice is expanding to attend to algebraic reasoning opportunities.

The second technique is the *spiraling of algebraic themes over significant periods of time* across the school year. Whether this spiraling was planned or the result of her own developing mathematical knowledge, June built the complexity of algebraic activity in the classroom over time. For example, at the beginning of the year, students generated function tables (t-charts) and identified simple recursive patterns in sequences of data. Identifying recursive patterns progressed to analyzing the relationships between two quantities whose values were represented in the table and describing more complex relationships, such as in the Trapezoid problem (described here in Category H). June also continued to revisit themes addressing generalizations in arithmetic (e.g., generalizations about sums of evens and odds), symbolizing to represent unknown quantities, and varying task parameters in order to generate data for covarying quantities and to identify and describe a relationship in the data (e.g., the Handshake problem).

The third technique involves the *integration of multiple and independently valid algebraic processes*. For example, during one class June asked students to use base-ten blocks to solve equations with unknown values, a task that is in itself algebraic in nature. After a discussion with students in which they shared their strategies, she focused on the equation $14 = 6 + n$ and expanded it in the following way: First she asked students to solve $140 = 60 + n$ and then $1400 = 600 + n$. Thus, June turned solving this family of equations into a pattern-finding activity about similarities in the solutions of the equations, thereby superimposing a separate algebraic process on this activity.

The fourth technique we call *activity engineering*—that is, adapting or developing tasks to include algebraic reasoning. In our GEAAR professional development sessions, we encouraged teachers to develop their own tasks or to adapt other resources, because this activity supports the capacity to think beyond the finite resource base provided by professional development. In June's case, a few weeks after we introduced the Handshake problem to GEAAR participants, she adapted an activity in which students explored the number of gifts received, based on the words of the song "The Twelve Days of Christmas." (Her third-graders were learning the song for a school production.) In this song, one's paramour receives an accumulation of gifts over a twelve-day period. June's version of the task was as follows:

> How many gifts did your true love receive on each day? If the song was titled "The Twenty-Five Days of Christmas," how many gifts would your true love receive on the twenty-fifth day? How many total gifts did she or he receive on the first 2 days? The first 3 days? The first 4 days? How many gifts did she or he receive on all twelve days?

You probably notice that this problem is mathematically similar to other problems, such as the Handshake problem (see Blanton & Kaput, 2004, for other examples)—problems that are pattern-eliciting tasks that rely on sums left in their unexecuted form to find and describe general function relationships. The significance here was in June's capacity to capitalize on what her students were doing outside of her classroom (e.g., learning a song for a school program) and to incorporate it into her lessons. Thus, she was developing her own algebra "eyes and ears," learning to plan instruction independently of the resources provided in professional development.

Activity engineering is important because curricular materials often do not have a rich base for developing students' algebraic thinking. Teachers often need to find the potential in the tasks available to them. We summarize below some techniques that can build algebraic reasoning in your classroom (see also Blanton et al., 2011).

1. *Choose numbers strategically.* This technique might involve using numbers that students cannot operate on arithmetically and must, instead, think about structurally (e.g., Categories D, I). It might also involve choosing numbers within an arithmetic task that intentionally make a particular generalization or arithmetic property more visible.

2. *Write equations in nonstandard formats.* This technique will help children as they learn to think relationally about the equal sign. Equations for which the value or quantity to the right of the equal sign is always a single numerical value (e.g., $4 + 5 = 9$) can reinforce misconceptions in children's thinking (e.g., Category C).

3. *Make known information unknown.* Simple arithmetic tasks can be transformed into algebraic tasks by simply making a known value unknown. For example, the task

   > *If Jackson has 3 pencils and his teacher gives him 4 more, how many pencils does he have now?*

   can be transformed to an algebraic task by making the number of pencils Jackson has initially unknown:

   > *If Jackson has some pencils and his teacher gives him 4 more, how would you describe or represent the number of pencils he has now?*

4. *Vary a fixed quantity to create function tasks.* Create a function task by varying a fixed quantity. For example, the Handshake problem began as an arithmetic task for which students were to find the number of handshakes for a single group with four people. By varying the number of people in the group, students could explore a relationship between the number of people in the group and the number of handshakes (e.g., Category H).

Conclusion

The view that children would be better prepared for a formal study of algebra in secondary grades if the development of their algebraic reasoning began informally in the elementary grades came about as a result of a critical lesson learned: The historical "arithmetic-then-algebra" approach, where the exclusive study of arithmetic in the elementary grades was followed by a largely superficial treatment of algebra in secondary grades, led to the widespread failure and alienation of students in school and society (Kaput, 2008). One of the most important lessons learned since, in the abundance of early algebra research over the last several decades, is that children are far more capable of rich and varied forms of algebraic reasoning than previously thought. And another important lesson is emerging. Studies are beginning to show that early algebra instruction can significantly affect the growth of children's understanding of core algebraic concepts over time (e.g., Blanton, Stephens, Knuth, Gardiner, & Isler, 2013; Britt & Irwin, 2008). Such research holds much promise that early algebra instruction can ameliorate the well-known difficulties older students—students whose mathematical experiences in the elementary grades focused on arithmetic and did not include algebraic reasoning—historically have had.

This potential in young children's capacity for algebraic thinking brings into relief an important reality: Elementary teachers are critical to teaching and learning a new algebra. For us, this reality underscores the importance of a central lesson learned in the *JRME* article discussed here: Elementary teachers—even those who might see themselves as math phobic—have a tremendous capacity to build rich and engaging classroom environments in which children can learn to generalize, represent, justify, and reason with mathematical structure and relationships.

References

Blanton, M. (in press). Algebraic reasoning in Grades 3–5. In M. Battista (Ed.), *Reasoning and sense making in grades K–8*. Reston, VA: National Council of Teachers of Mathematics.

Blanton, M., & Kaput, J. (2000). Generalizing and progressively formalizing in a third grade mathematics classroom: Conversations about even and odd numbers. In M. Fernández (Ed.), *Proceedings of the XXII Annual Meeting of the North American Chapter of the International Group for the Psychology of Mathematics Education* (pp. 115–119). Columbus, OH: ERIC Clearinghouse.

Blanton, M., & Kaput, J. (2004). Design principles for instructional contexts that support students' transition from arithmetic to algebraic reasoning: Elements of task and culture. In R. Nemirovsky, B. Warren, A. Rosebery, & J. Solomon (Eds.), *Everyday matters in science and mathematics* (pp. 211–234). Mahwah, NJ: Lawrence Erlbaum.

Blanton, M., & Kaput, J. J. (2005). Characterizing a classroom practice that promotes algebraic reasoning. *Journal for Research in Mathematics Education, 36*, 412–446.

Blanton, M., Levi, L., Crites, T., & Dougherty, B. (2011). *Developing essential understanding of algebraic thinking for teaching mathematics in grades 3–5* [Essential Understanding Series]. Reston, VA: National Council of Teachers of Mathematics.

Blanton, M., Stephens, A., Knuth, E., Gardiner, A., Isler, I., & Kim, J. (2015). The development of children's algebraic thinking: The impact of a comprehensive early algebra intervention in third grade. *Journal for Research in Mathematics Education, 46*(1), 39–87.

Britt, M., & Irwin, K. (2008). Algebraic thinking with and without algebraic representation: A three-year longitudinal study. *ZDM, 40*(1), 39–53.

Falkner, K. P., Levi, L., & Carpenter, T. P. (1999). Children's understanding of equality: A foundation for algebra. *Teaching Children Mathematics, 6*, 232–236.

Kaput, J. (2008). What is algebra? What is algebraic reasoning? In J. Kaput, D. Carraher, & M. Blanton (Eds.), *Algebra in the early grades* (pp. 5–17). Mahwah, NJ: Lawrence Erlbaum.

Kaput, J., & Blanton, M. (1999, April). *Algebraic reasoning in the context of elementary mathematics: Making it implementable on a massive scale.* Paper presented at the annual meeting of the American Educational Research Association, Montreal, Canada.

National Council of Teachers of Mathematics. (2000). *Principles and standards for school mathematics*. Reston, VA: Author.

National Governors Association Center for Best Practices & Council of Chief State School Officers (NGA Center & CCSSO). (2010). *Common core state standards for mathematics.* Retrieved from http://www.corestandards.org.

Schifter, D. (2009). Representation-based proof in the elementary grades. In D. Stylianou, M. Blanton, & E. Knuth (Eds.), *Teaching and learning proof across grades: A K–16 perspective* (pp. 71–86). New York, NY: Routledge.

Soares, J., Blanton, M., & Kaput, J. (2006). Thinking algebraically across the elementary school curriculum. *Teaching Children Mathematics, 12*(5), 228–235.

Chapter 12

The Equal Sign *Does* Matter

Eric Knuth and Ana Stephens
University of Wisconsin–Madison

In our *JRME* article (Knuth, Stephens, McNeil, & Alibali, 2006) and its companion *MTMS* article (Knuth, Alibali, Hattikudar, McNeil, & Stephens, 2008), we shared research findings emphasizing the importance of students developing a relational understanding of the equal sign (that is, an understanding that the equal sign expresses an equivalence relation between two quantities). On the surface, it may seem that understanding the meaning of the equal sign is relatively trivial. After all, students use the symbol throughout their school mathematics experiences beginning in the early elementary grades, and beyond an initial introduction there is generally little time spent on the concept in later grades. Yet, as both articles highlight, many middle school students fail to develop a relational understanding of the equal sign—a failure that may lead to later difficulties in learning algebra.

Since the publication of both articles, we have continued to study the development of students' algebraic reasoning in grades 3 through 8 and have continued to find that—absent instruction explicitly addressing the concept—students have difficulty developing a relational understanding of the equal sign. To illustrate these more recent findings, consider the following series of three related tasks along with representative student responses. Task 1 was presented to approximately three hundred students in grades 3–5, and tasks 2 and 3 were presented to approximately four hundred students in grades 6–8. All students were from the same school district. (For more detail about the grades 3–5 work, see Stephens, Knuth, Blanton, Isler, Murphy Gardiner, & Marum, 2013). As you read each task, take a few minutes to think about how your students might respond.

Task 1

The following number sentence is true: $15 + 8 = 23$.

Is $15 + 8 + 12 = 23 + 12$ true or false? How do you know?

We found that only 10 percent of grade 3 students, 31 percent of grade 4 students, and 51 percent of grade 5 students correctly responded that the second number sentence is true. As might be expected (refer to the *MTMS* article as needed, available at nctm.org/more4u), the most common explanation—demonstrated by more than a third of the students at each grade level—was one that revealed students viewing the equal sign operationally (e.g., "$15 + 8 + 12 = 35$ and $35 \neq 23$, so the equation is false"). Among those students who correctly responded that the second number sentence is true, the most common approach was a computational

This commentary chapter is adapted from two sources: E. J. Knuth, A. C. Stephens, N. M. McNeil, & M. W. Alibali (2006), Does understanding the equal sign matter? Evidence from solving equations, *Journal for Research in Mathematics Education*, *37*, 297–312, and E. J. Knuth, M. W. Alibali, S. Hattikudur, N. M. McNeil, & A. C. Stephens (2008), The importance of equal sign understanding in the middle grades, *Mathematics Teaching in the Middle School*, *13*, 515–519. The *MTMS* article is available at nctm.org/more4u.

strategy (e.g., "15 + 8 + 12 = 35 and 23 + 12 = 35, so the equation is true"). While this response is mathematically correct, it essentially ignores the first number sentence. To recognize that 15 + 8 = 23 implies 15 + 8 + 12 = 23 + 12 and that no computation is necessary to verify the truth of the second number sentence requires an understanding that adding the same quantity to each side of an equation preserves the equivalence relation. Unfortunately, responses such as the two below were relatively rare among the students (none of the grade 3 students, 5 percent of grade 4 students, and 9 percent of grade 5 students provided such an explanation):

> The second equation is true because you are just adding 12 to both sides. (grade 4 student)

> If you do the same thing to both sides of an equation it is still true. (grade 5 student)

Now consider task 2.

Task 2

Ian says that because 37 + 10 = 47, he knows that

37 + 10 − 24 = 47 − 24.

Do you agree with Ian? Why or why not?

In contrast to the grades 3–5 students' responses to task 1, we found that 53 percent of grade 6 students, 62 percent of grade 7 students, and 81 percent of grade 8 students correctly responded that the second number sentence is true. As was the case with the grades 3–5 students' responses to task 1, many students (16 percent in grade 6, 27 percent in grade 7, and 32 percent in grade 8) computed each side of the second equation to determine the equation was true, essentially ignoring the first equation. Relative to the elementary school students' responses to task 1, we also found that a greater percentage of the middle school students used a strategy that suggested they viewed the equal sign relationally (37 percent in grade 6, 32 percent in grade 7, and 49 percent in grade 8). In this case, students recognized that the computation of the sum of each side of the equation is not needed in order to know that the two sides are equal. Representative student responses included the following:

> Yes because if you add something or subtract something on both sides the answer stays the same. (grade 6 student)

> Yes, because he subtracted 24 on each side to make it even. (grade 7 student)

> Yes he is subtracting 24 from both sides of the equation so it is balanced out. (grade 8 student)

Finally, consider task 3.

Task 3

Do these two equations have the same solution? How do you know?

$2 \times n + 15 = 31$ $\qquad$ $2 \times n + 15 + 9 = 31 + 9$

This task addresses the same underlying concept as tasks 1 and 2 but "looks" more like what we would expect to see in a traditional algebra course. In this case, we found that 49 percent of grade 6 students, 63 percent of grade 7 students, and 72 percent of grade 8 students (the same students who responded to task 2) stated that the two equations have the same solution. As might now be expected, a popular strategy among the students was to solve each equation and compare the solutions (or a slight variation: solve the first equation and substitute that value into the second equation); this type of strategy was used successfully by 37 percent of grade 6 students, 53 percent of grade 7 students, and 61 percent of grade 8 students. Despite students' relative success stating that the equations do have the same solution (especially considering the students had little, if any, formal algebra experience), very few students provided an explanation indicating they held a relational view of the equal sign and thus understood that performing the same operation on both sides of an equation preserves the equivalence relation (12 percent in grade 6, 8 percent in grade 7, and 11 percent in grade 8). The following responses are representative of the few who did provide such an explanation:

> Yes because it's the same question, but the second one has plus 9 on both sides of the equal sign. (grade 6 student)

> Yes these two equations have the same solution for *n* because with the one on the right because you added 9 on the left and right side of the " = " sign, it doesn't change any of the numbers. (grade 7 student)

> Yes because you subtract 9 from both sides, which cancels both 9s out, which makes both equations the same. (grade 8 student)

Collectively, the results from these three tasks (as well as results from other equal sign-related tasks used in our current work and the results presented in our *JRME* article) strongly suggest that many students are not developing a relational understanding of the equal sign; moreover, there is little evidence to suggest that their equal sign understanding significantly improves as they progress from elementary through middle school. In some respects this is not surprising, given that the equal sign typically receives little, if any, instructional attention after its initial introduction in the early elementary grades. However, as the title of our *JRME* article suggests, understanding the equal sign *does* matter, and it warrants greater instructional attention if we want students to be able to "meaningfully generate and interpret equations but also to meaningfully operate on equations" (Knuth et al., 2006, p. 309).

The recent publication and subsequent adoption by most states of the Common Core State Standards for Mathematics (National Governors Association Center for Best Practices & Council of Chief State School Officers [NGA Center & CCSSO], 2010) has significantly increased the "algebraic demands" being placed on students. Two of the Common Core's standards for mathematical practice (standard 6: Attend to precision, and standard 7: Look for and make use of structure) underscore the need for students to possess a relational understanding of the equal sign. Standard 6 states that students must "state the meaning of the symbols they choose, including using the equal sign consistently and appropriately" (p. 7). Standard 7 states that "mathematically proficient students look closely to discern a pattern or structure" (p. 8). Students who demonstrate this particular practice standard are able to recognize the structural similarity between equations such as those presented in the three tasks discussed previously and provide responses based on this recognition (e.g., in response to task 3, "Yes, because even though $2 \times n + 15 + 9 = 31 + 9$ is different than $2 \times n + 15 = 31$ they add 9 to each side so the value of *n* does not change.").

So how can elementary and middle school teachers help their students develop a relational understanding of the equal sign? First, consider the types of experiences your students have with this symbol. Elementary students who are asked to complete worksheets of "fill in the blank" number sentences such as $1 + 3 = \_\_$, $5 + 2 = \_\_$, and $9 - 4 = \_\_$ are only seeing the equal sign in one format—a format that may promote an operational view of the equal sign. It is critical, then, to ensure students have early exposure to the equal sign in a variety of equation formats. For example, true/false and open number sentences such as those shown in figure 12.1 are particularly good vehicles for engaging students in discussion about the meaning of the equal sign and challenging any existing operational conceptions. Equations with operations on both sides of the equal sign are particularly useful for eliciting a relational understanding in students (Carpenter, Franke, & Levi, 2003; McNeil & Alibali, 2005).

True/False number sentences	Open number sentences
$3 + 5 = 8$	$6 + 4 = \square$
$3 + 5 = 0 + 8$	$6 + 4 = \square + 3$
$8 = 3 + 5$	$\square = 9 + 3$
$8 = 8$	$5 + 3 = \square + 3$

Fig. 12.1. Equal sign use in a variety of equation formats

In our *MTMS* article (Knuth et al., 2008; available at nctm.org/more4u) we offered several suggestions for upper elementary and middle school teachers to help their students develop a relational understanding of the equal sign. These suggestions included posing equivalent equations problems in both arithmetic contexts (e.g., Is $3 + 5 = 8$ the same as $3 + 5 - 4 = 8 - 4$?) and algebraic contexts (e.g., If $x = 7$ is the solution to $3x + 10 = 31$, what is the solution to $3x + 10 - 8 = 31 - 8$?). We also advocated looking for everyday classroom opportunities to discuss appropriate use of the equal sign. For example, when a student keeps track of his or her calculations by writing a computational "string" such as $3 + 4 = 7 \times 2 = 14 - 8 = 6$, the opportunity arises to consider whether the representation makes sense and to discuss the equal sign's meaning.

We are not necessarily advocating that entire lessons be designed specifically to address the meaning of the equal sign nor that significant class time be spent discussing the equal sign (especially in the middle grades). We suggest, rather, that posing tasks such as the ones mentioned above as occasional warm-up problems or taking advantage of opportunities to discuss the meaning of the equal sign when they naturally arise (e.g., when students write computational strings) can go a long way toward helping students develop a relational understanding of the equal sign.

In our work with grade 3 students, for example, we have seen significant improvement in students' understanding of the equal sign after experiencing only two targeted lessons coupled with follow-up discussions as opportunities arose during the course of classroom activities (see Blanton et al., 2015). In particular, we administered a pretest at the beginning of the year and a posttest at the end of the year addressing students' understanding of the equal sign (as well as several other algebraic concepts). On the pretest, students overwhelmingly demonstrated an operational view of the equal sign (95 percent of the students!). On the posttest, however, only about 10 percent of these students continued to demonstrate an operational view of the equal sign—a significant improvement with little instructional time and effort. Of course we are not suggesting that with two lessons and an occasional conversation, voilà, students now have a relational understanding of the equal sign. What we are suggesting, however, is that a periodic focus on the meaning of the equal sign over the course of students' elementary and middle school experiences can help students overcome the often well-entrenched operational view of the equal sign and develop a relational understanding.

In our *JRME* article (Knuth et al. 2006) we documented the difficulty that middle school students seem to have developing a relational understanding of the equal sign as well as the importance of such an understanding to their ability to successfully work with equations. In our companion *MTMS* article (available at nctm.org/more4u), we illustrated the findings from the aforementioned work and provided suggestions for teachers. And finally, in the most recent work shared here (more than eight years after the *JRME* study was conducted), we show that students in both upper elementary and middle grades continue to struggle to develop a relational understanding of the equal sign. Yet our recent work also suggests promise regarding students' equal sign understanding, namely, that with a little bit of instructional attention students can learn to view the equal sign relationally and, in turn, learn to use that view to meaningfully generate, interpret, and operate on both arithmetic and algebraic equations.

References

Blanton, M., Stephens, A. Knuth, E., Gardiner, A. M., Isler, I., & Kim, J. (2015). The development of children's algebraic thinking: The impact of a comprehensive early algebra intervention in third grade. *Journal for Research in Mathematics Education, 46*(1), 39–87.

Carpenter, T., Franke, M., & Levi, L. (2003). *Thinking mathematically: Integrating arithmetic and algebra in elementary school.* Portsmouth, NH: Heinemann.

Knuth, E. J., Albali, M. W., Hattikudur, S., McNeil, N. M., & Stephens, A. C. (2008). The importance of equal sign understanding in the middle grades. *Mathematics Teaching in the Middle School, 13,* 514–519.

Knuth, E. J., Stephens, A. C., McNeil, N. M., & Alibali, M. W. (2006). Does understanding the equal sign matter? Evidence from solving equations. *Journal for Research in Mathematics Education, 37,* 297–312.

McNeil, N., & Alibali, M. (2005). Knowledge change as a function of mathematics experience: All contexts are not created equal. *Journal of Cognition and Development, 6,* 385–406.

National Governors Association Center for Best Practices & Council of Chief State School Officers (NGA Center & CCSSO). (2010). *Common core state standards for mathematics.* Washington, DC: Author. Retrieved from http://www.corestandards.org.

Stephens, A., Knuth, E., Blanton, M., Isler, I., Murphy Gardiner, A., & Marum, T. (2013). Equation structure and the meaning of the equal sign: The impact of task selection in eliciting elementary students' understandings. *Journal of Mathematical Behavior, 32,* 173–182.

Chapter 13

Students' Metaphors for Limit Concepts in Introductory Calculus

Michael Oehrtman
Oklahoma State University

Limit concepts are notoriously difficult for students in introductory calculus. As a result, limits are often de-emphasized in calculus textbooks and classes. Since the central concepts in calculus are defined in terms of limits, however, simply avoiding these foundational ideas is likely to have severe, unintended consequences. Furthermore, the techniques for finding limits that are covered in calculus courses often have very little to do with the ways limits are used to develop ideas such as the derivative, definite integral, or Taylor series. The combined result is a missed opportunity to build on the most significant unifying concept throughout calculus.

Little is known, however, about potentially powerful aspects of calculus students' reasoning about limits. The purpose of my study (Oehrtman, 2009) was to look beyond difficulties and misconceptions to detail the rich conceptual structure of calculus students' intuitive ways of reasoning about limits. At the very least, a deeper understanding of how students interpret what we present to them can inform our decisions about language, images, and activities that we introduce in class. Ideally, we as practitioners may also discover powerful aspects of students' intuitive reasoning that can serve as the foundation for instruction that is accessible, yet rigorous and coherent.

Table 13.1 summarizes eleven of the questions I asked students to collect the data for this study. The questions were designed to be challenging rather than routine so I could see how students attempted to make sense of difficult limit concepts. Before proceeding, I invite you to think about how you would respond to each of these questions. How do you think about the limits involved? What language would you use to help students understand these limits and the related calculus concepts? How would your students interpret your descriptions? Keep these questions in mind as you read about the study I conducted in this chapter and compare your answers with the results.

Research Method	I chose a theory of metaphorical reasoning to guide the data collection and analysis for this study because it frames the understanding of complex ideas in terms of more familiar experiences, much as students are likely to do when faced with limits in calculus. By transferring properties and inferences from a concrete "source domain" to a more abstract "target domain," we gain access to reasoning about complex structures. Of course, the productivity of such metaphorical reasoning is limited by the extent to which the domains are similar. Thus we often selectively focus on what we perceive to be the most applicable aspects of the source domain or even reconceive them in order to support productive reasoning. Black's (1962a, 1962b, 1977) theory of metaphor focused on new ways of understanding that may emerge through such a dynamic interaction between source and target domains. He labeled "strong metaphors" those that are rich in implications and to which the user is deeply committed.

This chapter is adapted from M. Oehrtman (2009), Collapsing dimensions, physical limitations, and other student metaphors for limit concepts, *Journal for Research in Mathematics Education, 40*, 396–426.

Table 13.1

Labels and brief descriptions of problems used to develop the metaphor clusters

Problem label (Data collection method)	Abbreviated problem statement
Limit of a function (Interview)	Explain the meaning of $\displaystyle\lim_{x\to 1}\frac{x^3-1}{x-1}=3$
Derivative definition (Interview)	Let $f(x) = x^2 + 1$. Explain the meaning of $\displaystyle\lim_{h\to 0}\frac{f(3+h)-f(3)}{h}$
$0.\overline{9}=1$ (Pre/Postcourse survey)	Explain why $0.\overline{9}=1$
Derivative definition (Pre/Postcourse survey)	Explain why the derivative $f'(x)=\displaystyle\lim_{h\to 0}\frac{f(x+h)-f(x)}{h}$ gives the instantaneous rate of change of f at x.
L'Hôpital's rule (Web writing assignment)	Explain why L'Hôpital's rule works.
Volume of revolution (Web writing assignment)	Explain how the solid obtained by revolving the graph of $y = {}^1/_x$ around the x-axis can have finite volume but infinite surface area.
Limit comparison test (Web writing assignment)	Explain why the limit comparison test works.
Taylor Series of sin x (Web writing assignment)	Explain in what sense $\sin x = 1 - \dfrac{1}{3!}x^3 + \dfrac{1}{5!}x^5 + \dfrac{1}{7!}x^7 + \cdots$
Sequence of sets (Web writing assignment)	Explain how the length of each jagged line shown below can be $\sqrt{2}$ while the limit has length 1.
Multivariable continuity (Web writing assignment)	Explain what it means for a function of two variables to be continuous.
Volume and area of a sphere (Web writing assignment)	Explain why the derivative of the formula for the volume of a sphere, $V = \frac{4}{3}\pi r^3$ is the surface area of the sphere, $\dfrac{dV}{dr} = 4\pi r^2 = A$.

In order to collect data on the metaphors students use to reason about limits, I recruited 120 subjects from a yearlong introductory calculus sequence at a large university in the southwestern region of the United States. I attended all classes during the year to take field notes and collected data from a series of interviews and written assignments in which students described their reasoning about challenging limit problems. The eleven tasks in table 13.1 generated the richest data and evidence for characterizing students' metaphorical reasoning. I analyzed the resulting data following Black's characterization of strong metaphors by requiring that (i) multiple students committed to similar metaphors (ii) applied to multiple problem contexts (iii) in ways that clearly influenced their perception of a problem, solution method, and results. I categorized metaphors that met all three criteria as strong metaphors, and those that lacked at least two of the criteria as weak metaphors. No metaphors satisfied only two of the criteria.

Students' Strong Metaphors for Limits

My data analysis resulted in five strong metaphors based on intuitions about (i) a collapse in dimension, (ii) approximation and error analyses, (iii) proximity in a space of point-locations, (iv) a small physical scale beyond which nothing exists, and (v) the treatment of infinity as a number. Each of these strong metaphors is discussed in the sections that follow.

Collapse metaphors

The students often reasoned about limits using an image of a multidimensional object decreasing in size along one of its dimensions and ultimately vanishing, resulting in a "collapsed" object of reduced dimension. The argument of the limit was conceived as measuring a property of the original object (e.g., volume), while the limit measured an analogous, lower dimensional property of the collapsed object (e.g., area). Even when geometric interpretations were not apparent, students often described limits in terms of an abstract collapse metaphor, for example arguing that $0.\overline{9} = 1$ "because the numbers [digits] you are adding eventually become zero at infinity."

More than one-third of the students used a collapse metaphor when considering the derivative defined as the limit of a difference quotient, as illustrated in figure 13.1. These students began with a standard illustration of slope and described a secant line augmented with the base and height of a right triangle. They imagined points x and x_0 in the domain or $(x, f(x))$ and $(x_0, f(x_0))$ on the graph being chosen closer together and viewed this process as "creating" a tangent line once the two points were moved to the same location. At that moment, the base and height of the triangle were imagined to collapse to a single point or to infinitesimal lengths so that "the slope of the secant line becomes the slope of the tangent." Students who used a collapse metaphor for the derivative were also consistent in its application in the context of the position of a car as a function of time.

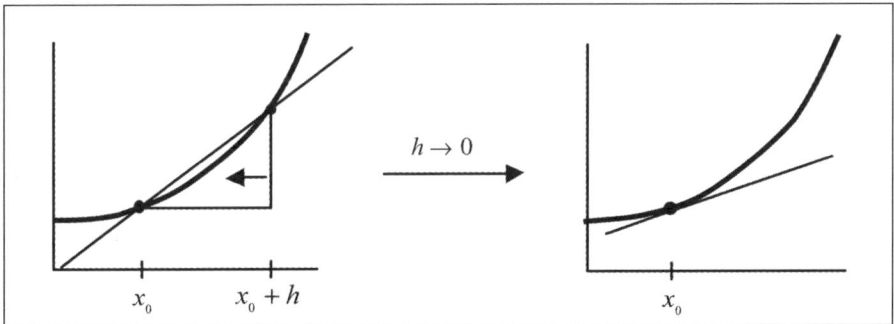

Fig. 13.1. A secant line between two points collapsing to a tangent line through "two points" at a single location

A second collapse metaphor emerged when students attempted to account for the seeming paradox of Torricelli's trumpet, the surface obtained by revolving the graph of $y = 1/x$ around the x-axis for $1 \leq x < \infty$, which has infinite surface area but finite volume. These students imagined the radius of a cross-sectional disk decreasing to 0 at some location along the axis of revolution so that the two-dimensional disk collapsed to a point or a line, as illustrated in figure 13.2.

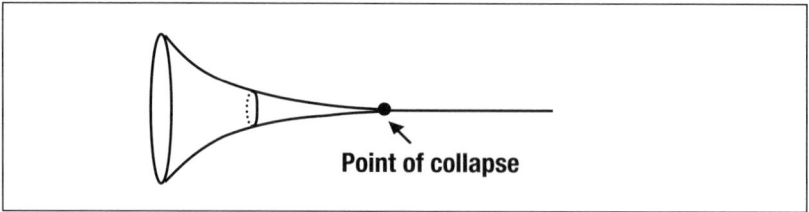

Fig. 13.2. A solid of revolution in which cross-sectional disks are imagined to collapse to a point, allowing finite volume

Nearly half of students used a collapse metaphor when attempting to explain the meaning of the Riemann integral or when justifying the fundamental theorem of calculus. Students described the fundamental theorem of calculus for a function, $\frac{d}{dx}\int_a^x f(t)\,dt = f(x)$ as saying that a final "rectangle" representing $\int_x^{x+\Delta x} f(t)\,dt$ becomes thinner as $\Delta x \to 0$ until it collapses to a line with height $f(x)$, as illustrated in figure 13.3a. Similarly, when asked why the derivative of the formula for the volume of a sphere $V = \frac{4}{3}\pi r^3$ is equal to the surface area of the sphere $\frac{dV}{dr} = 4\pi r^2 = A$, these students described an incorrect image of the fundamental theorem of calculus with thin concentric spherical shells where $\Delta r \to 0$ implies "the last shell of the sphere" gets thinner and thinner, eventually becoming the "last sphere's surface area."

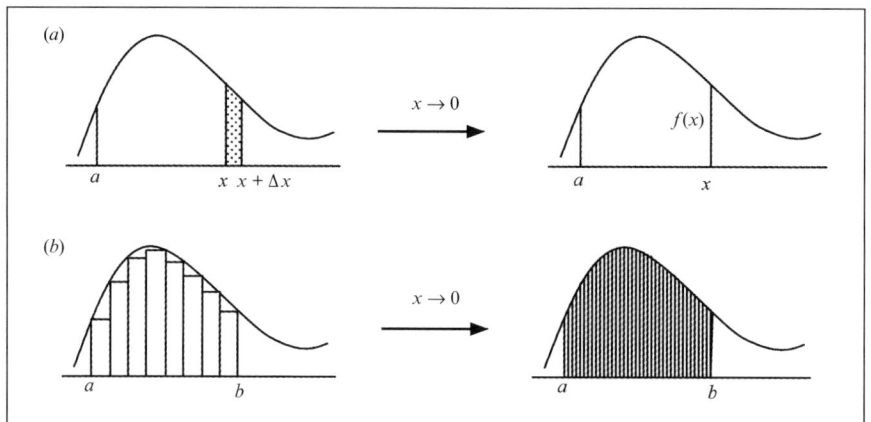

Fig. 13.3. Students' images of (a) the fundamental theorem of calculus where a final rectangle was imagined to collapse to a line of height f(x) and (b) a definite integral where a finite sum of areas of rectangles collapses to a sum of infinitely many one-dimensional heights

Students also imagined the Riemann integral $\int_a^b f(x)\,dx$ graphically as a "sum" of infinitely many one-dimensional vertical lines over the interval $[a,b]$ extending from the x-axis to a height $f(x)$ produced by a collapse of two-dimensional rectangles from the Riemann sum as their widths became zero, as illustrated in figure 13.3b. When asked to elaborate the meaning of the definite integral in the equation $\frac{4}{3}\pi R^3 = \int_0^R 4\pi r^2\,dr$, these students described a solid ball of radius R composed of concentric shells of small thickness Δr. They explained that a limit with $\Delta r \to 0$ meant all the shells became thinner and eventually became two-dimensional surfaces. Improperly transferring the idea of a sum of volumes of the shells with $\Delta r > 0$ to the limiting case with $\Delta r = 0$, these students argued that "the volume is just adding up the surface area of small [infinitely thin] spheres."

Collapse metaphors are almost always mathematically incorrect: Two points at the same location cannot define a tangent line, volumes of revolution do not have to shrink to a point to have finite volume, and you cannot add infinitely many lengths to obtain an area. Nevertheless, collapse metaphors were frequently productive for students. In many situations, students were unable to gain any mathematical perspective on the tasks presented until they invoked a collapse metaphor. Further, once they did, they often continued exploring important relationships and even resolved some of the inconsistencies introduced by the metaphor. For example, in the definition of the derivative, the collapse metaphor allowed students to see a connection between secant lines and the tangent, to recognize the need for a limit because the distance between the points is going to 0, to make sense of the slopes of these secants being the quotient of two quantities that both go to 0, and to see the connection to limits of functions with removable discontinuities. Rather than avoiding collapse metaphors, instructors can be aware of their frequent occurrence and help students ask and reason through questions that lead to a deeper understanding.

Approximation metaphors	The most common metaphor that emerged from the data analysis involved ideas about approximation to reason about a limit, which we will symbolize $\lim_{cv \to s} A = U$ to aid in description. Components of students' approximation metaphors were an unknown quantity, U, and approximations, A, close in value to U. Approximations could be made more accurate by making some controlling variable, cv, closer to a particular value, s (typically corresponding to a singularity in the approximation method). Each approximation has an associated error $= |U - A|$. A bound on the error, ε, was often used to restrict the range of values for the unknown, $A - \varepsilon < U < A + \varepsilon$. Approximations were contextually judged to be accurate if the error was small, and a good method of approximation allowed for improved accuracy to make the error as small as desired. Approximations were considered precise if there was not a significant difference among them.

Students described limits for repeating decimals and Taylor series as an unknown value to be approximated, the partial sums as approximations, and the difference between the two (i.e., the remainder) as the error. Discussions of accuracy were abundant in both cases, but students invoked more details of error analyses reflecting the structure of $\varepsilon - N$ definitions and arguments when talking about Taylor series compared to infinite decimals. Most students using an approximation metaphor for Taylor series indicated that errors could be made as small as you wanted by adding more terms, and several described specific methods for being able to bound the error using either the Lagrange formula or the fact for alternating series that "the maximum error . . . is the next term."

Students also applied an approximation metaphor to reason about the definition of the derivative, in which the unknown quantity was the slope of a tangent line and approximations were the slopes of secant lines. Many students discussed the resulting accuracy and a method for improving it, such as "The smaller you make your h, the better an approximation you would have, since the two points would be getting closer and closer."

These results suggest a unique potential power of approximation metaphors that allowed many students to reason about structures nearly equivalent to formal $\varepsilon - \delta$ and $\varepsilon - N$ arguments typically considered beyond the comprehension of students in introductory calculus. The intuitive nature of approximation ideas and error analyses combined with their central role in the mathematical structure of the calculus curriculum and their practical value in applied settings provide the basis for developing such reasoning as a coherent conceptual foundation throughout calculus. Results from subsequent instructional design research shows that ideas about approximation and error analyses can support students' powerful reasoning about limits and, subsequently, about the other major concepts in calculus (Oehrtman, 2008). This structure can be reinforced and generalized by repeatedly asking students to develop and work with multiple representations of the following five questions: (i) What is being approximated? (ii) What are the approximations? (iii) What are the errors? (iv) Given an approximation, how can you find a bound on the error? and (v) Given a desired error bound, how can you generate an approximation with that level of accuracy?

Proximity metaphors	Spatial representations of numbers are abundant in calculus, and such imagery supports metaphors for limits in terms of intuitive "closeness" and "clustering" in a space composed of point locations and distances between them. Students described points in space as having numerically measured properties (such as temperature) with small changes in physical locations resulting in small changes in the properties.

For the limit of a function and continuity, students often described separate spaces for the domain and the range with continuous functions preserving closeness. In the single-variable case "the y-points corresponding to an x-value that are sufficiently close to zero must also be sufficiently close to each other." For two variables, students often had difficulty reasoning about all possible paths approaching a point in the domain, and instead focused on a "circle around the point in question" and argued that the function "values within a sufficiently small radius around the origin must also be sufficiently close." Students used very physical language in their proximity metaphors: for example, when describing Taylor polynomials for sin x, "the closer the polynomials will wrap themselves around the original function," and further out the x-axis, "the polynomial becomes more and more loosely fitted around the curve."

Students using proximity metaphors often had difficulty reasoning quantitatively. For example, they thought that a limit was a point on a graph rather than a numerical value that the function outputs approach. They described points or lines being close to each other rather than conceiving of the relevant difference in values measuring a distance. These students also frequently misinterpreted the shapes of graphs as the shapes of physical objects in a context rather than representing a relationship between two quantities. When students begin to use very physical language, instructors can clarify such confusion by asking them to identify what quantity or quantities they are describing, how those quantities are seen and measured in the context, and how they are represented mathematically in the graph, table, or equation.

Infinity as a number metaphors	Students often treated infinity as a number that could be used in calculations or as inputs and outputs of functions, for example stating $\ln(\infty) = \infty$. Alternately, some students simply treated infinity as a "really big number." Dividing by infinity, students were led to consider infinitesimal quantities, often describing them as ambiguously nonexistent in size, yet not 0.

Several students represented $-\infty$ and ∞ as endpoints of the real number line. For example, as illustrated in figure 13.4, a student justified L'Hôpital's rule for functions f and g in an interval $[0,x]$ where $f(x) = g(x) = \infty$, beginning with, "By extending the Mean Value Theorem to the theoretical (x,∞) point, we can assume that there is a point on the domain where the slope of the line at that point is equal to the slope of the initial point and the (x,∞) point."

Infinity as a number metaphors often have correct corresponding limit statements; for example, $\ln(\infty) = \infty$ can be correctly reinterpreted as $\lim_{x \to \infty} \ln(x) = \infty$. It is helpful to remind students that infinity is not actually a number that can be plugged into a function or obtained as an output and to ask them how they can express the same idea using a limit. In other cases, infinity as a number metaphors are misleading: for example, incorrectly concluding that $\frac{\infty}{\infty} = 1$. In such cases, instructors may ask students to construct several easily analyzed scenarios with different results, such as $\lim_{x \to \infty} \frac{x^2}{x}$, $\lim_{x \to \infty} \frac{\sqrt{x}}{x}$, or $\lim_{x \to \infty} \frac{5x}{x}$.

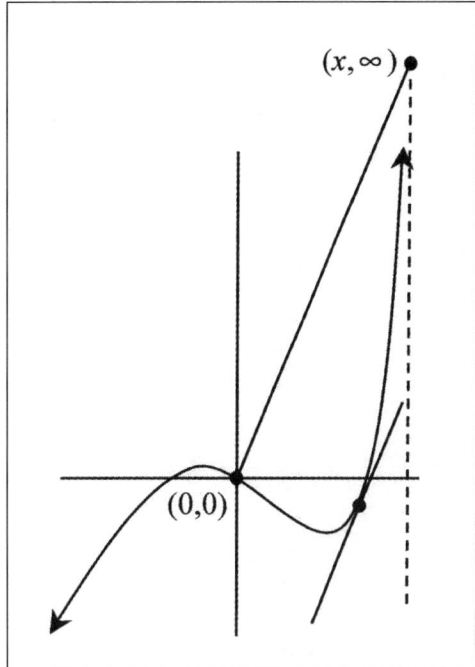

Fig. 13.4. A students' extension of the mean value theorem to a point at infinity

Physical limitation metaphors

Students used ideas about small-scale physical objects and phenomena to reason about limits, typically stating that there is a scale beyond which nothing can be observed, measured, or perhaps even exist. Such physical limitation metaphors consisted of an object representing the smallest physical size considered possible (e.g., a molecule, electron, or quark) and other objects composed of, interacting with, or measured against that "limiting" object.

The most striking application of a physical limitation metaphor occurred in students' interpretations of Torricelli's trumpet. The professor for the class presented the improper integral computations for this example and emphasized the apparent paradox by describing the surface as a paint can that could be filled (because it has finite volume), yet whose surface could not be painted (because it has infinite surface area). Making what he intended as a throw-away comment, the professor added, "Of course, you could never actually fill the can with paint, because at some point, the diameter gets smaller than a single molecule of paint." Over a third of the students responding to the writing assignment about this example invoked this compelling imagery, but with an interpretation critically different from the professor's, claiming that the volume was finite because, at some point, a single molecule would plug up the container, allowing the rest to fill, as illustrated in figure 13.5.

Students' physical limitation metaphors were often inspired by references to extremely small or large phenomena in fields such as physics, astronomy, or chemistry. Calculus instructors should exercise caution when drawing from these contexts to describe limits since they are fraught with alternate, and often conflicting, interpretations. As with infinity as a number metaphors, focusing students on the quantities of interest can also help distinguish the mathematical relationships from physical phenomena.

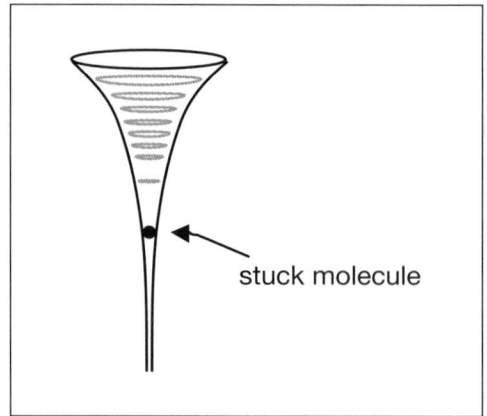

Fig. 13.5. An image of physical limitation
in a solid of revolution where a molecule of
water becomes stuck, allowing the container
to hold a finite volume

Weak Metaphors for Limits

In addition to the five strong metaphors for limits described above, three other domains were prominent in the data from the students' and professor's language to convey limit ideas: (i) experiences of motion, (ii) language about quantities being "sufficiently" and "arbitrarily" close or small, and (iii) images of zooming in on a graph. Language and images from these domains are frequently prescribed as ways to help students understand calculus and are commonly used by textbooks and instructors. Surprisingly, however, my data analysis categorized each of these as weak metaphors.

Zooming imagery and interpretations of local linearity

The professor for the course in which this study was conducted regularly described and demonstrated ideas of local linearity by magnifying the graph of a function with various technologies, such as a graphing calculator. Throughout both semesters of data collection, however, no students employed such zooming imagery to describe any of the limit situations presented to them. When directly asked about zooming, instead of interpreting the imagery in terms of local linearity as the professor intended, students described the result as a thick or blurry line, pixels on a display screen, or individual points on a theoretical graph. Imagery of zooming in on a graph was not categorized as a strong metaphor for the students in this study since they did not invoke these ideas on their own and did not apply them to reason about problems.

Motion imagery and interpretations of "approaching"

A dynamic conceptualization of quantities and relations is crucial to students understanding concepts in calculus and was regularly emphasized by the professor. Although the students frequently used words suggesting motion, such as *approaching* or *tends to,* the words were rarely accompanied by descriptions of something actually moving. In the few exceptions, students' reasoning relied on other features of their imagery (e.g., holes or cliffs in a surface to describe continuity). The addition of motion added visual effect or drama (e.g., a mouse falling to its death) but did not contribute observable conceptual structure or purpose. When asked directly about their use of words such as *approaches,* most students denied thinking of something moving and provided alternate dynamic interpretations, such as sequentially "selecting different points." Motion was classified as a weak metaphor in this study since nearly all students' use of related language was not intended to describe objects actually moving, and those who did imagine motion based their understanding and reasoning on other aspects of the situation.

Interpretations of "arbitrarily" and "sufficiently"

A third category of common language in the study involved the words *arbitrarily* and *sufficiently,* reflecting their frequent use by both the textbook and the professor to intuitively capture limit ideas. For example, the professor intended descriptions such as "You can make $f(x)$ arbitrarily close to L by choosing x sufficiently close to a" to make a rigorous understanding of a formal definition accessible. In this formulation, however, "arbitrarily close" and "sufficiently close" encode complex mathematical and logical ideas. Although students widely adopted such phrases, they interpreted them simply as modifiers of degree, such as "very close" and "very, very, very, very close." Use of the words *arbitrarily* and *sufficiently* did not meet the criteria for a strong metaphor since students were content to use equally intuitive, simple descriptions and these interpretations did not influence their reasoning and understanding.

Discussion and Implications

Determination of students' strong metaphors in this study did not include criteria for correctness, and in fact, much of what students said when applying these metaphors was mathematically incorrect. This does not mean, however, that all incorrect metaphors had negative implications for students' learning. The only metaphor cluster that demonstrated a consistent detriment to students' understanding was that of physical limitation metaphors. Although nearly every collapse metaphor is technically incorrect, students' extended exploration of their entailments often led to highly productive insights. Students' use of infinity as number metaphors and approximation metaphors were often productive despite numerous errors and inconsistencies. Although several metaphors were observed to be useful for students, most initial applications resulted in nonsensical ideas such as adding infinitely many heights to compute an area or applying the mean value theorem to an interval with one point at infinity. Breakthroughs were only observed after significant effort toward making sense of a problematic situation.

Despite focused effort by the professor of the course to develop students' reasoning about limits in terms of motion, zooming in on a graph, and informal language involving the words *arbitrarily* and *sufficiently,* students did not adopt these ways of reasoning beyond surface-level language inconsistent with the professor's intent. The weakness of these metaphors for students suggests that commonly assumed ways of making difficult calculus concepts accessible to students may only be leading students to parrot the language superficially. The one strong metaphor that appears to have been influenced by the professor was a physical limitation metaphor in which students focused on an extraneous aspect of his comments that led to a significant misconception.

Williams (1991) found that intuitive models for limits were extremely resilient despite direct efforts to influence them, citing factors of a lack of motivation to consider intricacies requiring more carefully formulated descriptions of limits and a belief that such descriptions are not practically valuable. Any attempts to influence students' metaphors (for example, by using approximation metaphors throughout a calculus course) must begin by directly engaging students in solving problems whose successful resolutions require the desired conceptual structure. Since students' intuitive reasoning is highly idiosyncratic and resistant to targeted influence, it is critical to begin with simple limit contexts with a focus on systematically structuring ideas about approximation and error analyses. Subsequently, engaging students in the same approximation questions about limits in multiple situations, such as difference quotients, Riemann sums, and Taylor polynomials, will eventually provide students with a general and powerful tool to develop many of these ideas largely on their own.

We must be careful to avoid treating students' incorrect metaphorical statements as mere misconceptions and instead look for the roots of growth toward a future, deeper understanding of the corresponding concepts. As the examples from this study illustrate, many of students' nonstandard interpretations are, at least, fertile sites for positive discussions. Recognizing this potential for development of mathematical reasoning requires an effort on the part of curriculum developers and instructors to see beyond students' errors.

References

Black, M. (1962a). Metaphor. In M. Black (Ed.), *Models and metaphors: Studies in language and philosophy* (pp. 25–47). Ithaca, NY: Cornell University Press.

Black, M. (1962b). Models and archetypes. In M. Black (Ed.), *Models and metaphors: Studies in language and philosophy* (pp. 219–243). Ithaca, NY: Cornell University Press.

Black, M. (1977). More about metaphor. *Dialectica, 31*, 433–457.

Oehrtman, M. (2008). Layers of abstraction: Theory and design for the instruction of limit concepts. In M. Carlson & C. Rasmussen (Eds.), *MAA notes: Vol. 73. Making the connection: Research and teaching in undergraduate mathematics* (pp. 65–80). Washington, DC: Mathematical Association of America.

Oehrtman, M., (2009). Collapsing dimensions, physical limitations, and other student metaphors for limit concepts. *Journal for Research in Mathematics Education, 40*, 396–426.

Williams, S. (1991). Models of limit held by college calculus students. *Journal for Research in Mathematics Education, 22*, 219–236.

Chapter 14

Infinitesimals in Student Reasoning

Robert Ely
University of Idaho

How would you answer this question: Is it possible to choose two different points on the real number line that are infinitely close to one another?

When I asked this question of 233 undergraduate students in first- and second-semester calculus, 83 percent said that this *is* possible to do. When I asked mathematicians this question, they invariably said it was *not* possible to do, and most were incredulous that so many undergraduates could have such a misconception. Any two points on the standard real number line are a specific finite distance apart; this distance cannot be infinitesimal.

I interviewed some of these students to try to figure out what they were thinking. Some of their responses accorded with prior research. For instance, one student gave an example that 1.001 and 1.002 are infinitely close to each other, but, say, 1.1 and 1.2 are not infinitely close. Apparently she understood "infinitely close" to mean something more like "very close," which is understandable given the way people often colloquially use the word *infinite*.

Another student provided the example 0.999... (repeating) and 1. As we talked it became clear that he was viewing the first number as actually moving closer to 1 as more 9s are added to its decimal expansion. He said it never reaches 1, but gets infinitely close to it. He was treating 0.999... as an unending process of adding 9s rather than as a finished stationary object; this is a way of thinking that students have displayed in other studies also (e.g. Dubinsky, Weller, McDonald, & Brown, 2005).

But another student, Sarah, responded very differently by saying some things that had not been discerned in my prior research. She appeared to be reasoning with numbers and distances that really were infinitesimal, in the sense of being larger than zero but smaller than any finite number or distance. As she pursued the implications of her reasoning further and further through the interview, I began to realize that her conceptions were not really "misconceptions" at all. I now call them *nonstandard* conceptions: pieces of knowledge that significantly differ from the standard way of thinking but that could potentially be built into a system (of the real number line, in this case) just as mathematically consistent and powerful as the standard system. Her thinking also showed considerable parallels with historical ideas about the mathematical continuum. My goal in this chapter is to show and explain these parallels and what they mean. I summarize these historical ideas about infinitesimal numbers and then describe Sarah's conceptions, finishing with a discussion of nonstandard conceptions and their significance.

A Brief History of the Infinitesimal in Mathematics

Many of the famous mathematicians of the 1600s and 1700s might have answered "yes" to the question at the beginning of the chapter. Descartes, Fermat, Wallis, Newton, Leibniz, the Bernoullis, Euler, Legendre, and others all used infinitesimal numbers in their analysis (Bair et al., 2013). What we now just call calculus was known in the 1700s as the *infinitesimal calculus*, developed independently by Isaac Newton and Gottfried Wilhelm Leibniz.

This chapter is adapted from R. Ely (2010), Nonstandard student conceptions about infinitesimals, *Journal for Research in Mathematics Education, 41*, 117–146.

It is Leibniz's notation for how to calculate with infinitesimal quantities that we still use today, including dx and $\int$. Leibniz's system of infinitesimal calculus is fundamentally based on treating any curve as an infinite number of straight "sides" that are infinitesimal in length, smaller than any finite number but not zero. A derivative is the slope of one of these straight sides, and an integral is an infinite sum of infinitesimal polygonal areas bounded by them.

To calculate these derivatives and integrals, Leibniz developed an algebra of operations on infinitesimal, finite, and infinite numbers (Struik, 1969, pp. 272–280). This system includes infinitely many infinitesimals and infinite numbers, and there is no smallest one. For example, one can get smaller and smaller infinitesimals by raising an infinitesimal to higher and higher powers or by dividing it into smaller and smaller numbers. Indeed, in every neighborhood of a given number, there is a microcosmic world of numbers that looks like the larger continuum. In Leibniz's words: "Since the continuum is divisible to infinity, any 'atom' will be of infinite kinds like a sort of world, and there will be worlds within worlds to infinity" (Leibniz, 1663–72/1966, p. 241 [my translation]). Leibniz's algebra of operations on infinitesimal numbers included dozens of rules such as this: If i is infinitesimal and f is finite, then $i \times f$ is infinitesimal, $f \div i$ is infinite, and $f + i \approx f$. More aspects of Leibniz's system are summarized in table 14.1.

Although there was plenty of argument about what infinitesimals actually *were* (see Grabiner, 1981), mathematicians for more than a century happily used them to do calculus because they were relatively intuitive and wildly useful. These "useful fictions," as Leibniz called them, produced an explosion of mathematical discoveries in mechanics, calculus of variations, probability theory, astronomy, and more.

By the early 1800s mathematicians were encountering some unintuitive and even contradictory results, such as with the convergence of trigonometric series, that an informal reliance on the intuition of infinitesimals could not help resolve (Bressoud, 1994). By trying to resolve these issues, Weierstrass and other nineteenth-century mathematicians ultimately advocated eliminating infinitesimals entirely and grounding calculus in a formal theory of limits (Grattan-Guinness, 1970). Nonetheless it was a bit strange that infinitesimals could be so useful for discovering calculus-based results but had to be jettisoned when it came time to check one's proofs. The reason became clear only a century later: It was not infinitesimals that were the problem, but rather that the field of mathematical logic had not yet been sufficiently developed for people to be able to make infinitesimals rigorous.

By 1961 mathematical logic had developed sufficiently for Abraham Robinson to make infinitesimals rigorous. Robinson (1996) created the nonstandard real numbers (or "hyperreal" numbers), which include the standard real numbers and infinitesimal and infinite numbers also, as in Leibniz's system. Unlike Leibniz, he was able to carefully define what infinitesimal and infinite numbers were, in terms of infinite sequences of finite numbers. He proved a set of formal rules for consistently working with nonstandard real numbers, which mirrored Leibniz's informal rules almost exactly. Most important, Robinson proved that this system was just as consistent and powerful as the standard real numbers for doing calculus and advanced calculus. Some features of Robinson's nonstandard analysis are summarized in table 14.1.

The nonstandard real numbers provide a fascinating vindication of Leibniz's infinitesimal calculus and have created a new field of research: nonstandard analysis. Although the nonstandard model of the real numbers marks a powerful, coherent, and mathematically correct mode of thought, it differs substantially from the standard real number system taught in today's classrooms. This made it all the more intriguing when the student I was interviewing, Sarah, began talking about infinitesimal quantities in ways that shared surprising parallels with Robinson's and Leibniz's systems.

Table 14.1

Summary of the similarities between Sarah's system, Leibniz's system, and the system of nonstandard analysis

Leibniz's foundational system (c. 1690)	Nonstandard analysis (c. 1961)	Sarah's conceptions
There exist *infinitesimal, finite,* and *infinite* numbers.	There exist *infinitesimal, finite,* and *infinite* numbers.	There exist *infinitesimal, finite,* and *infinite* numbers.
There are *infinitely many* infinitesimal and infinite numbers. Operating on them produces other numbers: for example, squaring an infinitesimal produces a smaller infinitesimal; squaring an infinite number produces a larger infinite number. This can be done infinitely. (Here Leibniz disagrees with Nieuwentijt [Mancosu, 1996].) Any number, even an infinitesimal, can be divided infinitely many times.	There are *infinitely many* infinitesimal and infinite numbers. The nonstandard real numbers are a field extension of the real numbers, and are thus closed under the arithmetical operations, which gives rise to infinitely many infinitesimal and infinite numbers. An infinitesimal is the multiplicative inverse of an infinite number.	Sarah's is quite similar to Leibniz's. One can keep dividing any length into infinitely smaller and smaller segments. Infinitesimal numbers can be operated upon: squaring an infinitesimal produces a smaller infinitesimal, and the reciprocal of an infinitesimal is an infinite number.
An infinitesimal is not strictly defined or represented (other than as, say, *dx*); an infinite series has an infinitesimal term at its end.	An infinitesimal can be precisely represented by an equivalence class of sequences of positive rational numbers $\{a_n\}$ that converges to 0.	An infinitesimal can be represented by a decimal expansion that has digits extending past the "infinityth" decimal place. For instance, 0.0000… (infinite 0s) with a 1 at the end.
One can find two different numbers infinitely close to one another. In particular, $f + i \approx f$ (f is a finite number, i is an infinitesimal).	One can find two different numbers infinitely close to one another. Any nonstandard number is infinitely close to but not necessarily equal to its standard part ($n \approx s(n)$).	One can find two different numbers infinitely close to one another. For instance, 0.000…1 and 0.000…01 are both "infinitely close" to 0.
If one zooms in enough on part of a thing, one will see that this part looks like a microcosm of the thing itself. There are worlds within worlds. This can only be done finitely in the real world, but in the mathematical one it can be done infinitely, or at least it is a "useful fiction" to pretend that it can. On the other hand, *monads* are indivisible and spiritual units, as different from infinitesimals (which are infinitely divisible) as the limitlessness of God is from mere infinite quantities.	Keisler's "infinitesimal microscope" allows one to zoom in infinitely. A *monad* is defined to be a set of all numbers infinitely close to one another (this is not definable in first-order logic). This is sort of a microcosm of the standard real numbers. (Note that this differs dramatically from Leibniz's use of the word *monad*.)	Sarah's conception is similar to Leibniz's, at least with respect to the number line: "even in that tiny infinitely small space [between 0.999… and 1] you can still cut that into infinity too and put numbers in there," and this "would look exactly like the big number line except they wouldn't be called the same numbers."
I found no reference to rationality of infinitesimal numbers by Leibniz.	Rational and irrational infinitesimals exist.	There exist rational numbers in every small space, including infinitesimally small spaces.
The purpose of infinitesimals is to provide a method for doing calculus.	The purpose is to produce a rigorous model for analysis that uses infinitesimals.	The purpose of infinitesimals is sense-making, not some additional functional purpose.
Gradually discarded in the nineteenth century in favor of the limit-based formulation of the standard real numbers.	Established in 1961 (Robinson) as logically equivalent to the standard real numbers.	She refers to her conceptions as her own strange way of thinking.

Sarah's Thinking about Infinitesimals

A summary of Sarah's thinking about infinitesimals and the number line also appears in table 14.1. This section details her ideas and how I ascertained them.

Infinitely close

As Sarah started talking about her idea of why it is possible to find numbers that are infinitely close to each other, I began to suspect that she thought in terms of infinitesimals. I actually interviewed Sarah again a week later to continue probing her thinking about infinitesimals, and she continued to elaborate her conceptions along the same lines. It all started with her example of two numbers that were infinitely close to each other: 3.999... and 4. I asked if she thought there were other numbers infinitely close to both of these, and she said, "three point nine nine nine to infinity plus *[pause]* an infinitely small number." This suggested that she was not really conceptualizing "infinitely close" as meaning just "very close" or "getting arbitrarily close to," like the other students had been. It seemed that she was thinking of fixed numbers on the number line that were actually infinitely close together. Shortly after this I asked her about what was between 0.999... and 1, and she replied a similar way. Thinking I was talking about the distance between them, she said:

S: Um . . . I mean, this isn't really a number. *[She writes 0.000...1]* This isn't real, so . . .

I: Point 0 repeating with a 1?

S: Yeah. But um . . . I don't know. I mean . . . There *is* numbers, because no matter how small you get it there's still going to be some kind of space, and even in that tiny infinitely small space you can still cut that into infinity too and put numbers in there. So I would say yes but I don't know how to express that.

The fact that she acknowledges that this new number 0.000...1 "isn't real" suggests that she is not thinking of a very small number with a large, but finite, number of zeroes, but that she is envisioning an infinite number of zeroes. In the second interview she explicitly says that there is "an infinite number of zeroes" in this number.

Infinitely divisible number line with infinitely many infinitesimals

In the excerpt above, Sarah not only is describing that her number line has infinitesimal numbers and distances (although she never uses the word *infinitesimal*), she is also indicating that it is infinitely divisible even at the infinitesimal level. She confirms this in the second interview, and even begins to develop notation for these infinitesimals:

I: Okay. Um, so now I ask you, are there, are there other numbers between here and here [between 1 and .9 repeating]? Like . . .

S: Um, maybe an even smaller one, like . . . hahaha *[writes 0.000...1]*

I: Okay . . .

S: If you can do that.

I: No, that, how many numbers are between here and here? *[points to 0.999... and 1]*

S: An infinitely small, an infinite amount of infinitely small numbers.

Sarah is developing an emerging notation for infinitesimal numbers: 0.000...1 and 0.000...01 are different numbers because the latter has an extra 0 after the infinite string of zeroes. Sarah is generating this notation, and the conclusions that it affords, through the course of the interviews. At the beginning she clearly thought that there are infinitely small distances and numbers; by the end of the second interview, she has developed a notation for such numbers and has explored their properties extensively. This notation is suggestive to her. By simply adding a different digit at the end of the infinite decimal expansion, she is able to describe another infinitely small number. This enables her to generalize that there must be *infinitely* many such numbers, presumably because she could notate arbitrarily many different expansions like this.

When I ask her what the numbers in this spot would look like, she says, "Well, they would look like, they would look like the big number line except, they would look exactly like the big number line except they wouldn't be called the same numbers." This conception is not true for the standard real numbers, but it is true for Leibniz's system of infinitesimals and of Robinson's nonstandard real numbers. In particular, in both Leibniz's and Robinson's systems, infinitesimals are generative—once we posit that an infinitesimal number exists, we can generate, and order, infinitely many infinitesimal numbers from this one. In his correspondence with the early eighteenth-century mathematician Nieuwentijt, who believed that there existed only one infinitesimal number, Leibniz asserted that any consistent system of infinitesimals must contain infinitely many infinitesimals, and that these must be orderable and manipulable (Mancosu, 1996). The fact that Sarah's conceptions accord with Leibniz rather than Nieuwentijt on this point suggests that she may be motivated by a desire to be *consistent* with the affordances of her notation. The notational process by which she came to posit the existence of one infinitesimal number allows also for the existence of many other infinitesimals, so consistency forces her to acknowledge these infinitesimals as well.

Properties of and operations with infinitesimals

Sarah's emergent notation for infinitesimal numbers allows her to infer other properties of infinitesimal numbers:

1. Infinitesimal numbers can be *ordered*. She says that both 0.000...01 and 0.000...1 are "infinitely close to zero," but the latter is still "a little bigger than" the former.

2. Sarah says there must be *rational* numbers between 0.999... and 1, just as there are between any other two numbers on the number line.

3. Sarah is willing to *square* an infinitesimal number like 0.000...1. She says that doing this produces an even smaller infinitesimal, "infinitely more closer" to 0 than 0.000...1 is.

4. Sarah is willing to take the reciprocal of 0.000...1; she says you get something infinite. And when you take the reciprocal of 0.000...01, you get a "bigger infinity" yet. Even though she says, "yeah, I know that's really bad, heh," she explains, "I feel like if there's an infinity, then there has to be something *bigger* than the infinity, which is still infinity or, you know, small infinity."

In all of these cases, Sarah's views seem to be extensions of properties of the finite real numbers. This accords with Leibniz's stance that you can perform the same operations on infinitesimal numbers as you can on finite numbers, even though Sarah's notation allows her to describe only a few of the features of such operations, without the detail that one might use to describe such operations on regular finite numbers.

A resilient system

Sarah admits throughout the interviews that her ideas are wrong and off-the-wall, and clearly they are inconsistent with standard notions of the real number line. Yet she maintains and pursues her ideas in spite of being shown the "correct" conceptions. For instance, at the end of the first interview she says that she has seen a proof in high school that 0.999... equals 1, but she never believed the proof. She asks me why she was supposed to believe that 0.999... = 1. I oblige with a standard explanation that if $N = 0.999...$ then $10N = 9.999...$, so by subtracting the first equation from the second, you get $9N = 9$, so $N = 1$.[1]

Sarah is not persuaded by this argument at all, saying that if 0.999... really equals 1, then why do you even have 0.999...? When I ask her if she believes that 0.333... equals $1/3$, she says she does not believe this either, even though "it's what they make you memorize." She continues in the second interview asserting that 0.999... does not equal 1, so apparently my explanation does nothing to perturb her thinking.

Throughout the interviews, Sarah pursues the implications of her ideas about infinitesimals and never seems to experience contradictions within her system. Her system is at odds with, yet entirely resilient to, the standard system of real numbers taught in school. Indeed, she does not find it strange at all that there would be some inconsistency between what she actually thinks and what she knows she "should" think in math class. She says what she usually does is just to "memorize rules that someone made up," even though at heart she does not believe or understand them. It seems that no matter what anyone tries to tell her about the real numbers, her resilient system will remain intact and unperturbed. The reason for this fact is precisely that her conceptions are nonstandard conceptions, not misconceptions.

Nonstandard Conceptions	Sarah's conceptions about the real number line are nonstandard conceptions, not misconceptions. What is the distinction? When a teacher encounters a student with ideas that differ from hers, it is natural for the teacher to view these ideas as misconceptions and try to help the student correct them. One way to do this is to put the student in a situation where he experiences a perturbation. This could be (a) a situation where the student's conceptions lead him to anticipate a result that differs from the one he ends up actually experiencing; or it could be (b) a situation that forces the student to think through the implications of his conceptions and ultimately arrive at a contradiction. Either way, the perturbation requires the student to revise his conceptions to make them internally consistent or consistent with his experience. For example, suppose a student thinks additively about how to make a scaled-up replica (dilation) of a geometric figure in an additive way; she adds the same amount to each line segment rather than multiplying each segment by a common scale factor. The teacher may be able to devise a particular geometric figure that will look dramatically wrong when this additive strategy is applied, forcing the student to experience a perturbation between her anticipated outcome and her actual experience. Another kind of situation might be when the teacher does not suspect that the student's conceptions will produce a perturbation, but that the student is instead just limited in flexibility or generalizability. The teacher may try to show students (c) a situation that their conceptions cannot be generalized to but the standard ones can. For instance, if a student conceptualizes multiplication just as repeated addition, he may be limited in his ability to make sense of $\sqrt{3} \times \sqrt{5}$ until he is provided with a new meaning for multiplication as scaling.

When conceptions are nonstandard, they substantively differ from standard conceptions, just as misconceptions do. However, the interventions mentioned in the previous paragraph may not be possible for teachers to do, no matter how hard they try. The conceptions do not lead to (a) contradictions with the student's experiences, nor are they (b) internally inconsistent with one another, no matter how deeply their implications are explored and pursued. In addition, they are (c) no less powerful or applicable than the standard conceptions.

---

[1] Rudy Rucker (1983) points out that this proof is actually circular; the subtraction step actually presupposes that 0.999... = 1 rather than showing it. After all, if 0.999... is in fact some infinitesimal amount i smaller than 1, then $10N$ should be $10i$ smaller than 10, and the infinitesimal amount does not go away when the numbers are subtracted.

These are not just stubborn misconceptions. These conceptions are just as stable, viable, powerful, flexible, and consistent as the standard conceptions. Is it possible to get the student to abandon them? Should we try to get her to?

Sarah's conceptions are nonstandard, and it is precisely their parallels with historical conceptions about infinitesimals that suggests this is the case. The parallels between her system and those of Leibniz and Robinson can be seen in table 14.1. The similarities are extensive, but there is one enormous difference: Sarah has no "system"—she is exploring the entailments of her conceptions as the interview proceeds. Her ideas are emergent; yet because they are motivated by a need to maintain self-consistency, they develop along the lines of Leibniz's and Robinson's systems. This does not mean that she would readily be able to use these emergent ideas for a particular purpose, such as for doing calculus or analysis, as Leibniz and Robinson did. But because Robinson *proved* his nonstandard system to be as logically coherent, consistent, and powerful as the standard system, there is no reason to suspect that Sarah would ever run into a perturbation if she continued to reason courageously along these same lines.

This also explains why Sarah's underlying intuitions have remained unaltered in spite of her classroom mathematical experiences and even in the face of proofs she has been shown in order to dislodge them. Without an understanding of Leibniz's system of infinitesimals, or of Robinson's formalization of this system, it would be tempting to dismiss Sarah's ideas as being quirky but stubbornly wrong. With this understanding, it becomes evident that Sarah is developing a coherent framework and an emergent notation for thinking about infinitesimal quantities, one that she could continue to pursue without perturbation.

The idea of nonstandard conceptions gives us a reason to make sense of why there should be such significant parallels between a student's reasoning and the historical thinking of people she has never heard of. The parallel does not exist because she learned about infinitesimals in her classrooms—she said that she had not, and indeed her classroom experiences seem to have opposed rather than inspired her ideas. Instead, it seems that the reason for these parallels is the relationship between (a) the internal mechanisms for maintaining cognitive consistency that guide how Sarah's knowledge is constructed and (b) the external mechanisms for maintaining mathematical consistency that guide how public mathematical knowledge is constructed. In other words, cognition is a locally self-regulating system that resolves perturbations, a "rational game" that maintains semantic links and avoids contradictions (von Glasersfeld, 1990), and mathematical systematization is a sociocultural game with rules for establishing and maintaining deductive consistency. What is the relationship between the internal rules and the external rules? Is it that our culture's criteria for mathematical consistency and coherence are an externalization of our innate internal cognitive criteria? Or perhaps our personal cognitive mechanisms for constructing consistent knowledge are an internalization of the criteria for mathematical consistency (or more generally, of the norms of consistency in our discursive social practices). Or perhaps our internal cognitive mechanisms and our public mathematical rules influence one another in complex and iterative ways.

In any case, nonstandard conceptions like Sarah's confirm how students can have access to deep mathematical ideas that they are not explicitly taught, even ideas that are incommensurable with the ones they are taught. This case also provides an extreme example of how examining the functionality and structure of a student's conceptions, rather than dismissing these conceptions as misconceptions, reveals a meaningful structure to the student's conceptions that otherwise might have been overlooked.

References

Bair, J., Błaszczyk, P., Ely, R., Henry, V., Kanovei, V., Katz, . . . & Shnider, S. (2013). Is mathematical history written by the victors? *Notices of the AMS, 60*(7), 886–904.

Bressoud, D. (1994). *A radical approach to real analysis.* Washington, DC: Mathematical Association of America.

Dubinsky, E., Weller, K., McDonald, M. A., & Brown, A. (2005). Some historical issues and paradoxes regarding the concept of infinity: An APOS-based analysis, Part 1. *Educational Studies in Mathematics, 58*, 335–359.

Ely, R. (2010). Nonstandard student conceptions about infinitesimals. *Journal for Research in Mathematics Education, 41,* 117–146.

Grabiner, J. V. (1981). *The origins of Cauchy's rigorous calculus.* Cambridge, MA: MIT Press.

Grattan-Guinness, I. (1970). *The development of the foundations of mathematical analysis from Euler to Riemann.* Cambridge, MA: MIT Press.

Leibniz, G. W. (1966). *Philosophische Schriften: Sechste Reihe, zweiter Band. Sämtliche Schriften und Briefe* [Philosophical writings: Ser. 6, Vol. 2: All essays and letters] (W. Kabitz & H. Schepers, Eds.). Berlin: Akademie-Verlag. (Original works published 1663–1672)

Mancosu, P. (1996). *Philosophy of mathematics and mathematical practice in the seventeenth century.* New York, NY: Oxford University Press.

Robinson, A. (1996). *Princeton landmarks in mathematics and physics: Non-standard analysis* (Rev. ed.). Princeton, NJ: Princeton University Press.

Rucker, R. (1983). *Infinity and the mind.* New York, NY: Bantam.

Struik, D. J. (1969). *A source book in mathematics, 1200–1800.* Cambridge, MA: Harvard University Press.

von Glasersfeld, E. (1990). An exposition of constructivism: Why some like it radical. In R. B. Davis, C. A. Maher, & N. Noddings (Eds.), *Constructivist views on the teaching and learning of mathematics* (pp. 19–29). Reston, VA: NCTM.

Chapter 15

Modifying and Constructing Diagrams in Calculus

Elizabeth George Bremigan
Ball State University

Students encounter visual representations throughout their study of mathematics. Teachers often include diagrams, figures, or graphs as components of instructional explanations. Many mathematical problems presented in curriculum materials or encountered on tests include an accompanying visual representation that students must interpret and may modify as they seek to understand a problem situation or determine a solution. For some problems where no accompanying visual representation is given, constructing a visual representation is often a helpful and at times even an essential component of the solution process. Diverse types of mathematical problems have solution processes that require or are aided by reasoning with visual representations, and there are also many ways that students can use visual representations when solving problems. While teachers may take great care in choosing the symbols and words they use in teaching and while they may look carefully at the symbols and words that students write when solving mathematical problems, less attention may be given to the ways that students modify or construct visual representations. Yet teachers can gain important insights into students' mathematical understanding and thinking by observing and analyzing the visual representations students create when solving problems.

Reasoning with visual representations is particularly important in learning the theories and applications of calculus. Many of the problems traditionally included in the study of single-variable differentiable calculus initially draw upon students' abilities to interpret, modify, or construct a diagram representing some situation geometrically, with the application of calculus concepts and procedures occurring later in the problem-solving process. In particular, both optimization and related rates problems rely heavily on the interpretation and/or construction of diagrams. Optimization problems often involve reasoning from a specific diagram that represents a general case. In solving related rates problems, students must impose motion on static diagrams, recognizing which quantities are constant and which vary with respect to time.

Applications of integral calculus also heavily depend on the use of visual representations. Typically regions located on the coordinate plane are described, and the areas of regions and volumes of solids created by rotating such regions about an identified axis are computed by setting up and solving definite integrals. When solving these types of problems, students must carefully interpret the coordinate plane and be able to visualize three-dimensional solids based on two-dimensional representations. Perhaps it is in these aspects of solving applied calculus problems—the interpretation and construction of visual representations that represent the problem situation—that students face unrecognized challenges.

As a high school calculus teacher, I had noted over time the extent to which problems on the free-response section of the Advanced Placement (AP) Calculus examinations involved reasoning with visual representations. I was also aware of the gender gap in overall AP Calculus test scores, with males typically outperforming females (Morgan, 1996). Some researchers had proposed that differences in spatial visualization skills accounted for some

---

This chapter is adapted from E. G. Bremigan (2005), An analysis of diagram modification and construction in students' solutions to applied calculus problems, *Journal for Research in Mathematics Education, 36*, 248–277.

of the gender differences in mathematics achievement, particularly in high school geometry (Battista, 1990). This led me to wonder if there were interesting relationships between the ways students used visual representations in calculus and their achievement. Thus, I designed and conducted the following research study described next.

My Research Questions

The purpose of my study was to determine the extent to which high school calculus students modified or constructed diagrams in their written solutions to applied calculus problems and to describe the relationships between the nature of the diagrams they created and their problem-solving success. Specifically, I posed two research questions:

1. What was the **frequency** of diagram modification and construction evident in students' written solutions to selected free-response problems on the 1996 BC-level Advanced Placement Calculus Examination, and to what extent was gender or performance level on the examination associated with a difference in the frequency of diagram modification or construction?

2. What was the **nature** of the diagrams that students modified or constructed in their written solutions to selected free-response problems on the 1996 BC-level Advanced Placement Calculus Examination, and to what extent was gender, performance level, or problem-solving success associated with differences in the nature of the modified or constructed diagrams?

Research Question 1: Frequency of Diagram Use

What I did

In order to determine how frequently students modified or constructed diagrams, I analyzed students' written solutions to three of the six free-response problems. (See the appendix for the problems.) These problems were chosen because the presentation of the problem situation included a diagram, thus giving students the opportunity to modify the given diagram or to construct a new diagram. Students' understanding of the problem situation presented and the subproblems posed may either have been aided by or actually required reasoning directly from the given visual representation. In BC-1 and BC-6, the given diagram provided relational information that was not stated in the verbal presentation of the problem. No additional information other than that stated in the verbal presentation of the problem was shown in the diagram given in BC-5.

From the approximately 21,000 high school students who took this examination, I chose a random sample of 600 students. Table 15.1 shows the distribution of these 600 students according to gender and performance level, with students' overall AP scores determining performance level. High scorers achieved scores of 4 or 5, moderate scorers achieved a score of 3, and low scorers achieved a score of 1 or 2.

First, I coded each solution based on whether there was evidence of diagram use in the written solution. Any markings made on the given diagram (e.g., lines, arrows, numbers, variables, words, or highlighting of a cross section or of an embedded geometric figure) were considered a modification of the given diagram, and any new diagram drawn was considered to be a construction. For each free-response problem I determined the percentage of the 600 students whose written solutions contained modified or constructed diagrams and examined differences associated with gender and performance level. Statistically significant differences were identified using a two-way ANOVA test with an α-level of .01.

Table 15.1
Distribution of students in the sample by gender and performance level

	Males	**Females**	**Total**
High scorers	251	97	348
Moderate scorers	82	65	147
Low scorers	56	49	105
Total	389	211	600

What I learned

Overall, 70 percent of the 1,800 written solutions that the students produced showed evidence of visual representation use. Modified or constructed diagrams were found in 78 percent of students' written solutions to BC-1, 58 percent of written solutions to BC-5, and 73 percent of written solutions to BC-6. For all three problems, the written solutions of high scorers contained more evidence of diagram use than did solutions of low scorers, and this performance level difference was significant for BC-6. Females' solutions showed evidence of diagram use more frequently than did the solutions of males in all three problems; this gender difference was significant for BC-1 and BC-5.

While these results gave initial insight about the extent to which diagrams were used by various groups of students, there was much more to be learned from more closely analyzing the nature of the diagrams created and the relationship of particular types or features of the diagrams and students' problem-solving success. While knowing *whether* students used diagrams is interesting, learning *how* they used them is what provides important insights for teaching.

Research Question 2: Nature of Diagram Use

What I did

I randomly selected 45 students from each of the following four subgroups: high-scoring females, high-scoring males, low-scoring females, and low-scoring males, for a total of 180 students. I examined their written solutions to each free-response problem more closely with respect to both the types and the features of the diagrams created and their problem-solving success.

Students' written solutions were analyzed using several coding schemes I created. First, four categories were used to describe patterns of diagram modification and construction: (1) only modify the given diagram, (2) only construct a new diagram, (3) both modify the given diagram and construct a new diagram, and (4) no modifying or constructing. Second, I coded diagrams by identifying descriptors of the diagrams themselves; these descriptors were specific to each of the free-response problems. Examples of these descriptors are arrows showing direction of motion, adding pictorial elements, drawing a cross-sectional slice, highlighting an embedded geometric figure, and adding three-dimensional perspective on a two-dimensional representation. The third way written solutions were coded measured the degree of success students achieved in setting up an equation or integral that, when solved or evaluated, would lead to the problem's solution. These categories were (1) no success in set-up, (2) partial success in set-up, or (3) success in set-up. A solution was considered successful in set-up regardless of whether the student subsequently produced a correct solution to the problem.

What I learned

Examining patterns in the ways that students modified and/or constructed diagrams revealed some gender and performance level differences. More females than males tended to only modify the given diagrams. No diagram modification and construction was evident in more solutions of males than of females. For BC-5 and BC-6, high scorers both modified the given diagram and constructed a new diagram in their solution more often than did low scorers. For BC-1, it was the action of only constructing a new diagram that characterized high scorers from low scorers.

The most striking observation that emerged from analyzing the diagram descriptors was the diversity of the diagrams that students produced within each free-response problem, particularly since all students were presented with the same given diagram. Seldom were two identical diagrams found in students' solutions. For instance, of the 88 modified diagrams produced by students for BC-5, no two diagrams were coded alike. (See figures 15.1 and 15.2 for examples of diagrams created in solving BC-5.) In general, the diagrams that students modified were more diverse than were the new diagrams students constructed, and the modified diagrams tended to contain more detailed markings than did the constructed diagrams.

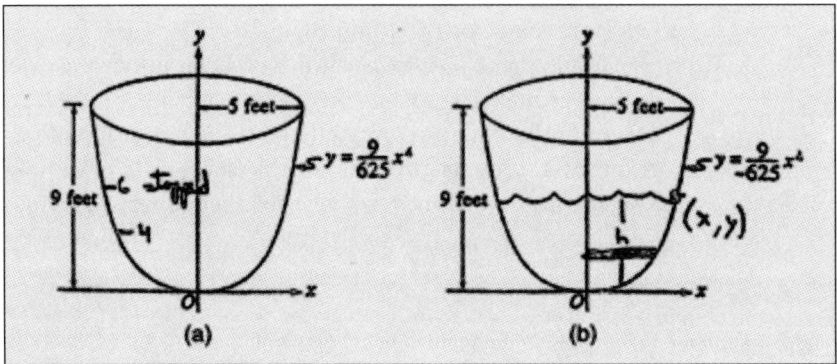

Fig. 15.1. Examples of modified diagrams contained in students' solutions to BC-5

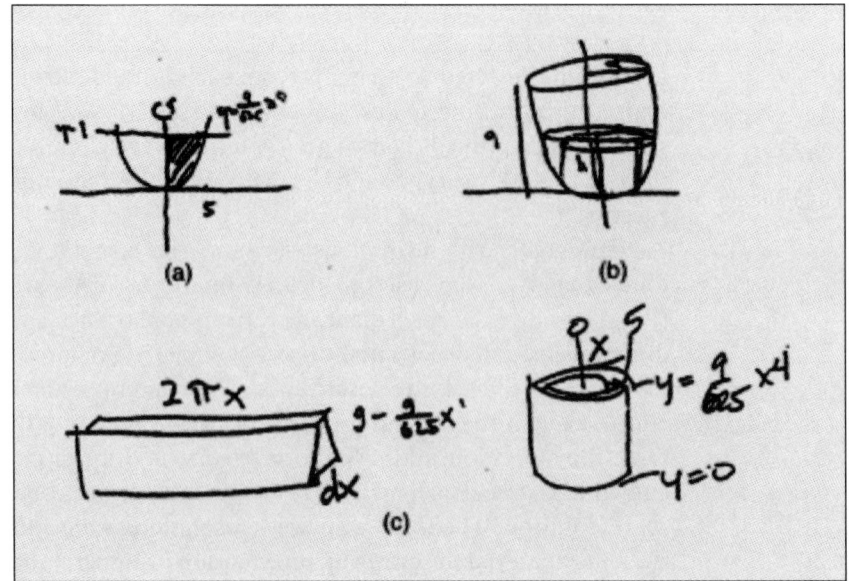

Fig. 15.2. Examples of constructed diagrams contained in students' solutions to BC-5

Some gender differences emerged when analyzing the diagram descriptors. Females' diagrams contained more markings than did males' diagrams, most notably the diagrams drawn by high-scoring females. Actions frequently taken by females included the highlighting of cross sections, particularly adding three-dimensional perspective along with pictorial elaboration and algebraic labeling. The diagrams produced by males generally focused on the graph of a function or simply highlighted an embedded two-dimensional figure. At times these geometric figures were drawn on a coordinate plane, but often males redrew the embedded figures as isolated figures. The diagrams produced by males were less frequently labeled with constant or variable quantities than the diagrams produced by females.

No consistent pattern in the nature of the diagrams produced by high and low scorers emerged across the free-response problems. This result may be explained, in part, by recognizing the varying visual demands that each problem situation presents. Rather than consistently modifying or constructing diagrams in some prescribed way or producing particular types of diagrams, high-scoring students appeared to be flexible in creating different types of diagrams, recognizing and using several different solution paths for a particular problem. This flexibility of thinking may explain one interesting result of this study, namely, that more high scorers than low scorers both modified the given diagram and constructed a new diagram. High scorers may have recognized the limitations of modifying a given diagram and chosen to restructure their thinking by constructing a new diagram, perhaps a simpler diagram than the given diagram they had begun to modify.

Note that the previous analyses were conducted using students' overall AP scores to categorize them as high scorers or low scorers. The following results are based on levels of problem-solving success within each free-response problem, thus allowing a closer look at the connection between the diagram created and students' success in setting up an equation or integral to solve the related problem.

The actions of modifying the given diagram and/or constructing a new diagram appear to be related to problem-solving success. Table 15.2 shows the percentage of students achieving each of the three degrees of success whose solutions to selected subproblems contained a modified or constructed diagram. For each of the five subproblems examined, more than 60 percent of students who achieved success in set-up and over 50 percent of students who achieved partial success in set-up either modified or constructed diagrams, whereas fewer diagrams were found in the written solutions of students who achieved no success in set-up. In addition, a greater percentage of students whose solutions included a diagram achieved success in set-up on most subproblems than students whose solutions did not include a diagram.

Table 15.2
Percentage of students achieving various degrees of success based on modification or construction of diagrams in selected subproblems

	BC-1a	BC-1b	BC-5a	BC-6a	BC-6b
Success in set-up	70	63	79	87	89
Partial success in set-up	65	57	54	95	91
No success in set-up	20	25	44	43	49

Although the production of a diagram appears to be related to success in problem solving, results of my study also show that the action of modifying or constructing a diagram is not a sufficient condition for problem-solving success. For instance, although nearly 60 percent of students modified or constructed a diagram in solving BC-5 and over 70 percent did so in solving BC-1 and BC-6, fewer than 20 percent of students achieved the highest degree of success—success in set-up—for BC-5a and BC-6b. Approximately 50 percent of students reached success in set-up for BC-1a, BC-1b, and BC-6a. Thus, many students only achieved no success or partial success in set-up, despite the fact that they had modified the given diagrams or constructed new diagrams.

Errors identified in the solutions of students who achieved partial success in set-up appeared to be, at times, related to their diagrams. For instance, students often confused the direction in which a cross section was drawn and the related choice of the appropriate method of cylindrical shells or disks when computing the volume of a solid. Reversing the conventional use of the variables x and y on the coordinate plane and assigning a constant to represent a variable quantity were other errors noted in students' diagrams that negatively influenced the set-up of an equation.

Recognizing and highlighting particular features in diagrams appeared to be a key to successful problem solving more often than did the production of complete diagrams. In solving BC-5a, identification of a cross section in either a modified or a constructed diagram characterized the diagrams created by students who achieved success in set-up. Drawing right triangles, when either modifying the given figure or constructing isolated figures, was the action most closely associated with success in set-up of the trigonometric ratios used in solving BC-6a and BC-6b. Students who achieved success in set-up when solving these problems typically focused on the embedded geometric figures located within the given diagrams.

In solving BC-1a, many students produced a graph of the function and annotated their graphs with indication of motion and/or drawing of a cross section, but descriptors of the diagrams were not associated with success in set-up. Approximately half of the students who produced simple graphs of the function in BC-1 achieved success in set-up, as did nearly the same proportion of students who drew diagrams containing a cross-sectional slice. Labeling the diagrams produced in BC-1a and BC-1b with variables was not a sufficient condition for success; students whose diagrams were characterized by algebraic labeling were as likely to only achieve partial success as full success in set-up. It is interesting to note that BC-5 and BC-6 were presented in the context of a real-world, dynamic setting and required the problem solver to impose motion on a given diagram. Problem BC-1 was not presented in a real-world, application context.

A closer look at male and female students who reached success in set-up provides additional insight into gender differences. More females than males achieving success in set-up produced diagrams; yet for most problems, particularly those involving motion over time, fewer females than males whose written solutions included diagrams were successful in set-up. So whereas more females successful in set-up had modified or constructed diagrams, fewer female students who modified or constructed diagrams were successful in reaching the highest degree of success.

One major conclusion of my study is that the relationship between diagram use and problem-solving success is indeed complex. Females, who achieved significantly lower overall AP scores and mean scores on each free-response problem, consistently produced more diagrams than did males, who tended to be higher achievers on the examination. High-scoring males recorded evidence of diagram use less frequently than high-scoring females, but may have used nondrawing visualization strategies without physically modifying the given diagram or constructing a new diagram. Their ability to use nondrawing visualization strategies may also explain why males produced simpler diagrams than did females, focusing narrowly on the graph of a function or the geometric figure on which the problem situation was based. In contrast, females may be more dependent on the use of drawing strategies and thus produce more, and more elaborate, diagrams in their written solutions. Overall, these calculus students used given and created diagrams in ways that were very problem-specific and not easily generalizable.

Suggestions for Teachers

So what are we as teachers to do with the results of this study? Based on what I have learned, I am more conscious of my own use of visual representations in planning, teaching, and assessment, and I incorporate more focused discussion about visual representation use in my classroom. I also look more closely at what students draw and listen to what they say when using visual representations, and I provide more verbal and written feedback on the drawings they produce. The following suggestions are offered so you can reflect on your own use of visual representations as you teach and so you can lead your students to be more successful mathematical problem solvers by acknowledging and encouraging their use of visual representations.

- Look carefully at the diagrams that accompany the verbal statement of problems you and your students encounter in your curriculum materials or assessments. Draw students' attention to the relationship between the information contained in the verbal statement of the problem and the given visual representation. Orchestrate discussion about the features of the given diagram and model alternative ways in which the given diagram could be drawn.

- When you are creating or choosing problems to use in class discussion, on assignments, or on tests, consider when to provide a visual representation as a component of the problem statement and when not to. Recognize situations in which a visual representation is necessary, when it is helpful, and when it might be distracting or unnecessary. Consider the advantages and disadvantages of leaving the construction of a diagram as part of the solution process, as opposed to providing a diagram that students can then modify.

- Consider whether a rough sketch will suffice or whether a more exact, detailed visual representation is necessary. Is it important, or even possible, that a diagram be drawn to scale? Are there advantages to providing students with a completed diagram they must interpret versus letting them see and participate in the various decisions one makes in the process of constructing a diagram?

- Encourage students to abandon working with an initial diagram and construct a new diagram, particularly if a diagram they are modifying becomes too complicated or if embedded figures can be "removed" and redrawn easily. If a diagram becomes too muddled, it may be less useful. Perhaps one first uses a diagram in order to understand the given problem situation, but then needs a different diagram to label and use in the process of determining a solution.

- Draw a series of diagrams illustrating different moments in time for problem situations involving motion, in particular for related rates problems. Help students discern which quantities remain constant and which quantities vary with respect to time.

- Recognize the advantages and limitations of using a single diagram to represent a general case, as is often done in solving optimization problems. Encourage students to draw several possible cases, in order to avoid overgeneralization from a single diagram, or to draw to scale, as much as possible, a diagram that represents the optimum solution once it is found.

- Broaden the meaning of "show your work" to include work with visual representations. Students should not be hesitant or afraid to leave a written record of their work with visual representations. We can gain insight into students' thinking and understanding by examining their work with mathematical symbols, words, *and* visual representations.

- Provide constructive feedback to students on their use of visual representations, in addition to feedback on their use of mathematical symbols and words. Scaffold the tasks you ask students to complete by including aspects of interpreting and/or drawing visual representations. Give written feedback on their use of diagrams. Look for errors that are related to visual representation use. Recognize that diagrams don't have to be "complete" to be useful if their purpose is to help one achieve success in set-up of an equation or inequality.

- Appreciate and use the diversity of ways that students can create and use visual representations when representing problem situations and solving mathematical problems. Allow students to share with each other the diagrams they created. Perhaps by seeing the various ways that diagrams can be modified and constructed, more students will use these drawing strategies and achieve success in problem solving. Or perhaps students will learn to visualize without drawing. Flexibility in using visual representations might be the result of more exposure to the various ways that visual representations can be used.

- Continue to ask and attempt to answer interesting questions that arise about visual representation use in mathematics. There is much more research that can be done in this area, and much of it can be accomplished at the classroom level. Then look for ways to share what you have learned with others in the mathematical education community.

Reference

Battista, M. (1990). Spatial visualization and gender differences in high school geometry. *Journal for Research in Mathematics Education, 21,* 47–60.

Bremigan, E. G. (2005). An analysis of diagram modification and construction in students' solutions to applied calculus problems. *Journal for Research in Mathematics Education, 36,* 248–277.

Morgan, R. (1996). *1996 Advanced Placement test analysis* [College Board Statistics Report No. 97–23]. Princeton, NJ: Educational Testing Service.

Appendix

Three Problems from the 1996 BC Level Advanced Placement Calculus Examination

BC-1

1. Consider the graph of the function h given by $h(x) = e^{-x^2}$ for $0 \le x < \infty$.

    (a) Let R be the unbounded region in the first quadrant below the graph of h. Find the volume of the solid generated when R is revolved around the y-axis.

    (b) Let $A(w)$ be the area of the shaded rectangle shown in the figure to the right. Show that $A(w)$ has its maximum value when w is the x-coordinate of the point of inflection of the graph of h.

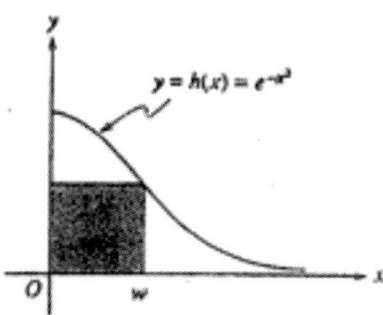

BC-5

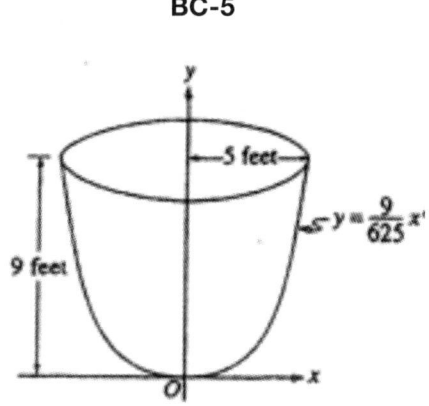

5. An oil storage tank has the shape shown above, obtained by revolving the curve $y = \frac{9}{625}x^4$ from $x = 0$ to $x = 5$ about the y-axis, where x and y are measured in feet. Oil weighing 50 pounds per cubic foot flowed into an initially empty tank at a constant rate of 8 cubic feet per minute. When the depth of the oil reached 6 feet, the flow stopped.

    (a) Let h be the depth, in feet, of oil in the tank. How fast was the depth of the oil in the tank increasing when $h = 4$? Indicate units of measure.

    (b) Find, to the nearest foot-pound, the amount of work required to empty the tank by pumping all of the oil back to the top of the tank.

BC-6

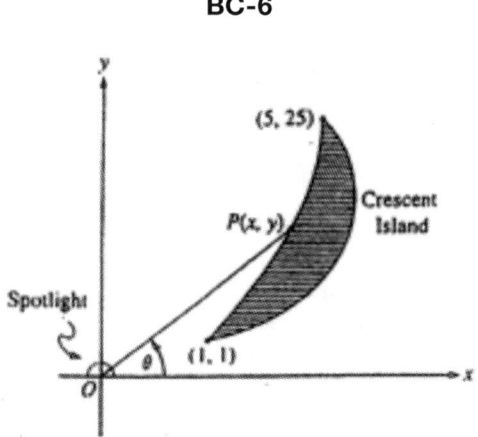

6. The figure above shows a spotlight shining on point $P(x,y)$ on the shoreline of Crescent Island. The spotlight is located at the origin and is rotating. The portion of the shoreline on which the spotlight shines is in the shape of the parabola $y = x^2$ from the point (1,1) to the point (5,25). Let θ be the angle between the beam of light and the positive x-axis.

    (a) For what values of θ between 0 and 2π does the spotlight shine on the shoreline?

    (b) Find the x- and y-coordinates of point P in terms of $\tan \theta$.

    (c) If the spotlight is rotating at the rate of one revolution per minute, how fast is the point P traveling along the shoreline at the instance it is at the point (3,9)?

Teaching and Learning Rational Numbers and Probabilistic Reasoning

Introduction

The final section of this volume includes six chapters that focus on research involving teaching rational numbers and three chapters that deal with teaching probabilistic reasoning. With respect to rational numbers, the NCTM *Principles and Standards* (2000) first mentions rational numbers—specifically, fractions—in the number and operations standards for grades 3–5, and the focus continues at the upper grades; rational numbers are subsumed in the real numbers at the higher grade levels. The Common Core State Standards for Mathematics (National Governors Association Center for Best Practices & Council of Chief State School Officers [NGA Center & CCSSO], 2010) mentions fractions first in the grade 3 overview category "number and operations—fractions"; by grade 6 fractions are subsumed under the more general category "number system."

In the case of probability, the NCTM Standards for pre-K–grade 2 and grades 3–5 include an introduction to data analysis and statistics. Formal mention of probability concepts begins with the grades 6–8 Standards, linking rational number concepts of proportionality to understanding of probability concepts through representations such as tree diagrams. By grades 9–12, the focus on probability expands to include sample spaces, random variables, and conditional/independent events, to name a few. The Common Core standards (NGA Center & CCSSO, 2010) begin by combining data with measurement for K–grade 5, with statistics and probability having its own category beginning with grade 6. By the high school grades, the Common Core standards focus directly on probability topics (e.g., conditional probability and rules of probability; using probability to make decisions).

As with the previous two sections in this volume, the placement and ordering of chapters within this final section was our decision as editors. We also chose what we believed to be an important "lesson learned" from each chapter. We encourage readers to think not only about the content of each chapter individually, but also about how these chapters could contribute to an understanding of teaching and learning of rational numbers and probabilistic reasoning. A brief summary of each chapter and at least one lesson that can be learned from its research follows.

Chapter 16. "Teaching Fractions for Understanding to Students Who May Experience Mathematics Difficulties: The Role of Classroom Interactions" by Susan B. Empson

Susan Empson's chapter is based on a case study of two first graders, Patrick and Pho, who experienced what many might refer to as mathematical difficulties. In particular, the author focuses on their knowledge of fractions and what their teacher, Ms. K, did to support their learning. The chapter begins with an overview of the conceptual framework based on difficulties students have with engaging in problem solving—here, *participation frameworks,* in which a teacher and students talk to each other during mathematics instruction.

Empson draws upon data from a five-week unit on fractions based on equal-sharing tasks (i.e., partitive division problems with fractional answers). With respect to equal-sharing problems, before instruction the two students could correctly solve only equal-sharing division problems without remainders (e.g., 12 cupcakes shared equally among 4 friends). But after instruction, each student could solve problems that involved a remainder. For comparison problems (e.g., "Who will get more pizza: a child at a table where 5 children are sharing a medium pizza, or a child at a table where 6 children are sharing a medium pizza?"), Patrick and Pho both answered problems like these correctly after the instruction. Likewise, they were able to correctly solve the transfer problems once the lessons were complete.

The author continues by providing an in-depth analysis of the participant frameworks during instruction that led to Patrick's and Pho's success with fraction tasks. She provides examples in the form of dialogues of how Ms. K, the boys, and the other students in the class resolved partial, unclear, or incorrect reasoning during instruction and how Patrick and Pho contributed productive mathematical ideas to group discussion.

Empson concludes by proposing three factors to explain how Patrick and Pho could participate profitably in instruction. First, the tasks allowed them to make use of their emerging understanding of partitioning. Second, the teacher created a variety of participant frameworks in which the boys were treated as competent in ways that counted for engagement in the classroom. Third, Patrick and Pho were afforded repeated opportunities to learn the values of their ideas.

Lessons learned

By creating an appropriate classroom environment, we as teachers have the capacity to reach students with mathematical difficulties and support their learning of new mathematical concepts. Empson also shows us how fraction concepts can be introduced in the early grades.

Chapter 17. "Fraction Notation and Reference in Children's Representations of Parts of Areas," by Geoffrey B. Saxe, Edd V. Taylor, Clifton McIntosh, and Maryl Gearhart

In their chapter, Geoffrey Saxe, Edd Taylor, Clifton McIntosh, and Maryl Gearhart highlight key findings from their research on developmental relationships between students' use of fraction notation and their understandings of part-whole relationships and the role of instruction in students' use of fraction notation to represent parts of area. The authors begin by introducing a framework based on two aspects of representing fractional parts of area: *notation* and *reference*. Notation consists of marks and symbols to represent fractions; reference is the conceptual work of using notational forms to represent relations between parts and whole—in particular, shaded regions of partitioned areas.

The authors used data from 384 students in nineteen upper-elementary classrooms during curriculum units on fractions to examine notation and area representations. They coded responses to a set of equal and unequal area problems to investigate three questions: (1) When students produce part-whole reference on area problems, are they treating the parts as discrete or as continuous quantities? (2) How do notation and part-whole reference develop (e.g., independently, concurrently, or sequentially)? and (3) How is fraction notation development affected by kind of instruction (e.g., inquiry vs. skills)?

With respect to part-whole references, most students referred to part-whole relations on both the equal and the unequal area problems but did not treat area as a continuous quantity. The authors found evidence of some independence between notation and reference; that is, student used either conventional notation or part-whole notation for both types of problems. For the question regarding fraction notation development between high-inquiry and low-inquiry (skills-based) classrooms, the authors focused on the unequal area problems (the more challenging of the two types) and found that the kind of instruction was not related to students' shifts to conventional notation. However, with respect to reference categories (part-whole: continuous; part-whole: discrete), students in the high-inquiry classrooms showed the greatest shift to part-whole: continuous.

The authors conclude with two implications for instruction. First, *tasks matter*. Non-routine tasks such as unequal area problems are valuable methods for investigating and assessing student understanding. Second, *instruction matters*. The students in the study who progressed in their understanding of notations and references were supported by high-inquiry instruction.

Lessons learned

We could not say it better than the authors did above: Tasks matter and instruction matters. We as practitioners need to ensure that students have access to nonroutine mathematical situations that evoke discussion; and the classroom environment should be structured to support mathematical inquiry.

Chapter 18. "Partitioning and Iterating When Teaching and Learning Fraction Addition on Number Lines," by Andrew Izsák, Erik Tillema, and Zelha Tunç-Pekkan

Andrew Izsák, Erik Tillema, and Zelha Tunç-Pekkan continue on the topic of fraction learning by focusing on what the Common Core standards recommend for teaching this topic. In particular, the authors find a difference in how fractions should be taught because of the increased emphasis on number line representations of fractions and decreased emphasis on circle representations. How can we help students represent fractions on number lines in ways consistent with CCSS-M? In the chapter, they present some key insights from their research study.

The authors present the experiences of how a sixth-grade teacher (Ms. Reese) and her students interpreted lessons about fraction addition in lessons based on using number lines, partitioning lengths, and iterating unit fractions. Ms. Reese's teaching strategy was in three steps: (1) determine how many whole numbers to represent (e.g., for $\frac{1}{4} + \frac{1}{8}$ the interval needed is 0 and 1); (2) partition unit intervals created in the first step (i.e., splitting the interval into eight equal parts); and (3) draw arrows for each interval added and circle the sum. When the authors interviewed students, it was apparent that they had different interpretations of the three steps; unintended consequences emerged. Izsák and his coauthors devote the remainder of this section to a case study of one student (Sonya) who interpreted the lessons in an unintended way.

The chapter concludes with a discussion of what teachers can do to strengthen students' understanding and use of number line representations for fractions and fraction addition. These include not changing the size of the whole after it has been established on the number line; having explicit discussions about partitioning intervals into equal length sub-intervals; and providing students with experience in partitioning a partition. Subtle nuances are important for success with visual representations such as representing fractions on number lines, and how teachers create and use such representations can have significant consequences for the way in which students interpret and understand instruction and for their future understandings.

Lessons learned

The authors point out that teaching students how to use visual representations—in this case, number lines as a way to work with fractions—is an important component of mathematical understanding. The example of Sonya's unintended interpretation of how to add fractions on a number line underscores the importance of teachers' devoting time to their instructional practice.

Chapter 19. "Developing Understanding of Multiplication of Fractions by Putting Research into Practice," by Nancy K. Mack

Nancy Mack shares her research on how students can develop a solid understanding of multiplication of fractions during instruction. In this chapter, the author focuses on how fifth-grade students used informal knowledge of fractions and their knowledge of partitioning quantities to solve word problems involving finding a portion of a part of a whole. She found that working with multiplication of fractions was not always straightforward for the students; however, as they became more flexible in their thinking about units and meanings of fractions, they were able to identify units in the problems and to successfully solve problems by partitioning units in appropriate ways.

Mack continues the chapter by sharing two research results that have been particularly effective in helping students think flexibly about units and ways to partition units. The first result involves the importance of *how the problems were worded* with respect to the unit for each fraction. Explicitly stating these units (e.g., $2/3$ of students in a survey said they rode in a vehicle to school; $3/4$ of the students who ride in a vehicle said they rode the bus) helped students understand the problem and solve it successfully. The second finding was that the structure and sequencing of word problems affected students' success with multiplication of fractions. The author recommends a progression of four kinds of problems, beginning with multiplication of fraction word problems like $1/2 \times 1/2$ and progressing to problems of the form $a/b \times c/d$ where b and d are relatively prime. Examples of interviews with students and their work on the four kinds of word problems illustrate student thinking about multiplication of fractions.

Mack concludes the chapter by referring to the Common Core State Standards for Mathematics and the emphasis placed on solving multiplication of fraction problems, and especially word problems. To assist teachers in their instruction of these important concepts, she includes a set of twenty-two word problems she developed from her classroom studies.

Lessons learned

Mack provides evidence that students' understanding of multiplication of fractions is likely not as straightforward as the Common Core standards imply. Attending to the wording, the structure, and the sequencing of multiplication of fraction word problems can help students develop a conceptual understanding and success with this important mathematical topic.

Chapter 20. "Helping Students Learn to Use Fractions," by Debra I. Johanning

Continuing on the topic of fraction learning, Debra Johanning shares findings from her research on a classroom practice referred to as "determining appropriateness." She proposes that when learning fractions, students typically explore how, why, and what—for example, how to determine if two fractions are equivalent, how to compare fractions, and what strategies can be used to operate with fractions. However, it is also important for student to investigate questions such as "Can I use a fraction if it is not in the problem context?," "Can I use a decimal instead of a fraction?," and "Do concepts that work for whole numbers also work for fractions?" The author details the methods used to carry out her investigation of determining appropriateness with a class of sixth and seventh graders and their teachers.

Results from the study allowed Johanning to create six subpractices of the *practice of determining appropriateness,* and she illustrates each subpractice with classroom examples.

The six subpractices are (1) representing fractions as decimals (e.g., the decimal approximation for $^1/_3$); (2) extending beyond whole numbers (e.g., scale factors applied to area); (3) drawing from multiple approaches and algorithms; (4) choosing to use fractions when fractions are not explicit (e.g., 5.2 represented as $5\,^2/_{10}$); (5) switching between equivalent fraction forms (e.g., $^{125}/_{100}$ and $1\,^{25}/_{100}$); and (6) using ideas about fractions with ratios (e.g., notationally $^a/_b$ and $a{:}b$).

The author concludes with the implications of her research for the classroom. For example, students need teacher support when trying to understand when it is appropriate to interchange fractions with decimals, such as in the case of $^1/_3$ and .33333... . Presenting students with tasks that use ideas they have previously learned can help students develop a deeper understanding of mathematics. As students look for and use connections, they come to view mathematics as a collection of connected, usable ideas rather than a set of arbitrary rules.

Lessons learned

Johanning presents important lessons for practitioners—that is, how important it is for students to determine appropriateness and make critical decisions about mathematical ideas. In this chapter, important connections between fractions and decimals are the focus, but the six appropriateness subpractices can be extended to other mathematical ideas as well.

Chapter 21. "Using Everyday Knowledge of Decimals to Enhance Understanding," by Kathryn C. Irwin

In the previous chapter, we were introduced to the connections between fractions and decimals. Here, Kathryn Irwin extends the discussion of decimal fractions to include misconceptions that students bring to this content topic. Some misconceptions include these: the longer the decimal, the larger the number; and putting a zero at the end of a decimal makes it ten times as large. The author posits that learning decimals should be anchored some way to students' existing knowledge—the knowledge they bring with them to school. This chapter is a summary of findings from her research study of New Zealand students (ages eleven and twelve) who came from a lower economic area. Students were classified as either high rank or low rank, based on prior performance in mathematics. Irwin devised problems based on three categories of decimal fraction problems—magnitude, addition/subtraction, and multiplication/division—both in context and in noncontextualized situations. Students provided written responses to the problems and were also interviewed.

When comparing pretest to posttest results, Irwin found that the lower ranked students who worked on contextualized problems improved at the highest rate. The author analyzed dialogues between pairs of students (one higher ranked student paired with one lower ranked student) and found that it was often the lower ranked student in the pairs who used everyday knowledge to make sense of the decimal problems. Dialogues between pairs are provided as examples in the chapter.

Irwin concludes with a summary and recommendations for practitioners. With respect to the misconceptions about decimals, she suggests that perhaps students had not had sufficient time to reflect on important decimal concepts. For example, students who believe that "one hundredth" can be written at 0.100 have not reflected on the incompatibility of this notion with principles of place value. She found that engaging students in discussions that involved some cognitive conflict (e.g., a fast food meal costing $95, based on a decimal misconception) gave them an opportunity to address such inconsistencies. Irwin closes by encouraging teachers to be aware of students' everyday knowledge and of any misconceptions developed on the way to achieving knowledge of decimal fractions.

Lessons learned

As we have learned from other authors in this section, the knowledge and experience that students bring with them to school provides fertile ground for mathematics learning. From Irwin's research, we learn how to recognize students' misconceptions and use real-world experiences to mitigate them.

Chapter 22. "Student Reasoning on Probability Tasks with Coins," by Laurie H. Rubel

We begin the section on probabilistic reasoning with a chapter by Laurie Rubel that focuses on how coin tasks can serve as models for important probability concepts such as randomness, sample space, and independence. The author is particularly interested in how students' ideas and beliefs about coins and other probabilistic devices can sometimes conflict with mathematics. She uses representativeness as a framework for analyzing probabilistic thinking and gives a list of student beliefs about coin tosses—such as, coins are unpredictable; a coin has a 50 percent chance or landing heads or tails; and there should be about equal numbers of heads and tails in a string of tosses. Rubel describes a study of 173 students in grades 5, 7, 9, and 11 in which each student completed a set of ten writing tasks. Based on the written responses, she chose a sample of 33 students across age groups and classes to interview.

Results from the interviews are reported for two tasks. The first task asked for the most likely fifth outcome for a coin that had heads on the first four tosses. Nearly three-fourths of students indicating there was no most likely outcome; about a third thought the most likely outcome would be tails because tails had not yet appeared in the tosses. The second task involved tossing a coin six times and predicting the pattern of outcomes (e.g., HTTTTT; TTTTHT; etc.). Most students chose the sequence HTHTHT as the most likely—a sequence that does not correspond to a belief that coin tosses result in random patterns. Through the student interviews, Rubel found an inconsistency between the tasks that involved the belief that there was not a most likely outcome on the first toss but that there was a most likely outcome on the second toss.

The chapter concludes with implications for instruction. Rubel highlights the importance of student explanations and justifications allowing teachers to understand misconceptions and beliefs. She recommends creating situations in which students solve problems and then are led toward confronting inconsistencies. Last, she poses a set of questions to be investigated with respect to different types of probability instruction in classroom materials.

Lessons learned

From Rubel's research we learn how valuable it is for teachers to create situations that evoke cognitive conflict in students' thinking—in her case, in the area of probabilistic reasoning. Additionally, paying attention to how students reason (either orally or in writing) in problem-solving situations can add to what teachers know about their students' understanding of mathematics.

Chapter 23. "Dice: Fair or Not Fair? That Is the Question," by Jane M. Watson

Jane Watson continues this section's focus on probabilistic reasoning by addressing the important topic of fairness—a topic that is often dismissed in classroom activities that deal with, for example, tossing dice. She notes that the rote answer of "$1/_6$" is the answer students give when asked what is the chance any side will come up when a six-sided die is tossed. Yet students often bring with them intuitions about dice and in the larger sense about probability that are resistant to instruction and do not improve across grade levels. The focus of her chapter is on research results from her studies on the development of students' beliefs about, and strategies dealing with, the fairness of dice. The results come from an interview study of 108 Australian students, ages eight to fifteen years.

The first area Watson examined was what students believe about fairness of dice. About half of the students simply stated, or agreed to the statement, that all numbers have the same chance on a fair die. There were many idiosyncratic or inconsistent responses to this question, such as believing that larger numbers (6s) come up more often than smaller numbers and that how you throw the die affects which numbers come up. Watson then examined how students would judge fairness of a specific pair of dice. Responses ranged from mention of nonstatistical issues, the assertion that the dice are fair without further investigation, references to physical features of the dice or a few unsystematic trials, and systematic trials of the dice. Last, the author looked at the association between beliefs about fairness and strategies for judging fairness, and found there was not a strong association.

Watson concludes with implications for teaching and suggests two instructional strategies to investigate issues of fairness. The first involves the importance of using concrete materials or computer simulations. The second strategy is the focused use of language, especially related to the word *chance*. Given the rise in prominence of statistics in today's curriculum, it may be time to move questioning the fairness of the random generator (like dice) to the forefront.

Lessons learned

Through her research findings, Watson makes us more aware of the importance of judging fairness within the larger realm of probabilistic thinking. Fairness is commonly assumed, but not tested.

Chapter 24. "Using a Computer Microworld to Make Sense of the Total of Two Dice," by Dave Pratt

Dave Pratt continues the discussion of randomness and chance in probabilistic reasoning with a focus on tossing two dice—this time, using an early study on how young learners (ages ten to eleven years) made sense of chance in a computer microworld. He is particularly interested in the equiprobability bias, a tendency to assume that different outcomes are equally likely in the case of the totals obtained by adding the dots on two dice. In the chapter, the author describes the Change-Maker microworld he developed and the responses of two students (Anna and Rebecca) as they tried to make sense of chance using "gadgets" (applications) within the microworld: two spinners and two dice.

Pratt found that the girls began their investigation of chance by using *local resources*—for example, using trial-by-trial variation that was immediately accessible in the microworld (setting the generator to 1,000 trials and examining the results). The girls progressed to using *global resources* that focused on an aggregated view of the stochastic, which he denotes as probability, large numbers, and distribution. For example, at first the girls did not attend to missing outcomes (e.g., no totals of 7 or 11) from the 1,000 trials of tossing two dice and adding the spots; they were depending only on the data they saw. However, with prompting from the interviewer, they generated a sample space that included all thirty-six totals, thus suggesting they were thinking globally about the activity and that some totals would be represented more than others (e.g., more totals of 7 than of 12).

Based on results from his study, the author concludes that the girls came to the activity dependent on local resources and with a strong equiprobability bias. As they worked through the two-dice microworld, they constructed new global resources. Pratt concludes by proposing that a microworld based on chance is beneficial because it allows for space to confirm or disavow prior knowledge, for setting up and testing conjectures, and for testing a large number of trials easily.

Lessons learned

More than a decade ago, Dave Pratt in his research showed how beneficial a computer microworld environment could be for students as they tested their probabilistic reasoning and formed new knowledge. With the advances in technology, we can benefit from continuing to use simulations in the classroom for probability topics and other mathematics concepts.

Reference

National Council of Teachers of Mathematics. (2000). *Principles and standards for school mathematics.* Reston, VA: Author.

National Governors Association Center for Best Practices & Council of Chief State School Officers (NGA Center & CCSSO). (2010). *Common core state standards for mathematics.* Washington, DC: Author. Retrieved from http://www.corestandards.org.

Teaching Fractions for Understanding to Students Who May Experience Mathematics Difficulties

The Role of Classroom Interactions

Susan B. Empson
University of Missouri–Columbia

> The reality of the self is relational.
> —*Varenne and McDermott (1998), 164*

Patrick and Pho were two first graders who experienced what many might refer to as mathematics difficulties. Out of all of their classmates at the beginning of a study on teaching fractions, they started out with the least knowledge, as measured by a problem-solving assessment administered by the researcher; and after five weeks of instruction, again compared to their classmates, they had the least knowledge of fractions (Empson, 1999). Nevertheless, the two boys engaged productively in problem solving and their understanding of fractions advanced in impressive ways. The analysis in this chapter (based on Empson, 2003) focuses on what the teacher did to support Patrick's and Pho's learning.

The mathematics teaching in Patrick and Pho's classroom centered on problem solving.[1] To teach fractions, their teacher, Ms. K, posed problems that allowed students to use informal strategies to partition items into equal shares. These strategies provided a context for the development of students' understanding of fractions. To illustrate the power of children's informal problem-solving strategies, consider the fact that before instruction on fractions began, the majority of children in Ms. K's first-grade classroom, fourteen out of seventeen children, were able to solve the following equal-sharing problem: "Four children want to share 10 cupcakes so that each child gets the same amount. Show how much 1 child can have." In a typical strategy, children dealt the whole cupcakes to each sharer and partitioned the remaining two cupcakes into halves. Ms. K posed problems such as this one to all of her students.

In this chapter I analyze Ms. K's interactions with Patrick and Pho and discuss how these interactions supported and extended the boys' understanding of fractions. Proof that their undertanding of fractions grew is offered in a comparison of their performance on a number of problems before and after instruction.

Conceptual Framework

Researchers disagree about why some children may have difficulty engaging in problem solving. Explanations often center on qualities that children lack that keep them from engaging productively in solving and discussing problems, such as a limited working memory or inability to handle complex information (Baxter, Woodward, & Olson, 2001; Woodward &

---

[1] The teacher was a participant in a longitudinal study of Cognitively Guided Instruction (CGI) (Fennema et al., 1996). See Carpenter, Fennema, Franke, Levi, and Empson (2014) for details about problem-type frameworks, children's strategy development, and instruction in CGI classrooms. See Empson & Levi (2011) for information about children's fraction thinking.

This chapter is adapted from S. B. Empson (2003), Low-performing students and teaching fractions for understanding: An interactional analysis, *Journal for Research in Mathematics Education, 34*, 305–343.

Baxter, 1997). However, focusing on limitations that are seen simply as part of the child stops short of considering the interactive nature of the child's engagement in problem solving and how a teacher can influence this engagement in profound ways (Ginsburg, 1997; Varenne and McDermott, 1998).

I used *participation frameworks* (Goffman, 1981) as a lens to analyze the teacher's interactions with Patrick and Pho during problem solving. A participant framework is created whenever a teacher and students talk to each other during mathematics instruction. It can be thought of as a set of social relationships involving roles centered on mathematics problems. Each person is positioned into a role by what participants are saying and doing with each other. For example, when a teacher asks a child to explain to the group how he or she solved a problem, that teacher is creating a participation framework that positions that child in the role of problem solver and solution reporter. When the teacher says, "Robert agrees with Sophia that you can split one-third in half and get two sixths," she is reinforcing Sophia's role as a claim-maker and Robert's role as someone who has evaluated and accepted the claim. The roles in participant frameworks can shift quickly, based on what the teacher or a child says. They differ from assigned roles in group work in that they are part of the fabric of discourse; teachers and their students are not necessarily aware of the participant frameworks created by how they talk to each other. *All* classroom interactions involve participant frameworks (O'Connor & Michaels, 1996).

In the study reported here based on my *JRME* article (Empson, 2003), I analyzed the types of participant frameworks that emerged in Ms. K's interactions with Patrick and Pho and how Ms. K's questions and comments positioned the boys into roles that supported their engagement in problem solving and the development of their understanding of fractions.

Background Information

Patrick was from a middle-class family and had been informally identified by school personnel as having difficulties focusing on academic tasks. Pho was from a family who had emigrated from south Asia and spoke English as a second language.

The data for this study came from a case study of a five-week unit on fractions in a first-grade classroom (Empson, 1999). The unit involved fifteen lessons, collaboratively planned by Ms. K and me. Instruction was based on equal-sharing tasks and related problems (Streefland, 1991). Equal-sharing tasks are partitive division problems with fractional answers. Like the cupcake problem at the beginning of the chapter, they involve sharing a given number of dividable objects, such as cupcakes, apples, pounds of flour, or liters of juice, among a given number of sharers or groups. Equal-sharing problems can be used to introduce and develop fraction concepts by engaging children in creating, reasoning about, and reflecting on fractional quantities in their solutions (Empson & Levi, 2011). For readers who might be interested in more information, the sequence and content of all lessons for the fractions unit can be found in Empson (1999, pp. 339–342).

Findings from the Study

In the next sections of this chapter, I share assessments of Patrick's and Pho's understanding. I conducted separate problem-solving interviews with Patrick and Pho before and after instruction. Selected problems appear in table 16.1. A before and after comparison of results showed that Patrick and Pho did indeed learn.

Table 16.1

Selected problems from the problem-solving interviews before and after instruction in fractions

Problem type	Problem
Before instruction	
Equal sharing	
(a) 5 ÷ 2	2 children want to share 5 cupcakes so that each child gets the same amount. Show how much 1 child can have. *If the child does not share the remainder:* They want to share this too. How could they do that?
(b) 10 ÷ 4	4 children want to share 10 cupcakes so that each child gets the same amount. Show how much can 1 child have. *If the child does not share the remainder:* They want to share these too. How could they do that?
Comparison	
(c) ¹/₅ vs ¹/₆	Who will get more pizza: a child at a table where 5 children are sharing a medium pizza, or a child at a table where 6 children are sharing a medium pizza? Why?

After instruction	
Equal sharing	
(a) 3 ÷ 2	2 children want to share 3 giant cookies so that each child gets the same amount. How much can each child have?
(b) 14 ÷ 6	6 children want to share 14 blueberry pancakes so that each child gets the same amount. How much can each child have?
Comparison	
(c) ¹/₆ vs ¹/₅	Who gets more pizza: a child at a table where 6 children are sharing a small pizza equally, or a child at a table where 5 children are sharing a small pizza?
(d) ²/₆ vs ²/₈	Some children are drinking lemonade. At one table, 6 children are sharing 2 cartons of lemonade. At another table, 8 children are sharing 2 cartons of lemonade. At what table does a child get more lemonade?
Subtraction	
(e) 3 − ¹/₂	Robert had 3 giant peanut butter cookies. He ate one-half of a cookie, and decided he didn't want to eat any more. How many cookies did Robert have left?

Equal sharing

Before instruction, Patrick and Pho could solve equal-sharing division problems with no remainders, such as 12 cupcakes shared equally among 4 friends, but neither boy generated a fractional quantity to equally share leftover items. For example, they were both given a problem that involved sharing 5 cupcakes between 2 children. Neither solved it for the fractional quantity 2 and a half. Using cubes, Pho dealt the 5 cupcakes one by one to each person, for a total of 2 cupcakes per person, and 1 cupcake left over. He said he could not break apart the extra cupcake.

After instruction, however, both children solved problems in which they partitioned leftover items into fractional quantities. For example, Pho shared 3 cookies between 2 children by halving each of the 3 cookies and giving each person 3 halves. To share 14 pancakes among 6 children, Patrick partitioned each pancake in half and distributed them one by one to the 6 people, saying "there sure is a lot to split . . . looks like they're gonna get more than

179

one half . . . because there's a lot of pancakes and not that many children." He continued to partition the leftover amounts into fractional amounts and described the final share as "a bunch of halves and one little piece," with each share consisting of four halves, one-half of a half, and one-sixth of a half.

Comparison

Before instruction, Patrick and Pho were given a problem that involved the simplest kind of comparison between two sharing situations: "Who will get more pizza: a child at a table where 5 children are sharing a medium pizza, or a child at a table where 6 children are sharing a medium pizza? Why?" Both answered that a child at the table of 6 children would get more because 6 is more than 5.

After instruction, they were given the same problem. Both boys answered correctly this time. For example, Patrick said a child at the table of five children would get more, "because if there's five it means you can get more pizza." The children were given another similar problem involving 6 children sharing 2 cartons of lemonade compared to 8 children sharing 2 cartons of lemondade, and again both answered correctly.

Transfer: Subtraction

To test the depth of children's understanding after instruction, all of the children were given problems to solve unlike any they had solved in instruction. This type of problem is considered a test of the "transfer" of students' understanding to new problems. Patrick and Pho were given a subtraction problem involving a half cookie taken away from three whole cookies. Both boys solved the problem correctly by drawing three giant cookies, partitioning one in half, and mentally separating one half cookie from the set. They described the remaining quantity as "one half cookie and two whole (or giant) cookies."

Participant Frameworks during Instruction

Instruction in Ms. K's classroom centered on children's thinking. I distinguished among participant frameworks on the basis of the nature of Patrick's or Pho's initial thinking about a problem as partial, unclear, incorrect, or valid and whether they were working one-on-one with Ms. K or involved in a discussion with the whole group (see table 16.2).

Table 16.2
Participant framework formats and their frequencies

	One-on-one with teacher	In group with teacher and other students
Reporting valid strategy or idea	5	18
Resolving partial or unclear strategy or idea	12	11
Exploring incorrect strategy or idea, not resolved	3	0

First I present instances in which Patrick's or Pho's reasoning was partial, unclear, or incorrect in some way, and how it was addressed either in group discussion or one-on-one with Ms. K. I begin with these because they capture some of the initial dilemmas Ms. K and the boys faced at the beginning of the fractions unit. Despite the tentativeness of the boys' reasoning in these episodes, Ms. K did not position the boys as deficient in their thinking. Second, I describe the most common type of participant framework: interactions in which Patrick or Pho were positioned to make productive contributions to group discussions.

Resolving partial, unclear, or incorrect reasoning during instruction

In the second lesson the children solved this problem: "There are 2 horses in the field. If there are 9 apples in a bushel, how many apples would each horse get?" Neither Patrick nor Pho partitioned the extra apple, which led to a predicament in the discussion that Ms. K resolved individually with each boy.

Interaction A begins after the first child explained his strategy to the group, including how he split the extra apple in half. Ms. K called on Pho to report his solution next. A predicament emerged when Pho said he could not give the horses the extra apple, and Ms. K asked him what he would do with it (see table 16.3).

Table 16.3
Interaction A

Line	Speaker	Dialogue
1	*Pho:*	8
2	*Ms. K:*	Each horse gets 8 apples? Tell me how you figured that out, Pho.
3 4	*Pho:*	*[Has 9 cubes for the apples, deals them one by one until each horse has 4 apples]* There's one more left, but I can't give them this.
5	*Ms. K:*	So what are you going to do with that one left?
6	*Pho:*	Uh.
7	*Ms. K:*	Anybody got an idea? What can we do with that other one?
8	*James:*	…. I know, I know.
9	*Ms. K:*	*[To Pho]* Do you know what you could do with this one?
10	*Pho:*	*[Shakes head no]*
11 12	*Ms. K:*	No. You're not sure. Could anyone share with him what he could maybe do with that other one? James?
13	*James:*	Cut it in half.
14	*Ms. K:*	Could you cut it in half, Pho?
15	*Pho:*	*[Inaudible]*
16	*Ms. K:*	And then how much would each horse get?
17	*Pho:*	It still wouldn't work.
18	*Ms. K:*	Why won't it work?
19	*Pho:*	'Cause, there still is one more left.
20	*Ms. K:*	OK. How much is this horse *[pointing to his cubes]* going to get?
21	*Pho:*	Four.
22	*Ms. K:*	How much is this horse *[pointing to second horse]* going to get?
23	*Pho:*	Four.
24	*Ms. K:*	…James, he says it is still not going to work. Would you please go down—
25	*Pho:*	*[Interrupting]* Because four and four is eight.
26 27	*James:*	*[To Pho]* You have eight so far. Just cut that [extra apple] in half, and then each one would get eight *[sic]* and a half.
28	*Pho:*	I still can't get it.

In this interaction, when Ms. K said to Pho, "Tell me how you figured that out" (line 2), she positioned him as a problem solver explaining his thinking. She reinforced this role again when she asked what he was going to do with the leftover apple (line 5). He held this role through line 25, when he gave his reason for why "cut it in half" would not solve the problem. In addition, Ms. K elicited help for Pho from the other children, initiating a shift in participant framework by distributing the role of problem solver, from Pho alone to Pho and the others. This move positioned the other children as sources for problem-solving ideas. Pho, by virtue of his role as problem solver in the participant framework, however, had the authority to accept or reject the idea offered by James, based on his understanding of that idea, and so he rejected a mathematically legitimate idea. His dilemma may have been that he was interpreting the idea of "cutting in half" in terms of operating on a set—here, the set of nine apples—instead of the single, leftover apple.

As the rest of the group solved a new problem, Ms. K worked with Patrick and Pho individually to help them arrive at a solution and verbalize how that quantity related to the whole apples. Because the interactions between Ms. K and the two boys were similar, only the interaction with Patrick is reproduced here. As shown in Interaction B, Ms. K began by holding a cube in her hand for reference and asking Patrick what the other children suggested be done with it (see table 16.4).

Table 16.4
Interaction B

Lines	Speaker	Dialogue
1	*Patrick:*	Hmmm.
2	*Ms. K:*	What did the other kids say we could do with this other apple?
3	*Patrick:*	Split it in half. But we can't do it.
4 5 6	*Ms. K:*	Well, let's pretend [this cube] is an apple. If it's an apple could we cut it in half? *[Patrick agrees.]* ... So how many apples is each horse gonna get in this now?
7	*Patrick:*	One half.
8 9	*Ms. K:*	OK. And they're gonna get these *[indicating four cubes each]* ... How much would they get?
10	*Patrick:*	Five.
11	*Ms. K:*	How did you figure out five apples?
12 13	*Patrick:*	Because if we cutted this [extra cube] in half they would each get five apples.
14 15 16	*Ms. K:*	Show me. *[Patrick counts four single apples each, and on the extra cube, two apples (one for each half).]* Is this last piece they're gonna get a whole apple or is that gonna be a half apple?
17	*Patrick:*	Half apple.
18	*Ms. K:*	Are these [four cubes] whole apples or are they half apples?
19	*Patrick:*	Whole.
20	*Ms. K:*	So how many whole, big apples are they gonna get?
21	*Patrick:*	Four.
22	*Ms. K:*	OK. And how many half apples?
23	*Patrick:*	One. *[Patrick writes "4 1" on his paper.]*

Patrick's dilemma may have been that the cubes could not be literally cut in half. In response, Ms. K suggested they pretend the cube is an apple, which Patrick easily took up in his solution. He did not differentiate between whole apples and half apples in figuring the total, however, so for each kind of unit he counted as one, Ms. K posed two options for quantification; Patrick chose appropriately for each and distinguished between them by noting separate totals ("4 1" in line 23).

Ms. K structured the interaction so that Patrick revoiced a key idea introduced by other children in the previous exchange. This move positioned Patrick in the role of evaluating for his own use a potentially useful mathematical idea. When he rejected the idea based on his understanding of how the physical materials worked, which was his right in this role, she invited him to consider a new idea, again giving him the option to agree or disagree (lines 4–5). Throughout the interaction, Patrick's responsibility for making sense of the solution was reinforced by Ms. K's "what?" and "how?" probing questions (lines 2, 11), while these same questions directed his attention to the mathematically critical aspects of the solution—such as the difference between whole apples and fractional apples as amounts and the use of physical materials as a support for thinking rather than as literal representations.

Ms. K's moves facilitated Patrick's and Pho's engagement in problem solving and taking responsibility for their thinking and built on the contributions each boy made. A significant feature of these interactions is that Patrick and Pho had ideas about how to solve all of the problems. Compared to some of the other children's contributions, their ideas were sometimes partially formed or unclearly stated. Nonetheless, Ms. K's interactions with the boys ensured their ideas were treated as legitimate contributions and were accepted or rejected based on mathematical reasons supplied by the other children. To enact these kinds of participation frameworks, Ms. K needed to be able to recognize the mathematical plausibility of the boys' ideas and know how to facilitate students' engagement with each others' ideas in group discussions (Webb et al., 2008).

| **Contributing productive mathematical ideas to group discussion** | In this section I present the participant frameworks in which Patrick and Pho were positioned as having mathematical ideas of use to the other children. Starting with the third lesson, Ms. K gave the children their first equal-sharing problem involving a partition into thirds, a partition that is harder to make than partitions into halves or fourths: "Three children want to share 7 candy bars so that everyone gets the same amount. How much would each child get?" Many children found this problem difficult because they could not figure out how to share all of the leftover candy bar. Pho, however, solved the problem using cubes by dealing them one by one to each of the three people, and deciding he would cut the last candy bar "in threes." One-on-one with Ms.K, he explained that each child got "three": "I split this [using a hand-cutting motion over extra cube]. One, two, three" (i.e., a whole, a whole, and a part from the candy bar cut in three pieces for the final share). Ms. K asked him to draw a picture, and he drew the circle partitioned into three parts shown in figure 16.1, along with the tallies. This activity set the stage for a participant framework during group discussion in which Pho was positioned as a problem solver making a valid mathematical claim with which other children disagreed. |

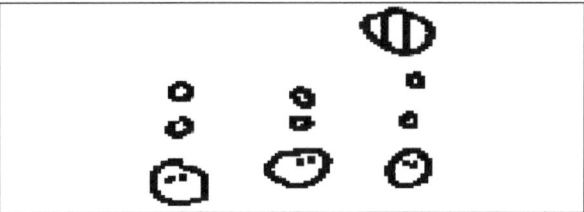

Fig. 16.1. Pho's representation of his solution for sharing 7 candy bars among 3 children

Ms. K called on Pho to report his strategy first. He described what he had done, and then Ms. K revoiced his strategy as she elicited reasons from Pho for his problem-solving moves. As he was explaining he had cut the extra candy bar "in threes . . . so each child can get one," another student, Marie, disagreed, saying, "You can split it in half, and quarters, but you can't split it in threes." Ms. K asked her why and, the conversation continued (see table 16.5, Interaction C).

Table 16.5
Interaction C

Line	Speaker	Dialogue
1 2	*Marie:*	Because there's—you can't leave one over *[i.e., if you make fourths, you will have an extra piece]*, so if you cut this one [extra cube] in half —
3 4 5	*Ms. K:*	But, Marie, look what Pho did. *[Marie looks at Pho's paper].* Pho, she says you can't split that in three. You think that's right? *[Inaudible answer from Pho.]* What do you think, guys?
6 7	*Tim:*	If you split in three, then you would get a half that remains, a half of a candy bar that's still there. [he may mean a fourth]
8	*Ms. K:*	If you split it in threes?
9	*Marie:*	That's what I'm talking about. You still get a quarter left.
10	*Ms. K:*	Pho, they said that you can't split it in threes.
11	*Tim:*	You can split it in threes but you have a half left.
12 13	*Ms. K:*	You'll have a half? Look at how he split it, Tim. Does he have a half left over? *[Lev goes over to look at Pho's paper; Tim looks too.]*
14	*Tim:*	No.
15	*Ms. K:*	*[to Pho]* Do you have a half left over?
16	*Kaitlin:*	No.
17	*Pho:*	No.
18	*Tim:*	That's because of the way you cut it.
19	*Ms. K:*	He cut it differently didn't he? Would that work, Tim, or not?
20	*Tim:*	It would work.
21	*Pho:*	I'll draw it over.
22	*Ms. K:*	Would you draw it bigger so people can see.

In this participation framework, Pho was positioned as a problem solver making a mathematical claim in opposition to another apparently reasonable mathematical claim. Ms. K created opportunities for Pho to respond directly to Tim's and Marie's claims about the impossibility of splitting the candy bar in threes by relaying their statements to Pho, focusing on the claims that she knew he could defend. With these moves, Ms. K created and helped sustain a role for Pho to express and defend his mathematical claim. Pho acknowledged the value his solution was positioned to have for others when he spontaneously suggested redrawing it to make it clearer (line 21). After this interaction, thirds become an acceptable partition to the other children.

In the eighth lesson, children solved an equal-sharing problem involving six children sharing eight pancakes: "Six children have ordered blueberry pancakes at a restaurant. The waiter brings 8 pancakes to their table. If the children share the pancakes evenly, how much can each child have?" After working one-on-one with Patrick and Pho to solve the problem (see fig. 16.2), Ms. K again called on them to report their solutions first, positioning them as problem solvers explaining their thinking. Patrick drew eight circles on the board to represent the pancakes. On the chalkboard both boys drew dots for people and arrows from each dot to one pancake. They continued as in the transcript of Interaction D, shown in table 16.6).

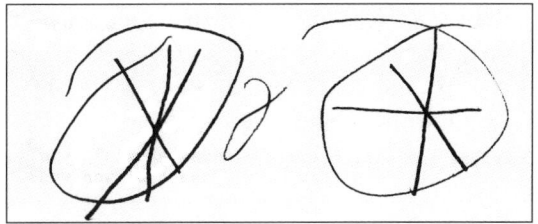

Fig. 16.2. Patrick's solution for how to partition
2 leftover pancakes to share among 6 children

Table 16.6
Interaction D

Line	Speaker	Dialogue
1 2	*Ms. K:*	*[to group]* Now listen to this. Do you think—*[to Patrick and Pho]* . . . What did you do so far? Tell 'em what you did so far.
3 4 5 6	*Pho:*	. . . There's six people, and we gave 'em each a pancake, and two were left, and we split 'em in six. *[Partitions one circle into six parts by making fourths, then a final partition through opposite fourths as if continuing to repeatedly halve, as shown in fig. 16.2.]*
7 8	*Ms. K:*	OK, you split your one, then let Patrick do another one. *[partitions his circle similarly]*
9	*Patrick:*	And we split 'em in six, and then we got *[counting pieces]* 1, 2, 3—
10	*Ms. K:*	Wait, why did you split 'em in sixths?
11	*Pho:*	So everyone can have one.
12	*Ms. K:*	So every child could have one?
13	*Pho:*	Yeah then—
14	*Ms. K:*	So wait, how much do they get of this one [partitioned pancake]?
15	*Pho:*	One.
16	*Ms. K:*	And how much do they get of this one [partitioned pancake]?
17	*Pho:*	One.
18	*Ms. K:*	One what?
19	*Pho:*	One piece.
20 21	*Ms. K:*	What do you call those pieces? He says you get one of these and one of these. What do you call those?
22	*Pho:*	Two [i.e., the total number of fractional pieces].
23	*Student:*	Fourths.
24	*Patrick:*	Ummm. Six[ths].
25	*Ms. K:*	Why do you call them sixths?
26	*Pho:*	'Cause there's six pieces.
27	*Ms. K:*	'Cause they're shared between six people.

This interaction shows Patrick's and Pho's ability to solve an equal-sharing problem involving fractional quantities beyond repeated halving. They were able to justify their problem-solving moves and show a way to partition the pancakes into six pieces.[2] Quantifying the fractional share continued to be problematic for them, suggesting that the relationship between partitioning actions and resulting quantities is not trivial to construct.

The episodes from the third and eighth lessons illustrate a common participant framework arrangement throughout the unit. Ms. K frequently asked Patrick or Pho to report their strategies first when they had successful solutions, which allowed Patrick and Pho to be positioned as problem solvers and to participate in explaining and justifying their thinking verbally. It also cast them in roles where they acted as mathematical authorities and got credit for their thinking.

Participant Frameworks and the Development of Understanding

There are few analyses of successful engagement in problem-solving based instruction on the part of students who may struggle in mathematics. A comparison of assessments before and after instruction clearly documents Patrick's and Pho's gains in understanding, and studying their participation in instruction suggests mechanisms of those gains. The participant frameworks orchestrated by Ms. K invited Patrick and Pho to make contributions to group discussions that positioned them in relation to the other students as problem solvers, solution reporters, and claim makers and defenders. In enacting these roles, Patrick and Pho "practiced" the content involved—they reconstructed it for themselves each time they were required to engage with it as part of their role.

I considered three main factors to explain how two boys who began instruction knowing considerably less than the other students could participate profitably in instruction. First, the problems posed to Patrick and Pho allowed them to make use of their emerging understanding of partitioning to generate new strategies. The equal-sharing problems in this study afforded a variety of strategies based on this understanding of partitioning (Empson & Levi, 2011). Despite knowing less about fractions, on average, compared to the rest of the class before instruction, Patrick and Pho had informal strategies they were able to use to solve problems from the very beginning, and Ms. K organized their engagement to support and extend these strategies, even when the strategies were unclear or seemed mistaken. Further, Patrick and Pho were never asked to use a strategy they did not understand. Instead they were expected to construct strategies based on their own ideas with mathematical potential and to listen to other children's ideas and use what made sense.

Second, Ms. K created a variety of participant frameworks in which Patrick and Pho were treated as competent in ways that counted for engagement in mathematics in this classroom. Both boys were positioned several times with the teacher and for their peers as problem solvers, claim makers, and solution reporters—roles that are central to doing mathematics. This positioning enhanced their identities (Cohen & Lotan, 1995).

Finally, the sheer frequency with which Patrick and Pho were positioned by Ms. K to engage productively suggests that they had repeated opportunities to learn the value of their ideas and to develop their understanding. These opportunities, offered regularly and consistently throughout instruction, outnumbered the times when either boy reached an impasse in thinking that could not be navigated and appeared to pay off in the long run as increased understanding of fractions.

---

[2] The pieces created by the partition were not, geometrically speaking, equal, but they were intended to be so; throughout the unit children explicitly expressed the need for equal pieces in their partitions and the difficulty of creating these pieces in models of continuous quantities. I have argued elsewhere (Empson, 1999) that this separation between the geometry of planar shapes and the mathematics of quantity (Schwartz, 1988) is reasonable in the early development of fraction understanding.

Patrick's and Pho's success depended on their interactions with Ms. K. Under other circumstances, such as working with a teacher who did not recognize the mathematical potential of informal strategies, one can easily imagine their difficulties getting worse. Just as productive engagement in mathematics depends on how the teacher interacts with students, so too may children's lack of engagement be explained in part in terms of teachers' interactions with students. If the participant frameworks that emerge in classroom interactions consistently position certain students as lacking or not knowing, one can expect disengagement as a consequence in the long run.

Although Ms. K. knew that Patrick and Pho did not, on average, understand as much about fractions as the other students, they were not framed as children who did not understand. Instead, they were positioned from the very beginning as children who engaged in problem solving, and, consequently, as children who understood mathematics. This positioning positively affected their learning.

References

Baxter, J., Woodward, J., & Olson, D. (2001). Effects of reform-based mathematics instruction on low achievers in five third-grade classrooms. *Elementary School Journal, 101,* 529–547.

Carpenter, T. P., Fennema, E., Franke, M., Levi, L., & Empson, S. B. (2014). *Children's mathematics: Cognitively guided instruction* (2nd ed.). Portsmouth, NH: Heinemann.

Cohen, E., & Lotan, R. (1995). Producing equal-status interaction in the heterogeneous classroom. *American Educational Research Journal, 32,* 99–120.

Empson, S. B. (1999). Equal sharing and shared meaning: The development of fraction concepts in a first-grade classroom. *Cognition and Instruction, 17*(3), 283–342.

Empson, S. B. (2003). Low-performing students and teaching fraction for understanding: An interactional analysis. *Journal for Research in Mathematics Education, 34,* 305–343.

Empson, S. B., & Levi, L. (2011). *Extending children's mathematics: Fractions and decimals.* Portsmouth, NH: Heinemann.

Fennema, E., Carpenter, T. P., Franke, M., Levi, L., Jacobs, V., & Empson, S. B. (1996). Mathematics instruction and teachers' beliefs: A longitudinal study of using children's thinking. *Journal for Research in Mathematics Education, 27,* 403–434.

Ginsburg, H. (1997). *Entering the child's mind: The clinical interview in psychological research and practice.* New York, NY: Cambridge University Press.

Goffman, E. (1981). *Forms of talk.* Philadelphia: University of Pennsylvania Press.

O'Connor, M. C., & Michaels, S. (1996). Shifting participant frameworks: Orchestrating thinking practices in group discussion. In D. Hicks (Ed.), *Discourse, learning, and schooling* (pp. 63–103). New York, NY: Cambridge University Press.

Schwartz, J. (1988). Intensive quantity and referent transforming arithmetic operations. In J. Hiebert & M. Behr (Eds.), *Number concepts and operations in the middle grades* (pp. 53–92). Hillsdale, NJ: Lawrence Erlbaum.

Streefland, L. (1991). *Fractions in realistic mathematics education.* Boston, MA: Kluwer.

Varenne, H., & McDermott, R. (1998). *Successful failure: The school America builds.* Boulder, CO: Westview Press.

Webb, N., Franke, M., Ing, M., Chan, A., De, T., Freund, D., & Battey, D. (2008). The role of teacher instructional practices in student collaboration. *Contemporary Educational Psychology, 33,* 360–381.

Woodward, J., & Baxter, J. (1997). The effects of an innovative approach to mathematics on academically low achieving students in inclusive settings. *Exceptional Children, 63,* 373–388.

Chapter 17

Fraction Notation and Reference in Children's Representations of Parts of Areas

Geoffrey B. Saxe, Edd V. Taylor, Clifton McIntosh, and Maryl Gearhart
University of California, Berkeley

In this chapter we highlight key findings from our 2005 *Journal for Research in Mathematics Education* (JRME) article (Saxe, Taylor, McIntosh, & Gearhart, 2005) that reported our analyses of (a) developmental relationships between students' uses of fractions notation and their understandings of part-whole relations and (b) the role of instruction in students' use of fractions notation to represent parts of an area. Researchers who were publishing at the time of our study suggested that a factor in the difficulties students confront with fractions may be the challenge of acquiring flexible use and understanding of written notation (e.g., Carpenter, Corbitt, Kepner, Lindquist, & Reys, 1980; Hiebert, 1988, 1989; Mack, 1990, 1995; Moss & Case, 1999). At that time there were few systematic studies of the ways that students develop understanding of fraction notation or the role that instruction played in fostering this development. Now, a decade later, it is a good time to revisit our work on representation of fractional notation in light of the major focus that the Common Core Standards for Mathematics (National Governors Association Center for Best Practices & Council of Chief State School Officers [NGA Center & CCSSO], 2010) places on fractions in grades 3 through 6.

Representing Parts of Area with Fractions Notation

Partitioning an area into fractional parts and using fractions notation to represent relations between parts and whole are common components of elementary fractions instruction. To organize our study, we introduced a framework that distinguished between two aspects of students' representations of fractional parts of area: *notation* and *reference*. *Notation* consists of marks and symbols (such as two numbers and a line) and rules for linking these marks (for example, one number is positioned above the other separated by a horizontal line). By *reference* we mean the conceptual work of using notational forms to refer to fractional relations between parts and wholes (such as shaded regions of partitioned areas).

Researchers have used longitudinal, qualitative studies to investigate developmental changes in students' understanding of part-whole relationships and the ways instruction may support change. These studies show that, in developing an understanding of fractions, students draw upon their knowledge of whole numbers and additive relations, and other, informal knowledge. We highlight the work of two researchers who focused on elementary students' understandings of fractional parts of area.

Mack (1995) engaged seven third- and fourth-graders in six individualized lessons over a three-week period, helping them to connect their informal knowledge with written notation by posing problems and asking question. She reported that these students faced two challenges coordinating written notation with an understanding of part-whole relations. On the one hand, the students tended to interpret written notation in whole number terms: For example, one student remarked that the notation for five-eighths of an area could be written

This chapter is adapted from G. B. Saxe, E. V. Taylor, C. McIntosh, & M. Gearhart (2005), Representing fractions with standard notation: A developmental analysis, *Journal for Research in Mathematics Education*, 36, 137–157.

as either "5" or "⁵/₈," explaining that "it doesn't matter. It's the same thing" (p. 435); this child's focus may have been on the five parts as discrete pieces without concern to represent the whole (eight pieces) in the notation. On the other hand, students tended to interpret whole number notation of mixed number expressions in terms of the language of fractions: For example, students consistently interpreted the 2 in the subtraction expression $2 - \frac{3}{8} = ?$ as two-eighths. Through her instructional methods of posing problems and asking questions, Mack was successful in supporting students' understanding of relations between fractional quantities of areas and conventional notation.

Another example of classroom-based research is the work of Ball (1993), who reported that her third-grade students initially made sense of notations for fractions in ways that were at odds with normative approaches. For example, one student interpreted the fraction expression $\frac{3}{4}$ as operations on discrete sets—make four groups of three objects and then take away one group. Other students invoked an area model, assuming that $\frac{1}{4}$ could only apply to one stereotypical shape, a quarter circle, thereby circumventing an analysis of parts and wholes. Ball designed her instruction to support students' emerging understandings of relations between fraction representations and fraction understandings by posing problems using various models (e.g., area, discrete) and guiding inquiry about these representational contexts. Ball reported evidence that students shifted from an inexplicit and tacit intuition of ideas like $\frac{1}{4}$ as a particular shape to a more principled understanding of fraction notations as expressions of part-whole relations.

Although Mack and Ball reported some success in supporting students' developing understandings, their studies demonstrate that students are challenged in their efforts to understand fractions notation as a means of representing part-whole relations of a continuous quantity like area (as distinct from a discontinuous or discrete quantity like objects). To add to the research on the development of fraction notation and reference for area, we created problems with two contrasting area representations. Examples of each problem type are shown in figure 17.1. *Equal Area problems* consisted of shapes partitioned into equal-sized parts, and *Unequal Area problems* had parts of unequal sizes, affording us an analysis of students' interpretation of size relations between parts of areas. We argued that students who interpret areas in whole number terms would respond correctly to an Equal Area problem by counting the discrete parts shaded without considering the relative sizes of shaded regions and represent their counts for the total number of parts and the number of parts shaded (e.g., use written notation such as "¹/₄" or "1-4" or "⁴/₁" for the left square illustrated in fig. 17.1). In contrast, students' notational responses to the Unequal Area problems would differentiate students who understand parts of areas as continuous quantities (e.g., the notation ¹/₈ for the right square in fig. 17.1) from those who conceptualize parts of areas as discrete quantities (e.g., the notation ¹/₅ for the right square in fig. 17.1). In the study, we examined students' reference to parts and part-whole relations for each problem type, as well as their use of conventional (e.g., ¹/₄) vs. unconventional (e.g., 1-4 or ⁴/₁) notation.

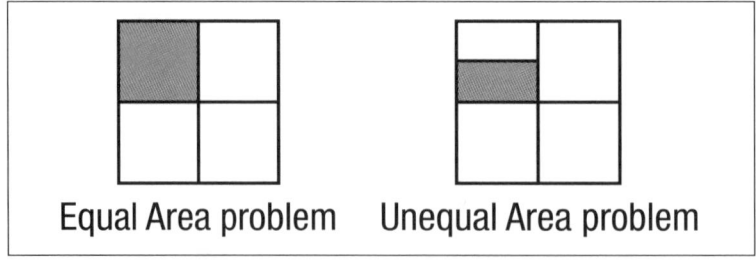

Fig. 17.1. Examples of Equal Area and Unequal Area problems: Write a fraction to show what part is gray.

**Details about
Our Study**

Our study (Saxe et al., 2005) made use of data collected in a research and development project entitled Integrated Mathematics Assessment. The study followed 384 students in nineteen upper-elementary classrooms during curriculum units on fractions. The classrooms were engaged with three categories of classroom practice: (a) use of inquiry curriculum in ways that were more aligned with inquiry principles (High Inquiry, eight classrooms); (b) use of inquiry curriculum in ways that were less aligned with inquiry principles (Low Inquiry, seven classrooms); and (c) use of traditional curriculum in ways that were less aligned with inquiry principles (Traditional, four classrooms). Judgments of alignment with inquiry principles were derived from video ratings of whole class lessons (Gearhart et al., 1999).

The study addressed three questions related to notation and reference. First, when students produce part-whole reference on Equal and Unequal Area problems, are they treating parts of area as discrete or continuous quantities? Second, do conventional notation and part-whole reference develop in concert, does one precede the other, or is there some independence? Third, for students who begin fractions instruction with similar notational approaches, how are their trajectories affected by participation in classrooms that stress inquiry as contrasted with classrooms that stress skills?

Data consisted of students' written notations based on four Equal Are problems (see fig. 17.2 a–d) and three Unequal Area problems (see fig. 17.2 e–g). Instructions to the students were: "For each picture below, write a fraction to show what part is gray." The data were collected just prior to initiation of instruction and again just after the completion of instruction.

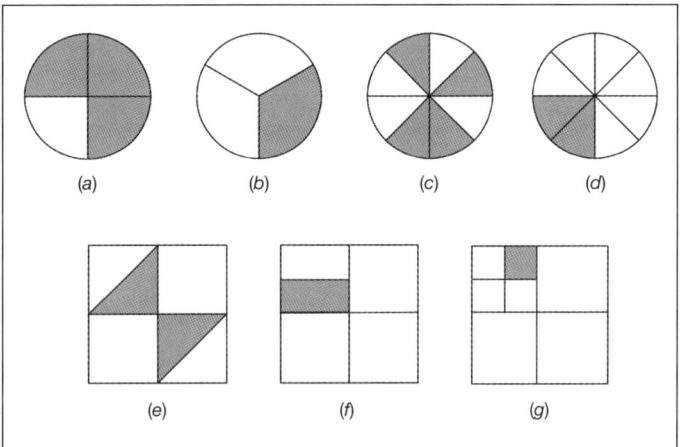

Fig. 17.2. Equal Area (a–d) and Unequal Area (e–g) problems

Table 17.1 contains brief descriptions and examples used in our coding scheme for the fraction problems shown in figure 17.2. For Equal Area problems, each answer received one of five codes: (1) Part-Whole (P-W); (2) Part-Part (P-P); (3) Integer (I); (4) Don't Know (dk) written as a response; and (5) Other (used for answers that did not match the number of gray or white parts or for which no numerical notation was given). For Unequal Area problems, the Part-Part, Integer, Don't Know, and Other categories were retained, but the Part-Whole category was separated into two types: (1) Part-Whole: Continuous (P-W:C) and (2) Part-Whole: Discrete (P-W:D). Part-Whole: Continuous reference accommodates the unequal area by treating area as a continuous quantity in a written representation, and thus the representation of the gray area of the Unequal Area problem example in table 17.1 would be $1/8$ or its fractional equivalent. In contrast, Part-Whole: Discrete reference treated the gray area as a discrete quantity, treating parts as equivalent regardless of size; the representation of the example square would be $^1/_5$ (or its equivalent), since 1 of 5 parts is shaded. We also coded students' responses for notation: (1) *conventional notation* was assigned to notations that contained a numerator and denominator separated by a horizontal line or a slash; (2) *unconventional notation* was assigned to answers such as single numbers and numbers separated by a hyphen, colon, or any mark other than a horizontal line or slash.

Table 17.1
Reference and notation codes for Equal Area and Unequal Area problems

Example of problem type	Reference code	Definition of reference code	Notation codes and sample representations	
			Unconventional	**Conventional**
Equal Area	Part-Whole (P-W)	Notation captures the number of gray parts and number of all parts in whole	3-4; 4-3; 3,4; 3 4; 6,8	$^3/_4$; $^6/_8$
	Part-Part (P-P)	Notation captures number of gray parts and number of white parts	3-1; 1-3; 3 1; 1 3	$^3/_1$
	Integer (I)	Notation captures the number of gray parts	3; 1	—
	Don't Know (dk)	"Don't Know" written as a response	—	—
	Other	None of above	7	$^7/_9$
Unequal Area	Part-Whole: Continuous (P-W:C)	Notation captures unequal size relations in the representation of part-whole relation	1-8; 8-1; 8 1; 8,1; 16,2; 2-16	$^1/_8$; $^2/_{16}$
	Part-Whole: Discrete (P-W:D)	Notation captures the number of gray parts and number of all parts in whole (without accommodating different size parts)	1-5; 5-1; 5 1; 1,5	$^1/_5$
	Part-Part (P-P)	Notation captures number of gray parts and number of white parts	1-4; 4 1; 4-1	$^1/_5$
	Integer (I)	Notation captures the number of gray parts	1; 4	—
	Don't Know (dk)	"Don't Know" written as a response	—	—
	Other	None of above	7	$^7/_9$

Following the coding, we assigned students to a category based on the consistency of their codes across tasks. For Equal Area problems, if students received the same code on at least three of four problems, they were assigned that code as their overall "Modal" code; all other students were assigned a code of No Modal. For the Unequal Area problems, we used a parallel procedure, but the criterion for the Modal categorization was consistency on at least two of the three problems.

Selected Results

Patterns of part-whole reference

We investigated how students who produced part-whole references on Equal Area problems interpreted the unequal-sized parts on the Unequal Area problems. We analyzed the pretest and posttest data separately. On the pretest, of those students who referred to part-whole relations on the Equal Area problems, many also did so on the Unequal Area problems but did not treat area as a continuous quantity in their notations: 25 percent treated Equal Area as Part-Whole but Unequal Area as discrete (P-W:D); only 9 percent interpreted Equal Area as part-whole and Unequal Area part-whole relations as continuous quantities (P-W:C); 35 percent treated only the Equal Area problems as part-whole (their responses to Unequal Area problems were Part-Part, Integers, or I don't know). This pattern was similar for posttest data, although there was substantial movement to part-whole reference: Almost all students produced part-whole relations for the Equal Area problems at posttest (92 percent), and 26 percent represented shaded parts in the Unequal Area part-whole relations as a continuous quantity (P-W:C).

Relations between notation and reference

To investigate whether students develop conventional notation and part-whole reference jointly, or whether there is some independence in the development of each, we produced a cross-tabulation of the use of part-whole reference and notation for both Equal Area and Unequal Area problems. The results in table 17.2 for the pretest data provide evidence of some independence between notation and reference. On Equal Area problems, 18 percent of the students used conventional notation without part-whole reference, and 7 percent used part-whole reference without conventional notation. On Unequal Area problems, 57 percent of the students used conventional notation without part-whole reference, and 4 percent used part-whole reference without conventional notation. A parallel analysis of the posttest data yielded similar patterns, and thus, across pretest and posttests, we found that a small to moderate percentage of students used either conventional notation *or* part-whole reference for both problem types.

Table 17.2

Percentage distribution of students' modal conventional and unconventional notation and reference categories based on pretest data

Notation	Reference	
	Not Part-Whole	**Part-Whole**
	Equal Area problems	
Unconventional	14	7
Conventional	18	61
	Unequal Area problems	
Unconventional	34	4
Conventional	57	5

Note: For the Unequal Area problems, the Part-Whole category includes only the Part-Whole: Continuous responses.

Additional longitudinal analyses focused on developmental relations between part-whole reference and conventional notation. If notation and reference influence one another in development, students' use of conventional notation *or* part-whole reference at pretest may predispose them to acquire the other at a more rapid rate than students who began instruction without using either conventional notation or part-whole reference. We illustrate one such analysis that we conducted to test these ideas.

We first identified students who had not created part-whole reference on either type of problem at the pretest; we then partitioned these students into two groups—those who did and those who did not use conventional notation on the pretest—and compared their posttest performance on the reference categories for Equal Area and Unequal Area problems. Figure 17.3 contains graphs of students' posttest reference categories based on their pretest notations category. The distributions of posttest reference categories are similar, with a large proportion (about 70 percent in each group) acquiring the use of part-whole relations only for the Equal Area problems; about 20 percent of each group (P-W, Both) used part-whole relations for Equal Area problems (P-W) and treated area as a continuous quantity for Unequal Area problems (P-W:C); and only a small percentage used no part-whole reference on the posttest (No Part-Whole). Thus this analysis yielded no evidence that accelerated acquisition of notational conventions advantaged students in their subsequent development of the part-whole reference.

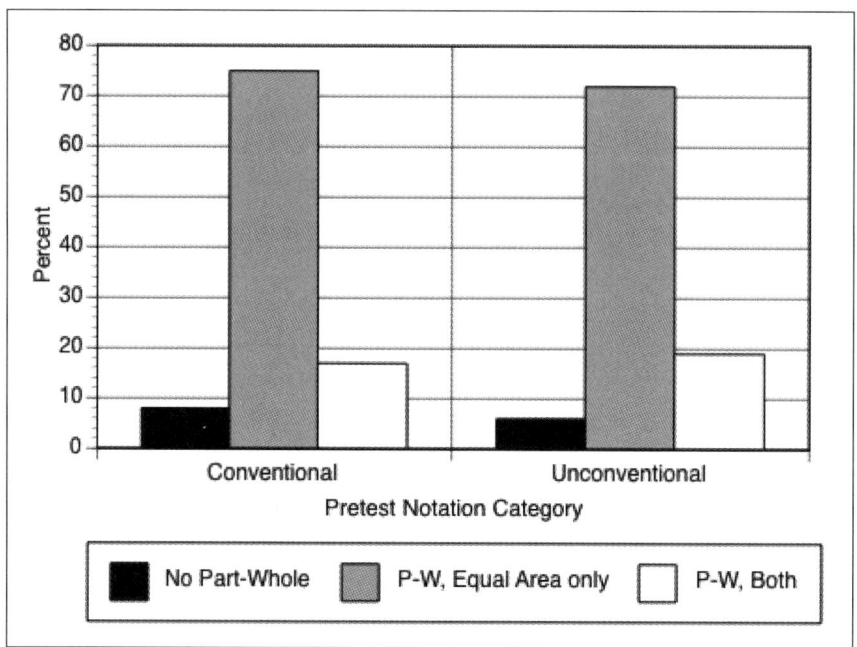

Fig. 17.3. Percentage distribution of students' use of part-whole reference categories at posttest based on notation used on the pretest

Trajectory of change from pretest to posttest focusing by type of classroom instruction

We conducted a final set of analyses to investigate students' trajectories of change from pretest to posttest as a function of different instructional conditions. Based on curricula and records of classroom activities, we classified the nineteen classrooms into one of three categories of instruction. The *Traditional* classification was assigned to seven classrooms in which students used textbooks and emphasis was on procedural skills; students were less likely to be engaged in conceptually rich discussions. The remaining twelve classrooms worked with an inquiry unit, *Seeing Fractions* (Corwin, Russell, & Tierney, 1991), which supported investigations of the quantitative meanings of standard notation in the context of area and linear models; we divided these classrooms into two categories of instruction, *High Inquiry* and *Low Inquiry*, based on evidence of students' engagement in more or less conceptually rich discussions.

We highlight here our findings regarding change over time with regard to performance on the Unequal Area problems, the most challenging problems for students in the study. We conducted two analyses of students' responses to these tasks, one for notation and one for reference. Results for the notation analysis indicated that classroom instruction (i.e., High Inquiry, Low Inquiry, or Traditional) was not related to shifts to conventional notation: In each classroom group, the proportions of students who shifted from unconventional to conventional notation were similar (67 percent in Traditional, 66 percent in Low Inquiry, and 71 percent in High Inquiry groups). Results for the reference analysis, however, revealed relationships between instruction and students' uses of reference categories (e.g., Part-Whole: Continuous, Part-Whole: Discrete, etc.). Regardless of students' uses of particular reference type at pretest, students in the High-Inquiry classroom group showed the greatest shift to the P-W:C form of reference. In addition, students in the Traditional classroom were far more likely to persist in their use of the P-W:D reference (62 percent of Traditional students) compared with students in the High Inquiry and Low Inquiry classrooms (40 percent and 36 percent, respectively). Stability in the Traditional students' use of Part-Whole: Discrete reference suggests that it provided them a coherent approach that was not challenged by instruction.

Implications for Instruction

Our research has two important implications for instruction. First, *tasks matter*. Non-routine tasks like Unequal Area problems are valuable methods for investigating and assessing student understanding. Second, *instruction matters*. The students in our study who progressed in their understandings of fractions notation and part-whole relations were supported by "high-inquiry" instruction focused on fractions concepts.

With regard to tasks, the inclusion of Unequal Area problems in our task set enabled us to contribute new findings on students' developing understanding of part-whole reference. We found that students who produced part-whole reference on Equal Area problems did not necessarily treat area as a continuous quantity in their notations on the Unequal Area problems. Thus if students disregarded size in their notations on the Unequal Area problems, their correct responses to Equal Area tasks were likely based on discrete counting models of part-whole relations. Inclusion of the nonroutine Unequal Area problems enabled us to identify properties of student understanding that would have remained hidden with only the routine Equal Area problems.

With regard to instruction, students in inquiry classrooms made greater progress than those in traditional classrooms in their representations of part-whole relations on the Unequal Area problems. This pattern was stronger in High Inquiry classrooms, where teachers were more likely to elicit students' thinking about fractions and engage students in mathematical discussions about part-whole relations. Our findings, furthermore, addressed a common concern about inquiry instruction—that it provides too little attention to facts and procedures. In fact, in the classrooms that were included in our sample, we found no difference between inquiry and traditional classrooms in the rate of improvement in students' performance on notational aspects of representation. Our findings were consistent with studies conducted from the 1980s through 2000 documenting that inquiry instruction, when compared with traditional instruction, is associated with greater gains in students' conceptual understanding and problem solving (e.g., Carpenter, Fennema, Peterson, Chiang, & Loef, 1989; Hiebert & Wearne, 1996; Riodan & Noyce, 2001; Saxe, Gearhart, & Nasir, 2001).

A "bottom line" message of our findings is that the traditional focus on the conventions of fractions notation is insufficient support for students' developing understandings of notation-reference relations. Our analyses of pretest to posttest shifts showed that students' use of conventional notation at pretest did not lead to any advantage in their development of part-whole reference at posttest, findings suggesting that teaching students notational skills may not support the development of an understanding of part-whole relations. As Mack (1995) and Ball (1993) argued nearly two decades ago and as we revealed through research conducted a decade ago, teachers need to support students in investigating relations between fraction reference and fraction notation in depth, and in multiple contexts. Across contexts, the "whole" may be a continuous or a discrete quantity, the continuous quantity may be a linear, area, or volume model, and part-whole relations for continuous quantities may be represented as equal or unequal parts. Tasks like Unequal Area problems can engage students with the core ideas of part-whole relations and appropriate fractions notation. Consistent with the National Council of Teachers of Mathematics Standards (2000) and now with the Common Core standards for mathematics (NGA Center & CCSSO, 2010), students need to make connections among mathematical representations both of the quantities and the notations for those quantities. And as teachers monitor their students' progress, they must recognize that an understanding of either notation or part-whole relations does not ensure understanding of the other.

References

Ball, D. L. (1993). Halves, pieces, and twoths: Constructing and using representational contexts in teaching fractions. In T. P. Carpenter & E. Fennema (Eds.), *Rational numbers: An integration of research* (pp. 157–195). Hillsdale, NJ: Lawrence Erlbaum.

Carpenter, T. P., Corbitt, M. K., Kepner, H. S., Jr., Lindquist, M. M., & Reys, R. E. (1980). National Assessment: A perspective of students' mastery of basic mathematics skills. In M. M. Lindquist (Ed.), *Selected issues in mathematics education* (pp. 215–227). Chicago, IL: National Society for the Study of Education; Reston, VA: National Council of Teachers of Mathematics.

Carpenter, T. P., Fennema, E., Peterson, P. L., Chiang, C. P., & Loef, M. (1989). Using knowledge of children's mathematics thinking in classroom teaching: An experimental study. *American Educational Research Journal, 26*, 499–531.

Corwin, R. B., Russell, S. J., & Tierney, C. (1991). *Seeing fractions: A unit for the upper elementary grades.* Sacramento: California Department of Education.

Gearhart, M., Saxe, G. B., Seltzer, M., Schlackman, J., Ching, C. C., Nasir, N., . . . Sloan, T. (1999). Opportunities to learn fractions in elementary mathematics classrooms. *Journal for Research in Mathematics Education, 30*, 286–315.

Hiebert, J. (1988). A theory of developing competence with written mathematical symbols. *Educational Studies in Mathematics, 19*, 333–355.

Hiebert, J. (1989). The struggle to link written symbols with understandings: An update. *Arithmetic Teacher, 36*, 38–44.

Hiebert, J., & Wearne, D. (1996). Instruction, understanding, and skill in multidigit addition and subtraction. *Cognition and Instruction, 14*, 251–283.

Mack, N. K. (1990). Learning fractions with understanding: Building on informal knowledge. *Journal for Research in Mathematics Education, 21*, 16–32.

Mack, N. K. (1995). Confounding whole-number and fraction concepts when building on informal knowledge. *Journal for Research in Mathematics Education, 26*, 422–444.

Moss, J., & Case, R. (1999). Developing children's understanding of the rational numbers: A new model and an experimental curriculum. *Journal for Research in Mathematics Education, 30*, 122–147.

National Council of Teachers of Mathematics. (2000). *Principles and standards for school mathematics.* Reston, VA: Author.

National Governors Association Center for Best Practices & Council of Chief State School Officers (NGA Center & CCSSO). (2010). *Common core state standards for mathematics.* Washington, DC: Author. Retrieved from http://www.corestandards.org.

Riordan, J. E., & Noyce, P. E. (2001). The impact of two standards-based mathematics curricula on student achievement in Massachusetts. *Journal for Research in Mathematics Education, 32,* 368–398.

Saxe, G. B., Gearhart, M., & Nasir, N. (2001). Enhancing students' understanding of mathematics: A study of three contrasting approaches to professional support. *Journal for Research in Teacher Education, 4,* 55–79.

Saxe, G. B, Taylor, E. V., McIntosh, C., & Gearhart, M. (2005). Representing fractions with standard notation: A developmental analysis. *Journal for Research in Mathematics Education, 36,* 137–157.

Partitioning and Iterating When Teaching and Learning Fraction Addition on Number Lines

Andrew Izsák, Erik Tillema, and Zelha Tunç-Pekkan
University of Georgia

The Common Core State Standards for Mathematics (CCSSM; National Governors Association Center for Best Practices & Council of Chief State School Officers [NGA Center & CCSSO], 2010) presented a perspective on teaching fractions that will be new for many teachers. One important difference is that the CCSSM presented a definition for fractions different from the part-whole meaning many teachers currently use with students (Norton et al., 2014). A second important difference is increased emphasis on number line representations of fractions and decreased emphasis on circle representations that are frequently found in instructional materials. These shifts will require teachers to rethink how they teach fractions and fraction arithmetic to their students.

We use the Candy Bar problem to illustrate the shift in meaning for fractions put forward by the CCSSM: Jamal and three of his friends are sharing one candy bar. Everyone gets the same amount. How much of the candy bar do Jamal's friends eat all together? Using the part-whole meaning for fractions, teachers might explain to their students that Jamal's friends eat 3 out of the 4 pieces, or $3/4$ of the candy bar. This meaning for fractions relies on counting the number of pieces in all and the number of pieces Jamal's friends eat. Thus, both counts are of the same thing: pieces of the candy bar.

In contrast, the CCSSM definition for fractions comes in two parts (NGA Center & CCSSO, 2010, p. 24). The first part says that just one share is $1/4$ of the candy bar, because all four equal-sized pieces create the whole candy bar. The second part says that 3 parts, each of which is $1/4$ of the candy bar, is $3/4$ of the candy bar. The critical difference is that the CCSSM definition first defines a unit fraction (a fraction whose numerator is 1) and then emphasizes combining copies of that unit fraction. As a consequence, the CCSSM definition makes explicit a distinction between the size of the parts, $1/4$ of the candy bar in this example, and the number of parts, 3. This distinction is not made as explicitly with the part-whole meaning discussed above. Attention to the size of parts is critical when comparing fractions and when performing arithmetic with fractions. A further advantage of the CCSSM definition is that it can support reasoning about improper fractions more readily than the part-whole definition: Interpreting $5/4$ of a candy bar as 5 out of 4 pieces does not make sense, but thinking of 5 pieces that are all the same size, $1/4$ of a candy bar, does, especially when fractions are represented as lengths. We explain this in the following paragraphs.

Representing fractions as lengths on number lines has not been common instructional practice in the United States but is attractive for many reasons. One reason is that using lengths that start at zero emphasizes the fact that fractions like $1/4$ are single numbers, not

Note: Tillema is now at Indiana University–Purdue University, Indianapolis; Tunç-Pekkan is at MEF University, Turkey.

This chapter is adapted from A. Izsák, E. Tillema, & Z. Tunç-Pekkan (2008), Teaching and learning fraction addition on number lines, *Journal for Research in Mathematics Education, 39,* 33–62.

pairs of whole numbers like 3 and 4. A second reason is that number lines provide a flexible medium with which to represent individual fractions and to develop meanings and numerical methods for all four arithmetic operations. A third reason is that a solid understanding of number lines that includes fractions provides an important foundation for subsequent topics, such as Cartesian graphs.

Although there are reasons for using lengths on number lines in fractions instruction, many U.S. teachers know that their students have trouble making sense of number lines, and some research also documents students' difficulties with fractions on number lines. For example, Tunç-Pekkan (2015) found that a sample of 656 U.S. fourth- and fifth-grade students were much more proficient using circles and rectangles to represent fractions than using number lines: 80 percent could partition circles and rectangles to show the fractions $3/4$ and $5/6$, respectively, but only 35 percent could partition a unit interval on the number line to locate $2/3$. These findings may reflect U.S. students' limited opportunities to work with lengths on number lines rather than persistent difficulties they might experience after instruction focused on number lines.

Evidence that students can learn to use number lines effectively can be found in international comparisons of student achievement (e.g., Gonzales et al., 2008). In these comparisons, students from Asian countries outperform U.S. students. Furthermore, students from several Asian countries experience systematic development of length-based representations of numbers, first with whole numbers and then with fractions, beginning in early elementary grades and continuing into upper elementary grades (e.g., Tokyo Shoseki, 2006). Furthermore, recent research has demonstrated that U.S. students can make significant gains in their understandings of integers and fractions as lengths or distances on number lines (e.g., Saxe, Diakow, & Gearhart, 2013) when offered instruction specifically designed to support such understandings.

If the CCSSM emphasizes treating fractions as lengths on number lines, and students in some countries that incorporate systematic development of length-based representations attain high levels of achievement relative to students in the United States, then it is natural to ask, How can we help students represent fractions on number lines in ways consistent with the CCSSM? In this chapter, we present some key insights from a study conducted by Izsák, Tillema, and Tunç-Pekkan (2008) for answering this question. The insights highlight the importance of how teachers discuss partitioning and iterating lengths. Partitioning involves subdividing a length into equal-sized sublengths. Researchers have long considered partitioning to be a key understanding in the domain of fractions, but one that is not straightforward for students (e.g., Kieren, 1980; Pothier & Sawada, 1983). Iterating involves concatenating copies of fixed length to create longer and longer lengths. As an example, one could start with one length of $1/4$, join a second length of $1/4$ to create a length of $2/4$, a third to create a length of $3/4$, a fourth to create a length of $4/4$, a fifth to relate a length of $5/4$, and so on. Thus, partitioning and iterating lengths can be used to make sense of both proper and improper fractions. In our *JRME* article (Izsák, Tillema, & Tunç-Pekkan, 2008), we examined ways that iterating and partitioning were important to developing students' understanding of fractions as lengths on a number line in one classroom. In the remainder of this chapter we summarize the study and derive three recommendations for teachers.

Ms. Reese Teaches Fraction Addition

In our *JRME* article, we examined how one experienced sixth-grade teacher, Ms. Reese, and her students (all names of teachers and students used here are pseudonyms) interpreted lessons about fractions and fraction addition in which they participated together. Although the study predated the release of the CCSSM by several years, Ms. Reese's instruction used number lines and emphasized partitioning lengths and iterating unit fractions. The lessons were based on a draft revision of the *Bits and Pieces II* unit from the Connected Mathematics Project (CMP) (Lappan, Fey, Fitzgerald, Friel, & Phillips, 2003). A key result from the study was that *how* Ms. Reese generated partitioned number lines in her demonstrated solutions to fraction addition problems had significant consequences for how her students made sense of the lessons.

The following solution to $1/4 + 1/8 = 3/8$ illustrates a consistent sequence of three steps that Ms. Reese and her students discussed for solving fraction addition problems. These three steps summarize Ms. Reese's practice for teaching fraction addition, which in many ways was consistent with the CCSSM. The first step was to determine how many whole numbers to represent. Ms. Reese and her students had worked on estimating sums of fractions in earlier lessons. In the present example, both addends were less than $1/2$, so their sum would be less than 1. Thus, for the current example, the interval from 0 and 1 was needed. In another case, students might estimate that a sum should be between 2 and 3. In such cases, intervals from 0 to larger whole numbers were needed. The locations of whole numbers were benchmarks that guided the subsequent location of fractions on the number line.

The second step was to partition unit intervals (intervals of length one) created in the first step. This required thought in the example of $1/4 + 1/8$, because fourths and eighths are different-sized pieces. Ms. Reese traced the interval from 0 to 1 with her finger and asked her students how to "divide up this amount." Her gestures were consistent with focusing on fractions as lengths. One student suggested a half; another suggested eighths. Ms. Reese took the second suggestion, saying that she needed "eight pieces that look about the same." She made seven tick marks from left to right (fig. 18.1a), labeling them $1/8, 2/8, \ldots, 7/8$ (fig. 18.1b).

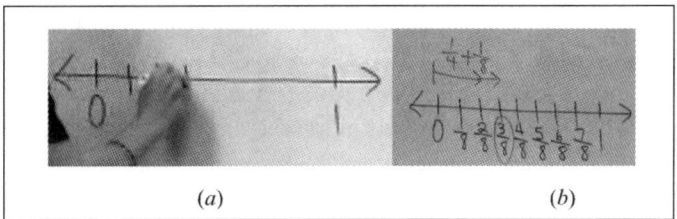

Fig. 18.1. Steps 2 and 3 in Ms. Reese's demonstrated method for adding fractions on number lines: (a) partition unit intervals; (b) draw arrows for each addend and circle the answer

The third step was to draw arrows for each addend and circle the sum (fig. 18.1b). Ms. Reese told the class that she could not see $1/4$ on the number line, but a student pointed out that $2/8$ was the same. Ms. Reese agreed and drew one arrow for each addend as she spoke: "OK. So start at zero and go over to two eighths and then go over one eighth more." Notice that this explanation is based on iterating unit fractions and thus is consistent with representing fractions as lengths on number lines and with the CCSSM definition for fractions.

Across examples, Ms. Reese consistently partitioned unit intervals by adding tick marks from left to right. She was skilled at estimating spacing so that the final tick mark corresponded to the 1. In cases where her final subinterval was slightly off, Ms. Reese moved the location of the 1 to create equal-length subintervals. To illustrate, during her solution to $2/5 + 1/10 = 5/10$, Ms. Reese pointed out that she was moving the 1 and told students:

Ms. Reese: I started putting pieces in here just eyeballing it and trying to space them out, and I didn't have enough spaces, so I moved the one over and made another mark because I need 10 spaces. . . . This is a space *[pointing with thumb and index finger]* and you need 10 of those to be called 10th-size pieces.

Although Ms. Reese consistently counted spaces, focusing on lengths, we learned that adjusting the location of the 1 had unintentional consequences for some of her students. Thus, deciding whether or not to move the 1 is an important piece of expertise when using lengths on number lines to represent fractions for students.

How Students Interpreted Ms. Reese's Instruction

As part of the study, the first author conducted sequences of interviews with pairs of students from Ms. Reese's class. Ms. Reese helped identify students with a range of success but who were not exceptional. During the interviews, the first author asked the students to work tasks similar to those used in the *Bits and Pieces II* lessons and to interpret short excerpts of Ms. Reese's video recorded lessons.

From the interviews we learned that students had different interpretations of moving the 1 when "eyeballing" was a little off. Students who had a strong understanding of a fixed whole unit interpreted the lesson in ways similar to Ms. Reese's intentions, as just a convenient adjustment so the tick marks did not have to be erased and redrawn. Students who did not have a strong understanding of a fixed whole unit, however, interpreted the lessons in ways that Ms. Reese did not intend.

Sonya was one student who interpreted the lessons in ways Ms. Reese did not intend. Ms. Reese identified Sonya and her interview partner, Jenny, as students who were often confused. During their third interview, Sonya and Jenny represented $2/3 + 3/4$ using a number line. Sonya's approach paralleled the three-step method that Ms. Reese had demonstrated several times by this point. First, Sonya drew a number line and put "0" on the left hand end, "1" in the middle, and "2" on the right hand end. Second she re-expressed $2/3 + 3/4$ first as $4/6 + 6/8$ and then as $8/12 + 12/16$. Thus, she doubled numerators and denominators to generate correct equivalent fractions but did not make progress toward common denominators. When asked to use the denominator of 12, the students generated $8/12$ and $9/12$ quickly. Sonya put 12 tick marks from left to right between 0 and 1, and Jenny did the same between 1 and 2. Thus, the students made the common error of placing one too many tick marks in each unit interval, indicating they were counting the tick marks rather than attending to the length. Sonya proceeded to label the tick marks, resulting in a number line that showed $12/12$ and 1 in two separate locations. (Sonya's original 1 is the longer tick mark labeled $13/12$ in figure 18.2a.) Third, the students drew arrows to represent the sum, although did so incorrectly.

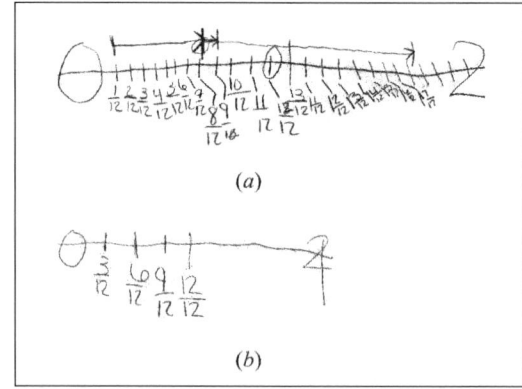

(a)

(b)

Fig. 18.2. (a) Sonya's number line when benchmarks are estimated locations; (b) Sonya's number line when benchmarks are exact locations

When the interviewer asked about the locations of 1 and $12/12$, Sonya explained that $12/12$ and 1 were the same because they were equal. The interviewer asked, "Does it make a difference if you have two different places where you've marked the same thing?" Sonya replied,

"No. Because both of them equals 1" and explained, "When you draw your number line you have to put 12 of them in each space because that is what the denominator is."

The interviewer then showed lesson video in which Ms. Reese demonstrated $1/4 + 1/8$ on the number line (see fig. 18.1). Sonya noticed that Ms. Reese collocated 1 and $8/8$. (During the lesson excerpt Ms. Reese asked where 8 eighths was on the number line, and several students commented, "On the 1.") When the interviewer returned the students' attention to their work and asked again about $12/12$ and 1 being in two different places, Sonya explained:

Sonya: You are trying, on the 0, 1, and 2 *[pointed to "0," "1," and "2"]*, you are trying to put where 8 over 12 and 9 over 12, *you are trying to guess* where, *like a estimate*, you trying to put where it goes here, and then when I drew these *[pointed to 12ths tick marks]*, when she drew these, you are like telling, *you know where 12 over 12 is at. [emphasis added]*

Sonya's explanation was strikingly similar to those cases where Ms. Reese adjusted the location of the 1, but instead of erasing the original 1, the students now had two 1s. Although at earlier moments during her interviews Sonya placed equivalent fractions at the same location and connected fractions of the form *n/n* with the whole unit, we saw no signs that separate locations for the 1 and $12/12$ created a contradiction for her here.

From the interviews, we learned not only that Ms. Reese's adjustment of the 1 unintentionally undermined Sonya's success, but also that Sonya could be considerably more successful with a seemingly small adjustment in instruction. When the interviewer asked the students to draw a new number line thinking of the original location of the 1 as exact, not an estimate, Sonya looked much more proficient. She drew a new number line with "0," "1," and "2" labels, added a "$12/12$" label under the "1," added a tick mark for $6/12$, and explained that "half of 12 is 6" (fig. 18.2b). She then located $3/12$ by partitioning the interval between 0 and $6/12$ in half and similarly located $9/12$ by partitioning the interval between $6/12$ and $12/12$ in half. Sonya continued to fill in additional tick marks, explaining that she could add two tick marks within each fourth to locate "7, and then 8, and then 9, and then 10, 11." Here her work indicated that she was focused on partitioning the fixed length from 0 to 1 into equal pieces. In addition, she accomplished her partition in stages: She first partitioned the length from 0 to 1 in half to locate $6/12$, then partitioned each half in half to locate $3/12$ and $9/12$, and then further partitioned the resulting fourths.

Discussion

A central question that Sonya's difficulties raise is, What can teachers do to strengthen students' understanding and use of number line representations for fractions and fraction addition? We present three recommendations. The first two recommendations are tied directly to the case of Ms. Reese, Sonya, and Jenny and may be more familiar to teachers than the last. The third recommendation reflects broader considerations when using number lines to support instruction in fractions.

The first recommendation is not to change the size of the whole (i.e., the 1) after it has been established on the number line. Like Sonya, many students have an emerging understanding that the whole should not be changed when solving fraction problems. Small actions that seem inconsequential to teachers can throw off students unintentionally. Students who had a strong understanding that the whole had to remain fixed in fraction addition problems did not misinterpret Ms. Reese when she adjusted the location of the 1. Sonya, however, needed more support than other students to maintain a fixed whole when working with a number line. In particular, she benefited from explicit instruction that the location of 1 should not be changed during the solution of a problem.

A second recommendation is to have explicit discussions with students about partitioning intervals into equal-length subintervals, because how to partition lengths on number lines is not self-evident to students, especially when they focus on counting tick marks instead of spaces. The case of Ms. Reese and Sonya demonstrates that how teachers produce number line representations can have significant consequences for students' attention to partitioning a fixed interval. Although Ms. Reese focused on equal-sized sublengths, Sonya interpreted left to right markings as a count of the number of tick marks (e.g., $^{12}/_{12}$ meant create 12 tick marks from left to right, not 12 equal-sized pieces). One instructional practice that can help students establish a fixed interval partitioned into sublengths is to have them partition a fixed whole on the number line and adjust the size of the pieces they create within that fixed length, rather than adjusting the whole itself. This activity focuses students' attention on the fact that the size of the whole should not change, and that the goal of the activity is to partition and make equal-sized pieces.

A third recommendation that builds on the first two, and that may be less familiar to teachers, is to provide students with experiences of partitioning a partition. By partitioning a partition, we mean first partitioning a fixed interval into a certain number of equal-sized pieces, and then partitioning each of those equal-sized pieces into still smaller equal-sized pieces. This recommendation supports the first recommendation discussed above, especially when partitioning an interval into a larger number of subintervals. Understandably, Ms. Reese had trouble "eyeballing" tenth-sized pieces when partitioning from left to right. An easier way to partition would have been to partition a fixed whole into halves first and then to partition each of those halves into five equal-sized pieces. Notice that such an approach might well have made sense to a struggling student like Sonya: She demonstrated some capacity to partition in stages when she first partitioned a whole into two equal-sized pieces, labeling her tick mark $^6/_{12}$, then partitioned each half into two more equal-sized pieces, labeling her new tick marks $^3/_{12}$ and $^9/_{12}$, and finally partitioning the resulting fourths (see figure 18.2b). Such activity can focus students' attention on equal-sized sublengths and provides opportunities to think about multiplication factors. To illustrate, Sonya's method creates opportunities to talk about two groups, each with six equal-sized pieces that create the whole unit ($2 \times 6 = 12$), and four groups, each with three equal-sized pieces that create the whole unit ($4 \times 3 = 12$). Connections between multiplication and partitioning can support understandings of equivalent factions, as discussed next.

By partitioning a partition students can use number lines to establish equivalent fractions (see Steffe, 2003; Steffe & Olive, 2010). To illustrate, Sonya created $^6/_{12}$ by partitioning the unit interval into two equal parts, and a discussion about equivalent fractions could follow—$^1/_2$ and $^6/_{12}$ name the same length, $^1/_4$ and $^3/_{12}$ name the same length, and $^3/_4$ and $^9/_{12}$ name the same length. In particular, students could begin by partitioning the whole into two halves and then continue by subdividing each of the halves in half again, creating four equal-sized pieces that make up the whole, or $^1/_4$. Using the fact that $4 \times 3 = 12$, students could partition each of the fourths into three equal-sized pieces. Then three of the twelfths are the same length as one of the fourths. This kind of activity and discussion can help students (a) understand how whole-number multiplication is involved in creating equivalent fractions, (b) use fixed lengths as the basis for understanding numerical notations for equivalent fractions, and (c) connect symbolic and length-based representations of number. Such understandings can form the foundation for studying subsequent topics, like multiplication with fractions, which is different than multiplication with whole numbers because it is based on equal-sized pieces that are smaller than the original whole.

The study reported in this chapter, and our recommendations that follow from it, illustrate subtle nuances that are important for successful instruction with visual representations. That is, how teachers create and use such representations in classrooms can have significant consequences for the way that students interpret and understand instruction. These consequences can either help or hinder students' future understanding. While Ms. Reese was certainly aware of some of the issues we raised in this paper (e.g., she focused her instruction on creating equal subintervals, not on counting tick marks), she may not have been aware how entrenched these interpretations can be for some students (e.g., counting tick marks). Therefore, it is essential for teachers to engage in instructional practices that provide opportunities for such students to continue working on these issues as they develop a more solid understanding of fractions and fraction addition. Careful attention to such issues, we think, is an important part of implementing the CCSSM successfully.

References

Gonzales, P., Williams, T., Jocelyn, L., Roey, S., Kastberg, D., & Brenwald, S. (2008). *Highlights from TIMSS 2007: Mathematics and science achievement of U.S. fourth- and eighth-grade students in an international context* (NCES 2009-001 Revised). Washington DC: National Center for Educational Statistics, Institute of Education Sciences, U.S. Department of Education.

Izsák, A., Tillema, E., & Tunç-Pekkan, Z. (2008). Teaching and learning fraction addition on number lines. *Journal for Research in Mathematics Education, 39*(1), 33–62.

Kieren, T. E. (1980). The rational number construct: Its elements and mechanisms. In T. E. Kieren, *Recent research on number learning* (pp. 125–150). Columbus, OH: ERIC/SMEAC.

Lappan, G., Fey, J. T., Fitzgerald, W., Friel, S. N., & Phillips, E. D. (2003). *Bits and pieces II: Using rational numbers* [draft of 2nd ed.]. Glenview, IL: Prentice Hall.

National Governors Association Center for Best Practices & Council of Chief State School Officers (NGA Center & CCSSO). (2010). *Common core state standards for mathematics*. Washington, DC: Author. Retrieved from www.corestandards.org.

Norton, A., Wilkins, J., Evans, M. A., Deater-Deckard, K., Balci, O., & Chang, M. (2014). Transcending part-whole conceptions of fractions. *Mathematics Teaching in the Middle School, 19*(6), 352–358.

Pothier, Y., & Sawada, D. (1983). Partitioning: The emergence of rational number ideas in young children. *Journal for Research in Mathematics Education, 14*(5), 307–317.

Saxe, G., Diakow, R., & Gearhart, M. (2013). Towards curricular coherence in integers and fractions: A study of the efficacy of a lesson sequence that uses the number line as the principal representational context. *ZDM: The International Journal on Mathematics Education, 45*(3), 343–364.

Steffe, L. (2003). Fractional commensurate, composition, and adding schemes learning trajectories of Jason and Laura: Grade 5. *Journal of Mathematical Behavior, 22*(3), 237–295.

Steffe, L., & Olive, J. (2010). *Children's fractional knowledge*. New York, NY: Springer.

Tokyo Shoseki (2006). *Mathematics for elementary school*. Tokyo, Japan: Author.

Tunç-Pekkan, Z. (2015). An analysis of elementary school children's fractional knowledge depicted with circle, rectangle, and number line representations. *Educational Studies in Mathematics, 89*(3), 419–441.

Developing Understanding of Multiplication of Fractions by Putting Research into Practice

Nancy K. Mack
Grand Valley State University

Take a moment to consider how your students would think about and solve the following problem.

> The fourth-grade class conducted a survey to learn how students get to our school every day. The results of the survey showed that two-thirds of the students get to school by riding in a vehicle. The results also showed that three-fourths of the students who ride in a vehicle ride the bus to school. What fraction of all the students at our school ride the bus to school?

Would the students understand the situation is a multiplication of fractions problem because it involves finding a portion (three-fourths) of a part (two-thirds) of one whole (all the students at our school)? Would the students comprehend that the "two-thirds" and the "three-fourths" refer to different wholes, or units? Would they partition each unit in an appropriate way and successfully solve the problem in a manner similar to one of the strategies shown in figure 19.1? (Note: These strategies are representative of ones my students have commonly used to solve this problem.)

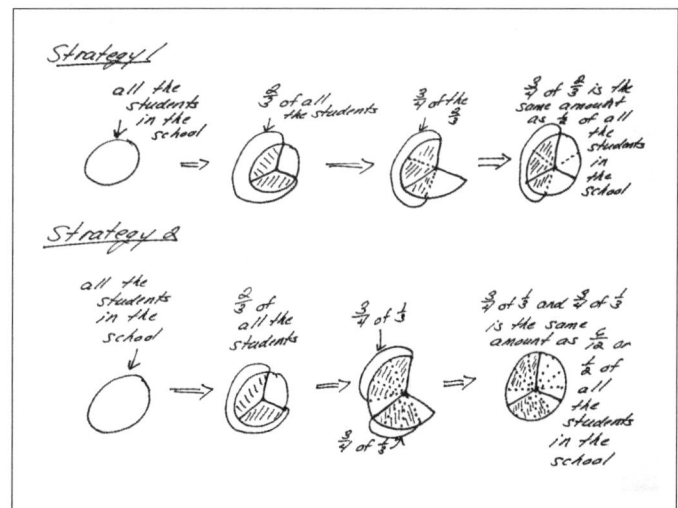

Fig. 19.1. Two strategies for finding three-fourths of two-thirds of all the students in the school

This chapter is adapted from N. K. Mack (2001), Building on informal knowledge in a complex content domain: Partitioning, units, and understanding multiplication of fractions, *Journal for Research in Mathematics Education*, 32, 267–295, and N. K. Mack (1998), Building a foundation for understanding the multiplication of fractions, *Teaching Children Mathematics*, 5, 34–38. The *TCM* article is available at nctm.org/more4u.

Multiplication of fractions is an important mathematical content area for intermediate and middle grades students to understand because it helps provide a foundation for proportional thinking (Behr, Harel, Post, & Lesh, 1992, 1994; National Governors Association Center for Best Practices & Council of Chief State School Officers [NGA Center & CCSSO], 2010). However, understanding multiplication of fractions can be quite challenging for students because this area may require students to think about units, or wholes, and the operation of multiplication in ways that differ from how they previously considered these ideas when operating on whole numbers and fractions. More specifically, understanding multiplication of fractions requires students to think flexibly about units and realize that a unit is not always one complete circle or rectangular strip. Students need to realize that quantities such as two-thirds of a circle, five-eighths of a rectangular region, and three-fourths of a set can also be considered a unit. Additionally, students need to comprehend that more than one unit is involved in multiplication of fraction problems, such as in the above problem where "all the students in the school" was the unit for the two-thirds of the students who ride in a vehicle and the two-thirds was the unit for the three-fourths of the students who ride a bus to school (Behr et al., 1992, 1994; Olive, 1999; Steffe, 1988). Furthermore, students need to view the operation of multiplication from a general perspective as a combining of groups of equal quantities. This view helps students understand that multiplication of fractions can be interpreted as "finding a portion of a part of a whole" (Behr, Harel, Post, & Lesh, 1993; Mack 1998, 2001). As shown in strategy 2 of figure 19.1, groups of equal fractional quantities (three-fourths of one-third) were combined to determine what fraction of all students ride the bus to school.

How can we help students comprehend these important mathematical ideas and develop a conceptual understanding of multiplication of fractions? A few years ago, I investigated this question through a research study that examined the development of students' understanding of multiplication of fractions during instruction. The study focused specifically on how fifth-grade students used their informal knowledge of fractions and their knowledge of partitioning quantities to solve word problems involving finding a portion of a part of a whole. Results of the study showed that understanding and solving multiplication of fraction word problems was not always simple and straightforward for the students. However, as students became more flexible in their thinking about units and meanings of fractions, they identified different units involved in problems and successfully solved problems by partitioning units in appropriate ways. Please see Mack (1998, 2001) for in-depth reports of this study. (Mack, 1998, is available at nctm.org/more4u.)

Since completing the multiplication of fractions study, I have applied results of this study to my own practice to help grades 4–6 students and preservice elementary teachers develop a conceptual understanding of multiplication of fractions. In this chapter, I discuss two results I applied that have been particularly effective in helping students think flexibly about units and about ways to partition units to solve multiplication of fraction word problems in meaningful ways.

Making Units in Problems Explicit

Results of the multiplication of fractions study suggested one primary factor that helped the students think deeply and flexibly about units when solving multiplication of fractions word problems was the way the problems were worded. When problems were worded in a manner where the unit for the multiplying fraction was implied rather than explicitly stated, students viewed the problems as being ambiguous and were unsure how to approach the problem. For example, when a problem was stated as

> You have three-fourths of a pizza. You give one-third to a friend. How much pizza did you give your friend?

students asked questions such as "Do you mean one-third of the whole pizza or one-third of three-fourths?" or they commented in a manner similar to Sam, who said, "This is confusing. I don't understand."

However, when problems were worded in a way that clearly stated the unit for each fraction, students correctly identified all the different units involved in the problem and attempted to solve the problem by partitioning each unit. The following protocol illustrates this manner of wording problems and how students solved problems. The protocol also shows how the student, Lee, readily comprehended that "four-fifths" referred to one whole cake and "one-fourth" referred to the four-fifths. Please note that although the problem context appears contrived, Lee himself suggested this context after viewing photos from a recent trip to the zoo.

NKM: During your trip to the zoo, you have four-fifths of the cake left to feed the camel [after feeding the meerkats one-fifth of the cake]. The camel is not very hungry. He is only going to eat one-fourth of what you have to feed him. How much does the camel eat of the whole cake?

Lee: *[Drew a circle and partitioned it into five equal-sized parts by drawing five radii one at a time. Put a dot on the one part fed to the meerkats. See fig. 19.2.]* That one *[points to one of the four unmarked pieces].* That's one-fifth *[of the whole cake]* . . . 'Cause there's five of these [pieces in the whole cake], five-fifths, and I gave one to him of these four there *[indicated the four unmarked pieces]* . . . You said one-fourth, so I need fourths. I need four pieces, and it's already cut into four, so that's four-fourths.

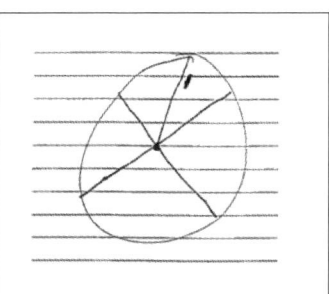

Fig. 19.2. Lee's strategy for finding one-fourth of four-fifths of a chocolate cake

After observing how the wording of problems influenced the research study students' understanding of the different units involved in problems, I began intentionally wording multiplication of fraction problems in ways that made the unit for each fraction explicit whenever I worked with grades 4–6 students or preservice teachers on this content area. The "What fraction ride the bus?" problem at the beginning of this chapter is an example of how I worded problems.

When I worded problems in this way during my own practice, students approached multiplication of fraction problems in ways that were similar to how the research study students approached solving the problems. Students first clearly identified appropriate units in the problem. They then partitioned each different unit in an appropriate way and stated the answer in terms of the original unit. This can be seen in strategies that two preservice teachers (India and Jacob) used for problems involving finding three-eighths of four-fifths of one whole garden (see fig. 19.3). Both preservice teachers first depicted four-fifths of the whole garden. They then clearly showed four-fifths was the unit they were partitioning to find the three-eighths. Last, their strategies showed how they stated the answer (three-tenths) in terms of the unit from which the four-fifths was derived.

Structure and Sequencing of Multiplication of Fraction Word Problems

Results of the multiplication of fractions study suggested another primary factor that influenced the students' ability to solve problems was the structure and sequencing of word problems that involved multiplying two proper fractions, $a/_b \times c/_d$ where $a < b$ and $c < d$. In brief, students did not view all problems involving the multiplication of two proper fractions as being the same nor did they solve all of these problems with the same ease. The numbers used in the problems for the multiplier, $a/_b$, and for the multiplicand, $c/_d$, appeared to influence the students' strategies and their thinking about multiplication of fractions.

Immediately prior to solving problems involving multiplying two proper fractions, the students solved equal-sharing problems and problems involving finding a fraction of a whole number amount, such as share ten cookies among four people and find one-third of twelve cookies, respectively. These two types of problems were used to help students begin to think broadly and flexibly about the various units involved in problems and what it means in general to partition a unit or quantity into a given fractional amount.

Students readily drew on their strategies for solving equal-sharing problems and problems involving finding a fraction of a whole number amount to solve word problems corresponding to $1/2 \times 1/2$ and $1/4 \times 1/2$. They solved these problems by identifying the different units in the problem: one whole "x" is the unit for $1/2$, and $1/2$ is the unit for the multiplier, $1/2$ or $1/4$. They also considered what it means to partition any unit or quantity into halves or fourths. For example, Sara solved a problem involving giving a friend one-fourth of one-half of a whole cookie. As shown in figure 19.4, Sara first drew one-half of a cookie and partitioned it into four equal-sized parts. She then related one of the four parts to the whole cookie and determined that she gave her friend one-eighth of the whole cookie.

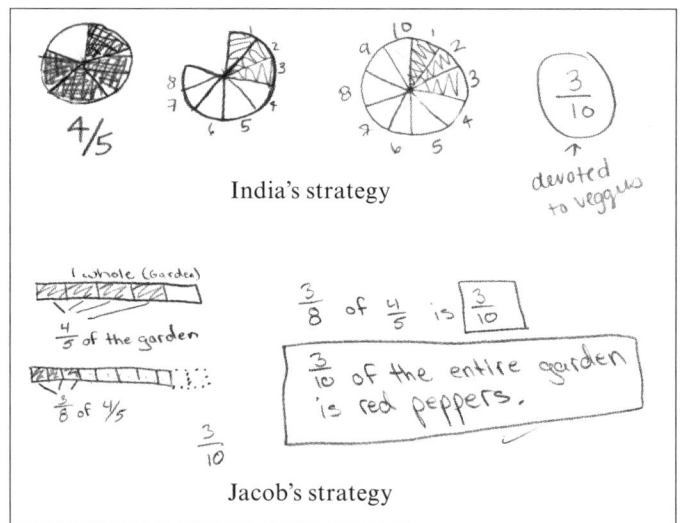

Fig. 19.3. India's and Jacob's strategies for finding three-eighths of four-fifths of one whole garden

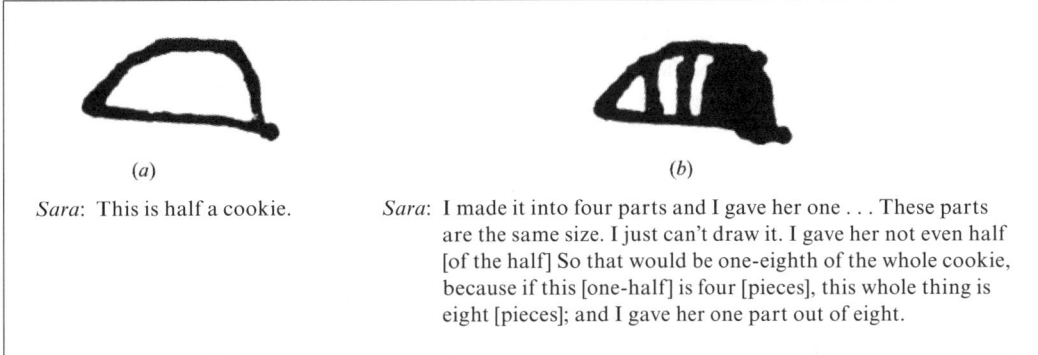

(a)

Sara: This is half a cookie.

(b)

Sara: I made it into four parts and I gave her one . . . These parts are the same size. I just can't draw it. I gave her not even half [of the half] So that would be one-eighth of the whole cookie, because if this [one-half] is four [pieces], this whole thing is eight [pieces]; and I gave her one part out of eight.

Fig. 19.4. Sara's strategy for finding one-fourth of one-half of one whole cookie

Other problems involving the multiplication of two proper fractions were more challenging for the students to solve. Students' ability to successfully solve word problems corresponding to $a/_b \times c/_d$ appeared to be related to numerical relationships the students perceived between the denominator of the multiplier (b) and the numerator of the multiplicand (c). Students' strategies also appeared to be influenced by the idea that any unit could be partitioned into a fractional amount and how the unit could be partitioned, such as how a unit of three-fourths could be partitioned into sixths. Table 19.1 summarizes the different numerical relationships the students perceived, the special cases of multiplication of fraction problems these numerical relationships created, and important ideas students needed to comprehend about units and partitioning units to successfully solve problems corresponding to each special case. The cases are presented in order of increasing difficulty for the students.

Problems corresponding to $a/_b \times b/_d$, where the denominator of the multiplier was the same as the numerator of the multiplicand, such as $1/4 \times 4/5$, were within students' reach after they solved problems involving finding one-fourth or one-half of one-half of one whole. Problems corresponding to $a/_b \times b/_d$ began to challenge students to realize that quantities other than one-half and one whole could serve as a unit and that these quantities could be partitioned. These problems also encouraged students to extend their thinking about what it means in general to partition any unit into a fractional amount, such as thirds, fourths, and so on. As students focused on these ideas, they realized the fractional unit was already partitioned appropriately in problems corresponding to $a/_b \times b/_d$. Thus, the students did not partition the fractional unit to solve the problem. This was previously shown in Lee's strategy for solving a problem involving finding one-fourth of four-fifths of a chocolate cake when Lee saw the four-fifths as a unit that was already partitioned into fourths.

As students considered what it means to partition any unit into a fractional amount, such as partitioning a unit of two-thirds into fourths, they began to realize they could repartition a fractional unit. Consequently, they were able to solve problems corresponding to $a/_{nb} \times b/_d$ where the denominator of the multiplier was a multiple of the numerator of the multiplicand, such as finding three-fourths of two-thirds of all the students in the school. Students viewed these problems as being different from ones they previously solved but often did not realize on their own that it was okay to repartition a fractional unit other than one-half when solving a problem. When students first tried to solve these problems, they commented in a manner similar to the following: "I don't know how to do this. It's already cut into two, but I need four pieces." I tried to help the students by asking them to think about how they had previously solved equal-sharing problems and other multiplication of fraction problems. For example, I asked questions such as these: "Remember when we shared cookies, did each person always have to get at least one whole cookie?"; "What did we do if each person did not get a whole cookie?"; and "If you need fourths of anything, how many pieces does your unit need to be partitioned into?" As the students considered my questions and the strategies they previously used when solving problems, they quickly realized they could repartition a fractional unit and solved the problems by using strategies similar to those shown in figures 19.1 and 19.3.

Table 19.1
Students' views of special cases of multiplication and fraction problems

Multiplication of fractions—algebraic representation	Numerical examples	Corresponding contextual situation example	Important ideas students need to comprehend about units and partitioning different units
$\frac{1}{b} \times \frac{1}{2}$	$\frac{1}{2} \times \frac{1}{2}$, $\frac{1}{4} \times \frac{1}{2}$	One-half of all the students in our class have on red shirts today. One-fourth of the students with red shirts have buttons on their shirts. What fraction of the whole class is wearing red shirts with buttons today?	• There can be more than one unit in a problem. • A quantity other than one whole can serve as a unit. • What it means in general to partition a unit into a fractional amount of one-half or one-fourth.
$\frac{a}{b} \times \frac{b}{d}$	$\frac{3}{5} \times \frac{5}{8}$ and $\frac{2}{3} \times \frac{3}{4}$	Three-fourths of the students in our class brought a sack lunch today. Two-thirds of the people who brought a sack lunch brought a banana in their lunch. What fraction of all the students in our class brought a banana for lunch today?	• What it means in general to partition a fractional unit other than one-half into any fractional amount, such as halves, thirds, fourths, fifths, and so on. • View $\frac{b}{d}$ as a unit that is already partitioned into the needed fractional amount, or b pieces.
$\frac{a}{nb} \times \frac{b}{d}$	$\frac{3}{4} \times \frac{2}{5}$ and $\frac{3}{4} \times \frac{2}{3}$	Our class planted a garden. We planted peppers in two-fifths of the whole garden. Three-fourths of the peppers we planted were red peppers. How much of the whole garden was planted with red peppers?	• View $\frac{b}{d}$ as a unit that can be repartitioned into a desired fractional amount. • Realize $\frac{b}{d}$ can be repartitioned by thinking b times what other number equals nb.
$\frac{a}{b} \times \frac{nb}{d}$	$\frac{2}{3} \times \frac{9}{11}$ and $\frac{2}{3} \times \frac{9}{10}$	Nine-elevenths of all the students in our school voted in the election for the school mascot. Two-thirds of the students who voted, voted for an animal as the school mascot. What fraction of all the students in our school voted for an animal as the school mascot?	• View nb as a unit whose pieces can first be grouped to form an equivalent unit of $\frac{b}{b}$. • Realize the unit composed of grouped pieces is now partitioned into the desired fractional amount.
$\frac{a}{b} \times \frac{c}{d}$ where b and c are relatively prime (share no common factors other than 1)	$\frac{3}{4} \times \frac{7}{8}$ and $\frac{3}{4} \times \frac{5}{6}$	Seven-eighths of the students in our class said their favorite snack food is cheese. Three-fourths of the students who like to eat cheese for a snack like to eat cheddar cheese for their snack. What fraction of our whole class likes to eat cheddar cheese for a snack?	• View $\frac{c}{d}$ as a unit that can be decomposed into c units of $\frac{1}{d}$. • Realize that each unit of $\frac{1}{d}$ can be partitioned into the desired fractional amount. • Realize that the pieces resulting from partitioning each unit of $\frac{1}{d}$ can be grouped to form a unit of the needed fractional amount.

Problems corresponding to $^a/_b \times {}^{nb}/_d$, where the denominator of the multiplier was a factor of the numerator of the multiplicand, further challenged students to extend their knowledge of what it means to partition a unit into a fractional amount. These problems also challenged students to think about grouping pieces to form equivalent units that could readily be partitioned into the desired amount, such as starting with a unit of nine-tenths, then making three groups of three-tenths that could be viewed as three-thirds, to find two-thirds of the nine-tenths. Problems corresponding to $^a/_b \times {}^{nb}/_d$ appeared to be more challenging for students than the other multiplication of fraction problems previously discussed. Students viewed these problems as different from ones they had previously solved but were initially unsure of how to approach solving the problems. To help the students, I once again asked questions that encouraged them to think about how they had solved equal-sharing problems and other multiplication problems. For example, I asked question such as "Remember when we shared cookies, could we give each person more than one whole cookie?" and "If we shared twelve cookies equally among six people, how many cookies would each person receive? What fraction of all the cookies would each person get?" These questions helped students realize they could group pieces to form equivalent units that could be partitioned into the desired amount.

The following example shows Sam initially struggling to solve a problem involving finding five-sixths of twelve-fifteenths of one whole can of dog food. The example also shows how Sam successfully solved the problem when he focused on what it means to partition a unit into a given fractional amount and on the idea of grouping pieces to form an equivalent unit that could be partitioned into sixths (see fig. 19.5).

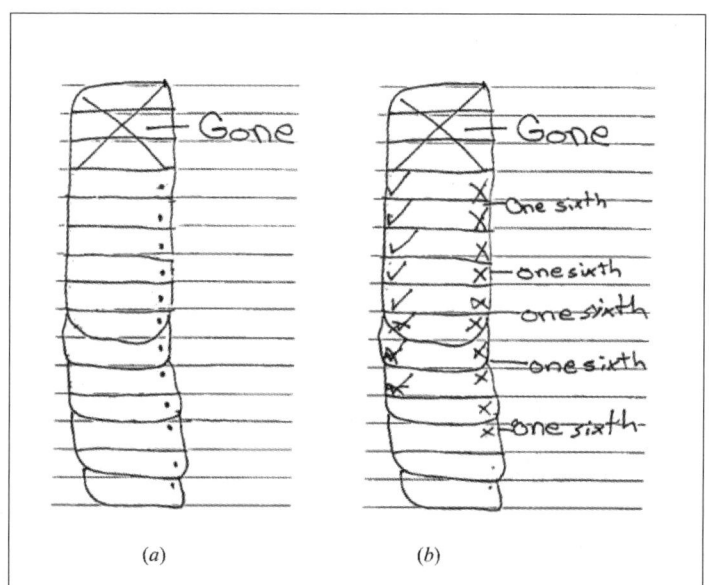

(a) (b)

Fig. 19.5. Sam's strategy for finding five-sixths of twelve-fifteenths of one whole can of dog food

NKM: You have twelve-fifteenths of a can of dog food. You're going to feed your dog five-sixths of the amount of dog food that you have. How much of the whole can of dog food do you feed your dog?

Sam: *[Drew a rectangle and partitioned it into fifteen parts by drawing horizontal lines one at a time, starting at the top of the rectangle. Marked the three-fifteenths that he did not have. See fig. 19.5a.]* This one's hard . . . I don't know.

213

NKM: Think about what five-sixths means. . . . If you needed to show sixths of any-thing, how many parts would it be partitioned into?

Sam: Six. Six times two is twelve. *[Formed six groups of two pieces each. Wrote "one sixth" by five of the groups of two. See fig. 19.5b.]* Ten-fifteenths . . . See this [twelve-fifteenths] is six-sixths, and you said I gave my dog five-sixths of this. That's these ten pieces. That's ten-fifteenths of the whole can of dog food.

Problems corresponding to $^a/_b \times {}^c/_d$, where the denominator of the multiplier and the numerator of the multiplicand were relatively prime (shared no common factors other than one), such as finding three-fourths of five-sixths of one whole, were particularly challenging for students to solve. These problems required multiple steps that could include decomposing fractional units into iterations of unit fractions, such as decomposing $^5/_6$ into five $^1/_6$ pieces, partitioning the unit fraction into the desired fractional amount (e.g., partitioning each $^1/_6$ piece into fourths), and grouping pieces resulting from the partitioning to form an equivalent fractional unit.

Initially, none of the students successfully solved these kinds of problems on their own. They used trial-and-error strategies that involved either repartitioning a fractional unit or they grouped pieces to form an equivalent unit, but not both. As with the other problems, I helped students by asking questions focused on what it means to partition any unit into a fractional amount, as well as questions about how they previously solved equal-sharing prob-lems and other multiplication of fraction problems. Such questions helped students realize they needed to both repartition a fractional unit and group pieces to form an equivalent unit that could be partitioned into the desired amount to solve the problems.

Repartitioning units and grouping pieces can be seen in Abby's strategy for a problem involving finding three-fourths of seven-eighths of one whole cookie (see fig. 19.6). Abby first partitioned a circle into eighths and clearly indicated seven-eighths of the entire cir-cle (see fig. 19.6a). Next, Abby partitioned each of the seven one-eighth pieces into fourths and found there were now twenty-eight small pieces (see fig. 19.6b). Abby communicated that she needed fourths, so she created four equal-sized groups from the twenty-eight small pieces and commented, "That's four-fourths and I need three of these" (see fig. 19.6c). Last, Abby partitioned the shaded one-eighth pieces into four equal-sized pieces and announced, "There's thirty-two pieces in the whole cookie . . . so [the answer is] twenty-one thirty-twos" (see fig. 19.6d).

The above results of the multiplication of fraction study have influenced my own prac-tice in important ways for the past several years. When working with intermediate-grade students and preservice teachers, I carefully considered the numbers used in multiplication of fraction problems. More specifically, I first thought about questions such as the follow-ing: What numerical relationships exist between the denominator of the multiplier and the numerator of the multiplicand? What particular numerical relationships do I want students to perceive between the fractions involved in the problem? How might students solve the problems if they do or do not perceive these numerical relationships? I then created word problems corresponding to the different numerical relationships I hoped students would per-ceive. The appendix contains examples of multiplication of fraction word problems I created and have used with intermediate-grade students and/or preservice teachers.

I also thought about how students in the research study viewed problems corresponding to $^a/_b \times {}^c/_d$ differently when particular numerical relationships existed between the denomina-tor of the multiplier and the numerator of the multiplicand. Additionally, I thought about how easily the research study students solved some problems, how challenging other prob-lems were for the students, and how the students in the study were able to use their strate-gies for previously solved problems to solve problems they found particularly challenging.

This helped me think about ways to sequence multiplication of fraction problems in my own practice. I intentionally began sequencing multiplication of fraction word problems in the following order:

(1) $\frac{1}{2} \times \frac{1}{2}$ and $\frac{1}{4} \times \frac{1}{2}$

(2) $\frac{a}{b} \times \frac{b}{d}$

(3) $\frac{a}{nb} \times \frac{b}{d}$

(4) $\frac{a}{b} \times \frac{nb}{d}$

(5) $\frac{a}{b} \times \frac{c}{d}$

As I presented problems to students in this order, I saw my students focusing on different units in problems and solving problems by using strategies similar to those used by students in the multiplication of fractions study and those used by India and Jacob in figure 19.3. In short, I observed my students developing a strong conceptual understanding of the multiplication of fractions.

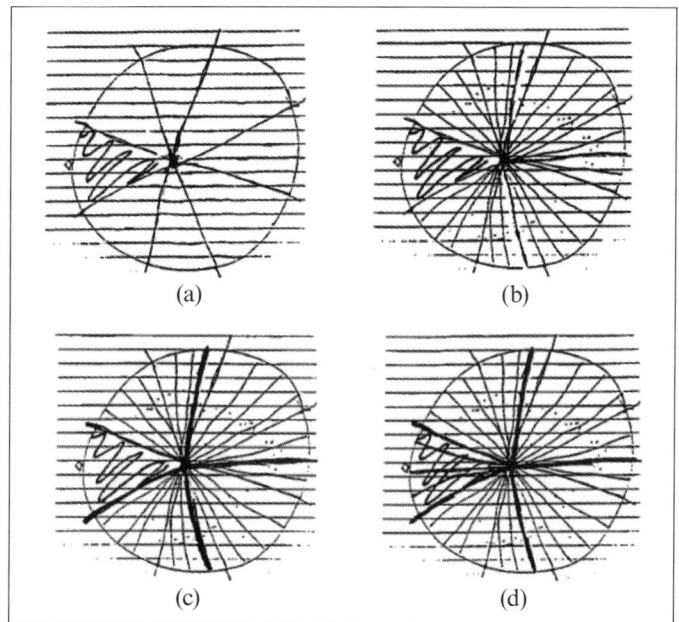

Fig. 19.6. Abby's strategy for finding three-fourths of seven-eighths of one whole cookie

Concluding Comments

The Common Core State Standards for Mathematics suggests that all students should be able to solve word problems corresponding to $\frac{a}{b} \times \frac{c}{d}$ and develop a conceptual understanding of multiplication of fractions (NGA Center & CCSSO, 2010). Results of the multiplication of fraction study suggested that guiding students' learning of this important content area may not be simple or straightforward. However, results of the study also provided insights into issues related to the wording, structure, and sequencing of problems that may be effectively applied to practice to help students develop a conceptual understanding of multiplication of fractions.

References

Behr, M. J., Harel, G., Post, T., & Lesh, R. (1992). Rational number, ratio, proportion. In D. A. Grouws (Ed.), *Handbook of research on mathematics teaching and learning* (pp. 296–333). New York, NY: Macmillan.

Behr, M. J., Harel, G., Post, T., & Lesh, R. (1993). Rational numbers: Toward a semantic analysis-emphasis on the operator construct. In T. P. Carpenter, E. Fennema, & T. A. Romberg (Eds.), *Rational numbers: An integration of research* (pp. 13–47). Hillsdale, NJ: Lawrence Erlbaum.

Behr, M. J., Harel, G., Post, T., & Lesh, R. (1994). Units of quantity: A conceptual basis common to additive and multiplicative structures. In G. Harel & J. Confrey (Eds.), *The development of multiplicative reasoning in the learning of mathematics* (pp. 121–176). Albany, NY: SUNY Press.

Mack, N. K. (1998). Building a foundation for understanding the multiplication of fractions. *Teaching Children Mathematics, 5,* 34–38.

Mack, N. K. (2001). Building on informal knowledge in a complex content domain: Partitioning, units, and understanding multiplication of fractions. *Journal for Research in Mathematics Education, 32,* 267–295.

National Governors Association Center for Best Practices & Council of Chief State School Officers (NGA Center & CCSSO). (2010). *Common core state standards for mathematic*s. Washington, DC: Author. Retrieved from http://www.corestandards.org.

Olive, J. (1999). From fractions to rational numbers of arithmetic: A reorganization hypothesis. *Mathematical Thinking and Learning, 1,* 279–314.

Steffe, L. P. (1988). Children's construction of number sequences and multiplying schemes. In J. Hiebert & M. Behr (Eds.), *Number concepts and operations in the middle grades* (pp. 119–140). Reston, VA: National Council of Teachers of Mathematics.

Appendix

Sample Multiplication of Fractions Problems

Please note that I write all fractions in words in contextual problems. Doing so helps students think deeply about what is happening to the quantities involved in the problems.

Problems Corresponding to: $\frac{1}{2} \times \frac{1}{2}$, $\frac{1}{4} \times \frac{1}{2}$, and $\frac{1}{a} \times \frac{1}{b}$

1. You walk one-fourth of a mile from home to school every day. Today, you walked one-half of the total distance from home to school then you stopped to eat a snack. How much of one whole mile did you walk before stopping to eat a snack?

2. One-half of our pizza has only veggies on it. One-fourth of the portion with veggies has only mushrooms on it. How much of our whole pizza has only mushrooms on it?

3. Today, one-half of the students in the class are wearing shirts with buttons. One-third of the students who are wearing a shirt with buttons on it are wearing a blue shirt. What fraction of the whole class is wearing a blue shirt with buttons on it today?

4. The zoo has many different animals. One-third of the animals at the zoo are bears. One-fourth of the bears are polar bears. What fraction of all the animals at the zoo are polar bears?

5. Three-fourths of the earth is water and one-fourth is land. One-half of the land is not habitable. Food can be grown on only one-fourth of the land where people can live. What fraction of the earth is land where food can be grown?

Problems Corresponding to: $\frac{a}{b} \times \frac{b}{d}$

6. Four-fifths of the books in the library are chapter books. One-fourth of the chapter books are non-fiction. What fraction of all the books in the library are non-fiction chapter books?

7. Three-fourths of the students in the class ate a piece of fruit for lunch today. One-third of the students who ate a piece of fruit ate an apple. What fraction of the whole class ate an apple for lunch today?

8. Three-fourths of the students in the class own a pet. Two-thirds of the students who own a pet own a cat. What fraction of all the students in the class own a cat?

9. Five-eighths of the plants in the garden sprouted two days ago. Last night, a rabbit ate four-fifths of the plants that had sprouted. What fraction of all the plants in the garden did the rabbit eat last night?

Problems Corresponding to: $\frac{a}{nb} \times \frac{b}{d}$

10. Two-thirds of the vehicles in the parking lot were cars. Five-sixths of the cars in the parking lot were red. What fraction of all the vehicles in the parking lot were red cars?

11. Four-tenths of the students in the class play an instrument in the school band. Three-eighths of these students play a brass instrument in the school band. What fraction of the whole class plays a brass instrument in the school band?

12. Five-twelfths of the students in the class have a sibling who is a boy. Seven-tenths of the students who have a brother have an older brother. What fraction of all the students in the class have an older brother?

13. You use a budget to manage your money. Last month, you designated three-eighths of your monthly income for school-related purposes. During the month, you spent five-sixths of the money you designated for school-related purposes on tuition. How much of your monthly income did you spend on tuition last month?

Problems Corresponding to: $^a/_b \times {}^{nb}/_d$

14. You had four-fifths of one pound of bird seed. You used one-half of the bird seed that you had to fill the bird feeder. How much of one pound of bird seed did you use to fill the bird feeder?

15. Nine-tenths of the students in the class were born in Michigan. Two-thirds of the students who were born in Michigan were born in Grand Rapids. What fraction of the whole class was born in Grand Rapids, Michigan?

16. You had eight-tenths of a pound of cheese. You used three-fourths of the cheese that you had to make a submarine sandwich. How much of one whole pound of cheese did you use to make the submarine sandwich?

17. Ten-twelfths of the students in the class entered a project in the school science fair. Three-fifths of these students won an award for their science project. What fraction of the whole class won an award for their science project in the school science fair?

Problems Corresponding to: $^a/_b \times {}^c/_d$

18. You had three-fourths of one yard of polka dot fabric. You used one-half of the polka dot fabric to make a quilt. How much of one yard of polka dot fabric did you use when making the quilt?

19. Five-sixths of the items that students in our class collected for the community food drive were canned goods. One-fourth of the canned goods were cans of soup. What fraction of all the items our class collected for the community food drive were cans of soup?

20. You teach at Learning Lane Elementary School. Seven-eighths of the students in your class are bilingual. Three-fourths of the bilingual students in your class speak Spanish as their first language. What fraction of all the students in your class speak Spanish as their first language?

21. There are lots of animals available for adoption at the Humane Society. Five-eighths of the animals available for adoption are dogs. Two-thirds of the available adoptable dogs have spots. What fraction of all the animals available for adoption at the Humane Society are dogs with spots?

22. You collected one and one-half bags of cans for recycling. Each full bag weighed three and three-fourths pounds. What is the total weight of the cans you collected for recycling?

Chapter 20

Helping Students Learn to Use Fractions

Debra I. Johanning
University of Toledo

The National Council of Teachers of Mathematics Standards (NCTM, 2000) recognize mathematics as more than a collection of concepts and skills to be mastered. Not only should mathematics focus on developing skills, it should be done in such a way that students see mathematics as usable. NCTM (2000) offers that students "should be responsible for what they have learned and for using that knowledge to understand and make sense of new ideas" (p. 64). A RAND Mathematics Study Panel report on mathematical proficiency states, "Simply knowing concepts does not equip one to use mathematics effectively" (RAND, 2003, p. 300).

There is a large body of research focused on what is involved in learning fractions when fractions are the focus of instruction. (A few examples include Behr, Harel, Post, & Lesh, 1992; Freudenthal, 1983; Kieren, 1976; Lamon, 1999; Mack, 1990; Moss & Case, 1999; Smith, 1995; Streefland, 1991). However, there is little research that explores how students learn to use what they have learned about fractions when they are presented in other mathematical contexts. The research findings shared here address this issue. In order to differentiate between when students "learn about" fractions and when they "learn to use" fractions, I use the phrase "learning about" to refer to what students typically study in formal curriculum units that are about fractions. In contrast, I use the phrase "learning to use" to refer to situations where a person has to use the fraction knowledge she has previously learned about when she comes upon problems or situations in other mathematical content areas where fractions are needed to solve a problem.

The research findings in this chapter reveal that what students learn when directly studying fraction content is different from what they learn when they have to use fractions as part of learning about other content. Broadly speaking, learning about fractions often involves learning how, why, and what. For example, when learning about fractions, students typically explore *how* to determine if two fractions are equivalent, *how* to compare and order fractions, *why* there can be multiple names for fractional quantities, and *what* strategies can be used to find equivalent fractions and to operate with fractions. In contrast, the data collected in this study revealed that when students were learning to use fractions, classroom conversations were focused on different issues. These conversations were related to making sense of what was appropriate in relation to the problem setting. These conversations were framed around determining what was appropriate for a specific situation. Questions that would be part of these conversations might include the following:

- Can I use a fraction if it is not in the problem context?

- Can I use a decimal instead of a fraction?

- Do concepts that work in whole number settings also work when fractions are used?

This chapter is adapted from D. I. Johanning (2008), Learning to use fractions: Examining middle school students' emerging fraction literacy, *Journal for Research in Mathematics Education, 39*, 281–310.

Questions such as these were the focus of a classroom practice referred to as "determining appropriateness." When students were solving problems that led them to use fractions in other content areas, they engaged in mathematical conversations where they considered what was appropriate for that situation involving fractions. In this chapter I present six different ways (referred to as subpractices) that the *practice of determining appropriateness* emerged in the classroom data that were collected and analyzed.

Methods Used to Carry Out the Study

All students in this study were exposed to the same mathematics curriculum from the Connected Mathematics Project (CMP2). During the first year of the study, a class of twenty-three sixth-grade students and their teacher were the focus of data collection. During the second year of the study, a seventh-grade class of twenty-three students and their teacher were observed in the same school. The sixth- and seventh-grade teachers were different teachers. Of the twenty-three original sixth-graders studied, eight were in the seventh-grade class. Four sixth-grade students were identified and followed in detail from sixth into seventh grade. These four students are referred to as focus students.

Data collection took place in two phases across two school years. During phase 1, I collected "learning about" fractions data. In the fall of the first year I collected data during the teaching of two sixth-grade fraction units. Phase 1 provided a record of the experiences students had and of the instructional tasks they worked when learning about fractions. These data also provided information about the experiences students brought to phase 2 of data collection.

Phase 2 is where "learning to use" fractions data collection took place. I examined curricular materials that were used after the two fraction units were taught (see table 20.1 for the data collection timeline) and identified instructional tasks where fractions were used. In the curriculum materials I identified five sixth-grade instructional tasks and nine seventh-grade instructional tasks where fractions were used. During phase 2 I followed students across sixth-grade into seventh-grade, collecting data when the identified instructional tasks were taught.

Table 20.1
Data collection timeline

Instructional unit: Mathematical context	Focus of data collection
Bits and Pieces I: Understanding Fractions, Decimals, and Percents (Lappan, Fey, Fitzgerald, Freil, & Phillips, 2002/3a)	Phase 1: Learning about fractions
Bits and Pieces II: Using Rational Numbers (Lappan et al., 2002/3b)	Phase 1: Learning about fractions
Covering and Surrounding: Two Dimensional Measurement (Lappan et al., 2002/3c)	Phase 2: Learning to use fractions
Data, Decimals, and Percents: Percents and Decimal Operations (Lappan et al., 2002/3d)	Phase 2: Learning to use fractions
Stretching and Shrinking: Similarity (Lappan et al., 2004a)	Phase 2: Learning to use fractions
Comparing and Scaling: Ratio, Proportion, and Percent (Lappan et al., 2004b)	Phase 2: Learning to use fractions

By collecting and analyzing field notes, copies of students' written work, video-recordings of classroom discussions, and data from three interviews with the four focus group students, I identified when the articulation of the *practice of determining appropriateness* and the six subpractices occurred. In particular, there were twenty-one different classroom conversations where the practice of determining appropriateness occurred in the data. Table 20.2 provides information regarding the content areas where each of these instances occurred.

Table 20.2
Frequency of subpractices associated with the practice of determining appropriateness by mathematical content in the unit

Subpractices	Area and perimeter	Decimal operations	Similarity	Ratio
Representing repeating decimals as fractions	1		1	
Moving from fractions to whole numbers and back	1	1	1	
Drawing from multiple approaches and algorithms	3			
Choosing to use fractions when fractions are not explicit		1	2	1
Does equivalence matter?	1	1		1
How are fractions and ratios related?			2	4

Results from the Study

As classroom episodes illustrating the six subpractices of the *practice of determining appropriateness* are presented, I encourage readers to note what the students knew about fractions as they engaged with the situations where they had to use what they learned about fractions. Consider the focus of these conversations, the questions students asked, and the questions asked of students. These conversations provided important opportunities for students to deepen their understanding of fractions.

Subpractice 1: Representing repeating decimals as fractions

The primary focus of these episodes was whether or not it was appropriate to use a decimal approximation for $1/3$. During the episodes it was clear that students knew $1/3$ could be represented with a repeating decimal. One of these episodes occurred during the sixth-grade area/perimeter measurement unit. Students were discussing a problem about building rectangular storm shelters with a floor area of 24 square meters. The walls of the rectangular storm shelter were constructed from 1-meter-wide panels that cost $125 each. Students were provided various measures for a storm shelter's wall length and were asked to find the width, resulting perimeter, and cost to buy the wall panels for each storm shelter. All but one of the measures given for wall lengths were in whole meters. One wall length was $5^1/_3$ meters. When the class was discussing what the width for a storm shelter with a length of $5^1/_3$ would be, various decimals were offered and then rejected as an accurate representation for $5^1/_3$.

Cathy: I did 5.3×4.6 and got 24.3.

Bryan: I did 5.3×4.5 and got an even 24.

Trevor: Well $5^1/_3$ is not equal to 5.3. It is 5.3 with a line over it. *[The teacher, Mrs. Kay, writes 5.33333 on board.]* So, let's say [5.33333] times 4.5. It equals 23.99999, which is pretty close to 24.

Next, a student named Katie asked how people found 4.5 meters as a solution for the width.

This led the conversation to change over to a discussion of division as an approach to finding the unknown width.

Katie: You divide 24 by $5^1/_3$.

Mrs. Kay: Corey, is that what you did?

Corey: Yeah.

Mrs. Kay: What did you get when you did that?

Katie: $4^1/_2$.

Corey: Four point 5 2 8.

Corey explained that he used a calculator and divided 24 by 5.333 to get 4.528. Now, for a second time, Katie asked how people found 4.5 as a solution.

Corey: Maybe like I did [using 24 ÷ 5.333].

Mrs. Kay: Okay, but you got 4.528. What would explain that?

Corey: I rounded off.

Mrs. Kay: That is the problem with sometimes switching to a decimal. If I don't [switch to a decimal] I can get the exact answer.

In this episode, approaches were questioned and even rejected because using a decimal approximation for $5^1/_3$ did not lead to an exact area of 24 meters. Shifting from fraction to decimal form represents knowledge commonly addressed when learning about fractions. In this situation, however, students had to step back and reflect on whether it was appropriate to use a decimal approximation rather then the fraction. This differs from learning how to move between forms. It involves determining appropriateness for the situation at hand.

Subpractice 2: Extending beyond whole numbers

A second form of the practice of determining appropriateness took place when students had to extend a concept they understood in whole number situations to fraction situations. These conversations raised the question of whether concepts that work in whole number contexts would also work when fractions are used. In a seventh-grade task students were asked to apply a scale factor of $2^1/_2$ to a 4-by-8 rectangle and then determine both the new area and the new perimeter.

During the discussion a student said that the rectangle was $2^1/_2$ times larger. The teacher (Mrs. Dew) pushed for an explanation of what that meant. Was the area or the perimeter becoming $2^1/_2$ times larger? Classroom discussions indicated that students understood that when each side length was multiplied by a scale factor of $2^1/_2$ it became $2^1/_2$ times longer, and the perimeter became $2^1/_2$ times longer. However, students struggled to figure out if the area was increasing by a scale factor of $2^1/_2$ or not. The class spent several minutes discussing how much larger the scaled rectangle was than the original. Students used the side lengths of each rectangle to determine its area. Next, they divided the area of the larger rectangle by the area of the smaller rectangle and found that the area of the larger rectangle was 6.25 times the area of the smaller rectangle. The teacher turned the discussion back to determining what the scale factor was.

Mrs. Dew: Is there any way I could have found [6.25] from where Bryan started this 10 minutes ago? Or do I have to know the area of this and the area of this and divide them to find out how many of these fit inside of this.

Janine:	For problem 3.1 *[see fig. 20.1[1]]* when you double it is 4, and when it is 3 it goes to 9 and 4 is 16, so if the scale factor is 2.5 wouldn't you square 2.5 or times it by itself to see the area change?
Mrs. Dew:	What is 2.5 times 2.5?
Class:	6.25.
Mrs. Dew:	So, Janine, is the same thing holding true?
Janine:	Yes.
Mrs. Dew:	Just because I went to [a scale factor] that wasn't quite as nice and pretty as a 2, a 3, and a 4, the rule we had for a week now didn't go away. The way you prove that, like Carl said [dividing area of larger rectangle by area of smaller rectangle], just helped me solidify that. Even with an ugly scale factor I still have the same relationship. The scale factor times itself will tell me how that area will change.
Amy:	So we spent almost a half an hour talking about this when we could have just done that?
Mrs. Dew:	Yeah.

Scale factor	Change in perimeter	Change in area
2	2	4
3	3	9
4	4	16

Fig. 20.1. Scale factor data table for prior triangle problem

The last student's comment is humorous yet enlightening. It points to the need to help students realize that ideas can carry from whole numbers to other forms of number. We see the teacher making connections between what the students were doing with fractions back to what they know about whole numbers. During discussion the teacher was prompting students to determine if it was appropriate to use an idea that worked with whole numbers in situations with fractions.

Subpractice 3: Drawing from multiple approaches and algorithms

The third subpractice involves situations in which the teacher is trying to establish that there are a variety of approaches used to solve a problem. In this case the teacher is pushing students to consider if it is also appropriate to use the standard fraction multiplication algorithm in which mixed numbers and whole numbers are converted to fractions and then numerators and denominators are multiplied.

In this episode, sixth-grade students were working on a task where they were finding the area and perimeter of parallelograms. Students were given a lab sheet with six parallelograms constructed on square centimeter grids, and were asked to determine the measures needed for calculating the area and perimeter. Five of the parallelograms on the grid were non-right parallelograms with whole number bases and heights. The sixth parallelogram was rectangular with a height of $3\frac{1}{2}$ centimeters and a base of $4\frac{1}{2}$ centimeters.

---

[1] When working on a prior problem in which whole number scale factors were used to determine the effect on the area and perimeter of a triangle, a data table was written on the board (see fig. 20.1). This is the data table referenced in the discussion.

During discussion of the rectangular parallelogram, it was established that the height was $3\frac{1}{2}$ centimeters and the length of the base was $4\frac{1}{2}$ centimeters. Two students offered approaches for finding area. One involved counting the number of square centimeters on the gridded diagram and the other approach used a combination of multiplying whole number dimensions and counting and adding the remaining fractional square centimeters. For example, one student multiplied 3 by 4 to find the area of the rectangular region formed with whole number square centimeters, and then counted the half and quarter square centimeter units to arrive at an area of $15\frac{1}{2}$ square centimeters.

Noting that students had used multiplication of the height and base to find area when a parallelogram had whole number dimensions, Mrs. Kay suggested that the standard algorithm for fraction multiplication developed in the previous fraction operation instructional unit was also a viable approach when a parallelogram had non-whole number dimensions. After using the algorithm to determine the area, she asked, "I am curious why no one used the algorithm with this problem. Is it that you don't like the algorithm or that it doesn't make sense or what?" One student offered that he "forgets" about it sometimes. During an interview with the focus-group students, a student offered that it did not make sense to use the algorithm since it was easy to count. "Why would you use the algorithm when the square is so small and you have all squares right there around the outside?"

The goal of the discussion initiated by the teacher was not to dismiss the other approaches. Rather, the goal was to have students consider the appropriateness of multiple approaches. She raised the possibility of using the standard algorithm because as the class continued to work on the area and perimeter unit, problems would shift away from presenting figures drawn on a grid. Although the students' partial product and informal counting approaches are useful in some cases, the standard algorithm that students had developed in the fraction unit was also appropriate as well as efficient.

Subpractice 4: Choosing to use fractions when fractions are not explicit	The sixth-grade unit on decimal operations began by having students explore the role of place value and the location of the decimal point when finding decimal sums and differences. The following problem was posed by the teacher:

> I ran 5.2 miles on Monday, 6.08 miles on Tuesday, and 2.455 miles this morning. How many miles have I gone so far?

A student presented a solution that involved lining up the digits by place value and lining up the decimal points. The class talked about why decimal points are lined up, the role of place value, and the meaning of the carry used when adding decimal numbers. They wrote out a general procedure for adding and subtracting decimals based on place value.

At this point, a student suggested that "maybe" there was another approach, one that involved fractions. "I think there might be another way. I think you can use fractions." Figure 20.2 shows the decimal form and the corresponding fraction form that was offered and recorded on the board. When the student first offered this approach he was not sure if it was reasonable. Together the class explored whether fractions could be used in place of decimals to operate and if the two forms were equivalent. They discussed how 5.2 could be represented as $5\frac{2}{10}$, 5.200, and $5\frac{200}{1000}$. In the episodes associated with this subpractice, fractions were introduced into the conversation as a tool for making sense of a mathematical concept. When the idea of using fractions to represent decimals was offered, students had to determine if using fractions was appropriate for the situation.

5.2	$5\,^2/_{10}$
6.08	$6\,^8/_{100}$
2.455	$2\,^{455}/_{1000}$

Fig. 20.2. Decimal/fraction representations
for the Running problem

Subpractice 5: Is switching between equivalent fractional forms appropriate?

This subpractice involved situations in which students were trying to determine whether using equivalent fraction forms was appropriate. One episode occurred when students were reworking decimal addition and subtraction problems they had previously solved using a place value approach. They were converting the given decimals into fraction form and then using fraction addition and subtraction as described in the previous section of this chapter. One problem read as follows:

> Emma signed up to clean 1.5 miles with the cross-country team. She stopped when it started to rain after 0.25 of a mile. How much did she have left to clean when the rain stopped?

As students worked in small groups, two focus-group students represented the solution to $1.5 - 0.25$ using different forms or representations. TJ wrote:

$$\frac{150}{100} - \frac{25}{100} = \frac{125}{100},$$

whereas Ali wrote:

$$1\frac{50}{100} - \frac{25}{100} = 1\frac{25}{100},$$

When TJ questioned Ali's solution, Ali pointed out that her solution was equivalent to his: "You did improper fractions and [I] did a mixed number." TJ replied, "Can mine be a mixed number? Is that okay?" TJ's concern was not whether the two forms represented equivalent quantities. He did not ask if $^{125}/_{100}$ and $1^{25}/_{100}$ were the same or different. Instead, he was trying to determine if mixed numbers were appropriate to use in this situation.

When learning about fractions, these students experienced instruction that focused on topics such as how to express fractions as decimals and decimals as fractions, fraction equivalence, and developing strategies from renaming mixed numbers as improper fractions and improper fractions as mixed numbers. However, some students were still not sure when to use these ideas in the decimal contexts that were presented to them. They wondered if it was appropriate to use equivalent forms when solving these decimal problems with fractions.

Subpractice 6: Can ideas about fractions be appropriately used with ratios?

Occurrences of this subpractice took place in seventh grade when students were formally introduced to ratio in a unit on similarity. Initially, students had been using same general shape and congruent angles as criteria to determine if figures were similar. Scale factor was then introduced, followed with ratio reasoning. In one situation using a rectangle whose length-to-width ratio was 4 to 1, Mrs. Dew introduced students to the ratio representation $^a/_b$ and $a{:}b$. When she introduced students to the representation $^a/_b$, students were unsure whether this representation of a ratio written in fraction form was a fraction or not. Students were trying to decide if part-whole ratios, which are also fractions, and part-part ratios written in fraction form behaved in the same way.

Next, students were given a task that included a diagram of three parallelograms with labeled measures (see fig. 20.3). Students were asked to find the ratio of the long side (base) to the short side (diagonal side length) and then use the ratio to decide if the parallelograms were similar. Most students eliminated parallelogram E right away because of the difference in angle measure. A majority of the students determined that parallelograms F and G were similar by showing that a scale factor of 1.25 existed between the corresponding side lengths of each parallelogram. If parallelogram G is used as the initial figure, applying a scale factor of 1.25 to each the base side length and the diagonal side length yields the base side length and the diagonal side length of parallelogram F.

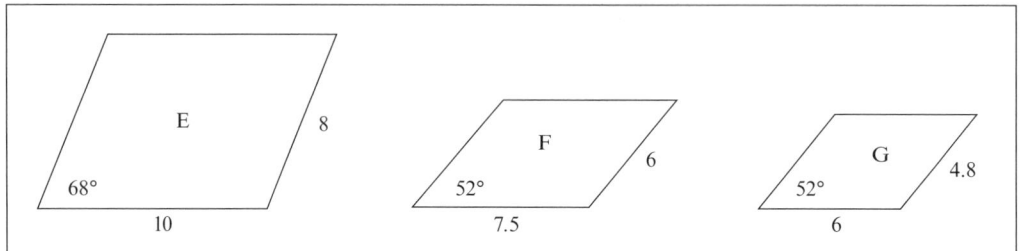

Fig. 20.3. Labeled parallelograms (Lappan, Fey, Fitzgerald, Friel, & Phillips, 2004a, p. 56)

During whole class discussion about the scale factor, Mrs. Dew wrote the long-side to short-side ratios for parallelograms F and G on the board (see fig. 20.4). At this point, a student asked about an approach that used the internal ratio, or the relationship between each individual parallelogram's base length and diagonal side length, to determine similarity rather than comparing the corresponding parts of two different parallelograms using a scale factor. In the discussion it became apparent that a student named Janine was also struggling with the placement of the two 6s in the problem. The teacher decided to use a different set of values to explore whether an internal ratio would also show that the relationship between each parallelogram's base length and diagonal side length are equivalent.

$$
\begin{array}{c}
\times 1.25 \\
\dfrac{7.5}{6} \longleftarrow \dfrac{6}{4.8} \\
\times 1.25
\end{array}
$$

Fig. 20.4. Long-side to short-side ratios of parallelograms

Mrs. Dew: What if we sort of did what you are saying? What if we divided this way? What if we took 7.5 and divided by 6?

Student A: 1.25.

Mrs. Dew: What is 6 divided by 4.8?

Student B: 1.25.

Mrs. Dew: Now that is interesting.

Janine: What if they are not the same? Which one would you use? Would you multiply it by the other one on the bottom, or over?

Mrs. Dew: [What] if we had ³/₄ and ¹²/₁₆? We were going across. The question is that we have been always going across. Janine is noticing that with both of these 6s here it got confusing. What if we did it vertically? What if you did 3 divided by 4?

Student C: 0.75.

Mrs. Dew: What is 12 divided by 16?

Student C: 0.75.

Mrs. Dew: Why are they coming out equal?

Class: They are the same.

Bryan: Because the fractions are equivalent.

Janine's question indicated that she was not clear about the appropriateness of applying ideas used with fractions in situations with ratios. She was not sure if part-part ratios could be compared in the same way part-whole ratios that are also fractions can be compared. Students had to reconcile how to use the fraction skills and concepts developed when they learned about fractions in these new contexts. As students learned to use fractions in their study of ratio, proportionality, and similarity, they had to integrate new ideas with their preexisting understanding.

Conclusions and Implications

The focus of my research was to identify common patterns in classroom discussions where students are posed problems that involve the use of fractions in other mathematical content areas. The data collected for this study revealed that classroom conversations students engaged in when learning to use fractions focused on different issues than those posed when learning about fractions. Using fractions did not come naturally to students. While the students did have a basic understanding of fraction-decimal relationships, equivalence, and operations, they had not yet developed a deep understanding of fractions. Students benefited from the opportunity to learn to use fractions in various contexts where fractions were used.

For example, connections among fractions and decimals were studied in the fraction units. However, students needed support understanding when it was appropriate to interchange fractions with decimals. They wondered if it was appropriate to use fraction form to represent quantities in decimal form when adding and subtracting fractions. Compare this to the Storm Shelter problem (described in subpractice 1) and whether to use ¹/₃ or 0.333. The conversations generated when determining what was appropriate for a given situation made it possible for students to learn that in some situations, such as when adding decimals, one could be flexible and choose either decimal or fraction form. In other situations, like the Storm Shelter problem, they learned that fraction form is preferable. Including fractional contexts in other mathematical content areas that students study helps students develop connections across various mathematical ideas.

The NCTM Standards (2000) point out that students need to develop the disposition to look for and use connections. Should we change the fraction to a decimal or do we need to operate with the given fractions? Can we use fractions in this situation? Which algorithmic approach would be most reasonable in this setting? How these teachers (Mrs. Kay and Mrs. Dew) engaged students in reasoning and problem solving by allowing them to wrestle with and talk about ideas, and not intervening too quickly in their conversations, supported the emergence of conversations where determining appropriateness was the focus. Presenting students with tasks that use ideas they have previously learned about—such as fractions or any other previously learned content—can help students develop a deeper understanding of either content. Furthermore, students are also supported in developing a disposition to use what they have learned to make sense of new ideas. As students look for and use connections, they develop a richer understanding of mathematics and they come to view mathematics as a collection of connected, usable ideas rather than a set of arbitrary rules.

References

Behr, M. J., Harel, G., Post, T., & Lesh, R. (1992). Rational number, ratio, and proportion. In D. A. Grouws (Ed.), *Handbook of research on mathematics teaching and learning* (pp. 296–333). New York, NY: Macmillan.

Freudenthal, H. (1983). *Didactical phenomenology of mathematical structures.* Boston, MA: D. Reidel.

Johanning, D. I. (2008). Learning to use fractions: Examining middle school students' emerging fraction literacy. *Journal for Research in Mathematics Education, 39,* 281–310.

Kieren, T. (1976). On the mathematical, cognitive and instructional foundations of rational numbers. In R. Lesh (Ed.), *Number and measurement* (pp. 101–144). Columbus, OH: ERIC/SMEAC.

Lamon, S. J. (1999). *Teaching fractions and ratios for understanding: Essential content knowledge and instructional strategies for teachers.* Mahwah, NJ: Lawrence Erlbaum.

Lappan, G., Fey, J. T., Fitzgerald, W. M., Friel, S. N., & Phillips, E. D. (2002/2003a). *Bits and pieces I: Understanding fractions, decimals, and percents.* Glenview, IL: Prentice Hall.

Lappan, G., Fey, J. T., Fitzgerald, W. M., Friel, S. N., & Phillips, E. D. (2002/2003b). *Bits and pieces II: Using rational numbers.* Glenview, IL: Prentice Hall.

Lappan, G., Fey, J. T., Fitzgerald, W. M., Friel, S. N., & Phillips, E. D. (2002/2003c). *Covering and surrounding: Two dimensional measurement.* Glenview, IL: Prentice Hall.

Lappan, G., Fey, J. T., Fitzgerald, W. M., Friel, S. N., & Phillips, E. D. (2002/2003d). *Data, decimals, and percent: Percents and decimal operations.* Glenview, IL: Prentice Hall.

Lappan, G., Fey, J. T., Fitzgerald, W. M., Friel, S. N., & Phillips, E. D. (2004a). *Stretching and shrinking: Similarity.* Glenview, IL: Prentice Hall.

Lappan, G., Fey, J. T., Fitzgerald, W. M., Friel, S. N., & Phillips, E. D. (2004b). *Comparing and scaling: Ratio, proportion, and percent.* Glenview, IL: Prentice Hall.

Mack, N. K. (1990). Learning fractions with understanding: Building on informal knowledge. *Journal for Research in Mathematics Education, 21,* 16–32.

Moss, J., & Case, R. (1999). Developing children's understanding of the rational numbers: A new model and experimental curriculum. *Journal for Research in Mathematics Education, 30,* 122–147.

National Council of Teachers of Mathematics. (2000). *Principles and standards for school mathematics.* Reston, VA: Author.

RAND Mathematics Study Panel. (2003). *Mathematical proficiency for all students: Toward a strategic research and development program in mathematics education* (DRU-2773-OERI). Arlington, VA: RAND Education & Science and Technology Policy Institute.

Smith, J. P. (1995). Competent reasoning with rational numbers. *Cognition and Instruction, 13,* 3–50.

Streefland, L. (1991). *Fractions in realistic mathematics education: A paradigm of developmental research.* Boston, MA: Kluwer Academic Publishers.

Chapter 21

Using Everyday Knowledge of Decimals to Enhance Understanding

Kathryn C. Irwin
University of Auckland

In 1887 Howard optimistically stated, "The system of decimal fractions is so eminently simplistic that when it is generally understood will entirely displace the clumsy system of common fractions" (Kerslake, 1991). But is this statement true? If you have taught decimal fractions in the elementary grades or worked with preservice elementary grade mathematics teachers on how they would structure lessons on decimal fractions, how many of the following misconceptions have emerged among your students?

- Longer decimal fractions are necessarily larger.

- Putting a zero at the end of a decimal number makes it ten times as large.

- Decimals as a "decorative dot" (see Bell, Swan, & Taylor, 1981): When you do something to one side of the dot you also do it to the other (e.g., 2.5 + 1 = 3.6).

- Decimal fractions are "below zero," or negative numbers.

- One hundredth is written 0.100

- $\frac{1}{4}$ can be written as 0.04 or 0.25.

These are among the misconceptions that I identified during interviews of elementary school students and that I included in my 2001 *Journal for Research in Mathematics Education* (*JRME*) article. (See also Irwin, 1995a, 1995b, 1996, for information on these interviews.) I also found, based on studies involving both school students and adults, that the system of decimal fractions was neither eminently simple to learn nor generally understood.

Decimal fractions are usually taught at school, but they need to be anchored in some way to students' existing knowledge. To test the idea of how everyday knowledge and its influence on how students—especially those from lower economic areas—learned decimals, I worked with sixteen students from a lower economic area in New Zealand. Details from this study and the results formed the basis for the *JRME* article mentioned above. In this chapter, I summarize this study and its results and offer ideas on how mathematics education practitioners might use the information in their classrooms.

Background

My study drew from a finding from a previous study (Britt, Irwin, Ellis, & Ritchie, 1993) that students from lower economic areas had more difficulty than did students from more affluent areas in understanding decimal fractions. Those researchers offered the example that 22 percent of thirteen-year-old students from one school in a lower income area understood the decimal concept of "hundredths" or more complex decimal relationships at the start of the

This chapter is adapted from K. C. Irwin (2001), Using everyday knowledge of decimals to enhance understanding, *Journal for Research in Mathematics Education*, *32*, 399–420. A brief summary also appears in K. C. Irwin (1999), *Difficulties with decimals and using everyday knowledge to overcome them: Set two* (Auckland: New Zealand Council for Educational Research).

school year, rising to 32 percent at the end of the year, whereas the comparable percentages from a similar cohort from a school in a middle-income area were 62 percent and 93 percent. I was also interested in research findings (e.g., Resnick, Bill, Lesgold, & Leer, 1991) suggesting that children from minority cultures—who may also reside in lower economic areas—are less likely than students from the dominant culture to spontaneously use the knowledge they have learned outside of school when learning new concepts in school. Researchers such as Boaler (1998) demonstrated the value of integration of everyday knowledge and school knowledge for students from lower economic areas.

The study I designed had as its premise that the usefulness of everyday contexts depends on the appropriateness of such contexts for that particular group of students and that problems from textbooks may not be appropriate. I also had to consider that integration of everyday knowledge and school mathematics can be problematic for reasons that have to do with the contexts of the problems. On one hand, when the problems are too closely tied to students' lives, factors other than mathematical ones can determine how problems are solved (see Lubienski, 2000). On the other hand, when the problems are too distant from their experiences, students either fail to associate problems with mathematics they know or they apply known mathematical skills without considering the appropriateness of the answer (e.g., Silver, Shapiro, & Deutsch, 1993).

Details about the Study

The purpose of the study reported in *JRME* (Irwin, 2001) was to investigate whether the understanding of decimals held by a group of students from a lower income area could be improved by asking them to solve problems set in everyday contexts. I was interested in how context contributed to increased understanding, the roles of peer collaboration, cognitive conflict, and any possible interaction among these factors. I structured the study around two groups of students: one group solving decimal-fraction problems set in a variety of contexts and another group solving similar problems but without contexts.

Participants

The participants were drawn from students from one class in an elementary school situated in a lower economic area of Auckland, New Zealand. The class was a combined Year 7 and 8 (ages eleven and twelve). Based on specific selection criteria (including parental permission), sixteen students participated in the study, and their teacher was asked to rank them by their general achievement in mathematics (not their specific understanding of decimal fractions). Students were then paired so that a higher ranked partner worked with a lower ranked one. More complete information about the participants appears in the article (Irwin, 2001, p. 403).

Pretest and posttest

I devised items for a pretest and posttest on decimal fractions. The purpose of these tests was to assess each student's understanding of decimal fractions before the intervention and to see if the intervention altered his or her understanding. The test questions appear in figure 21.1. These items came from existing research on misconceptions about decimal fractions. The test was not timed, but in general the students completed it in about ten minutes.

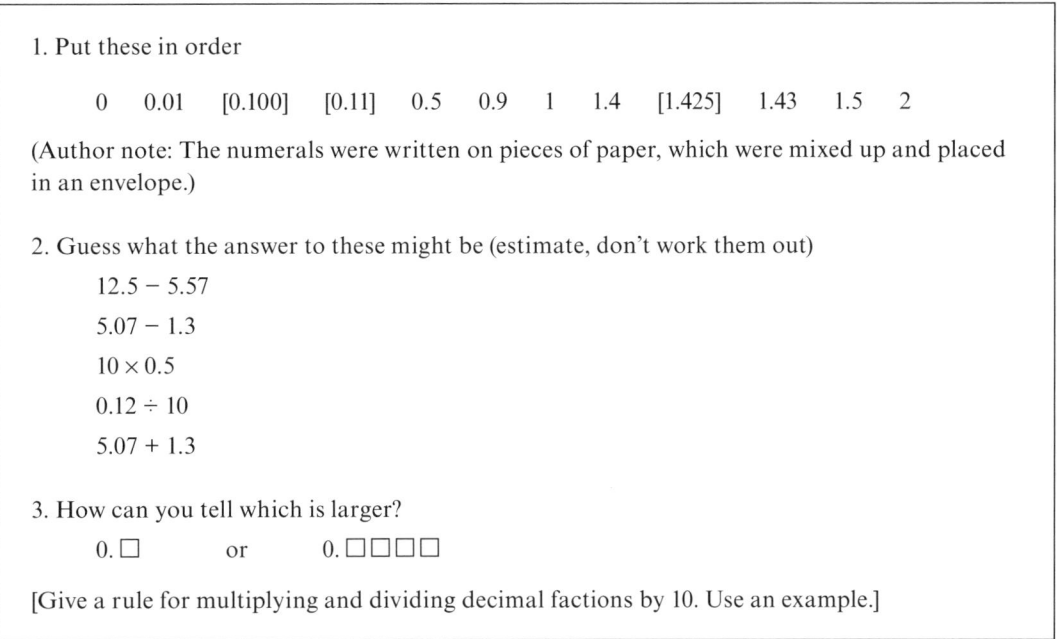

1. Put these in order

   0   0.01   [0.100]   [0.11]   0.5   0.9   1   1.4   [1.425]   1.43   1.5   2

(Author note: The numerals were written on pieces of paper, which were mixed up and placed in an envelope.)

2. Guess what the answer to these might be (estimate, don't work them out)

   12.5 − 5.57

   5.07 − 1.3

   10 × 0.5

   0.12 ÷ 10

   5.07 + 1.3

3. How can you tell which is larger?

   0. ☐   or   0. ☐☐☐☐

[Give a rule for multiplying and dividing decimal factions by 10. Use an example.]

Fig. 21.1. Pretest and posttest questions. Items in square brackets were on the posttest only.

Intervention decimal-fraction problems

I designed an intervention in which pairs of students solved problems, on the premise that this experience would increase their understanding of decimal fractions. The intervention problems consisted of two types: contextualized and noncontextualized. Contexts were selected from those offered by similar students from my other studies and included different sizes of soft-drink bottles, monetary exchanges between different countries, and metric measurements. There were three categories of decimal-fraction problems: magnitude, addition and subtraction, and multiplication and division. Within each category a problem was written to address each of these misconceptions:

1. The misconceptions that led students to treat whole numbers and decimal fractions as distinctly separate units separated by a "decorative dot"

2. Misconceptions that related the length of a number to its value

3. Misconceptions about the way in which quantities were represented in decimal fractions

4. Misconceptions about when a zero is important and when it can be omitted from a decimal fraction

Examples of the problems and the misconceptions they represent appear in table 21.1. All problems were designed to present students with a conflict that resulted from a misconception and a correct answer. For example:

> Teri said that 93¼ was written as 93.04 in decimals.
> Why did she say that? Do you agree?
>
> Peta said that 93¼ was written as 93.25 in decimals.
> Why did she say that? Do you agree?
>
> Who do you think is right?

There were an additional twelve problems similar to those above for each pair of students to discuss.

Table 21.1
Sample problems for the contextualized and noncontextualized conditions and the misconceptions

Category	Contextualized	Noncontextualized
Magnitude: Confusion of length and value	• A soft drink like Coke comes in different sizes. One is 1.5 liter and another is 355 ml. John says that 355 was bigger than 1.5 because that is a larger number. • Amoura says that 1.5 l is more than 355 ml because that is a bigger bottle.	• Alex said that .355 was more than .5 because 355 was more than 5. • Jasmine said that .5 was more than .355 because 5 tenths was more than 355 thousandths.
Magnitude: Different representations of decimal fractions	• The paper says that one New Zealand dollar = 0.9309 in Australian dollars. Susan said that would be 93.09 cents. • Andrew said it would be 9309 dollars.	• Teri said that 93¼ was written as 93.04 in decimals. • Peta said that 93¼ was written as 93.25 in decimals.
Addition and subtraction: Confusion of length and value	• How much do you think you will have left if you have a 1.5 liter bottle of drink and pour out enough to fill a 225 ml glass?	• If you subtract 0.225 from 1.5, what will you get?
Addition and subtraction Different representations of decimal fractions	• If you go on a trip and you buy 1 liter of petrol @ 90.9 cents, and a meal at McDonald's at $4.95, how much will it cost?	• If you add $90^9/_{10}$ and 4.95, what will your answer be?
Multiplication and division: Continuity of units across the decimal point	• Louise is making elastics for skipping and is buying 2 meters 30 cm for each. She needs to make up 10 for the class. She says that she will need 10 times 2.30 meters and that would be 20 meters and 300 cm. • Conrad says that 10 times 2 meters 30 cm would be 23 meters.	• Louise thought that 2.30×10 would be 20.300. • Conrad though that 2.30×10 would be 23.
Multiplication and division: Different representations of decimal fractions	• $1 New Zealand exchanges for 1.5989 Samoan tala. • How much would you get for $10 New Zealand?	• How much is 1½ × 10?

Procedure

Students took the pretest individually and then formed pairs and worked in a small room on the intervention problems on three separate days. The duration of this intervention was chosen to be similar to the time a teacher would normally spend reviewing decimal fractions in class. The posttest was given two months later; decimals were not part of the students' mathematics lessons in this intervening period. During the time students worked in pairs, I served as a clinical interviewer, making only statements about what the students were to be doing, answering requests for information, and facilitating collaboration. All sessions were audio-taped and then transcribed for data analysis purposes. (For information about how the data were collected, analyzed, and coded, please see pp. 407–408 of the *JRME* article.)

Results

Pretest versus posttest performance

The most important finding was that the lower ranked students who worked on contextualized problems improved at the highest rate between the pretest and posttest. In particular, these lower ranked students working on contextual problems showed the greatest rate of improvement from pretest to posttest (19 percent to 32 percent); the higher ranked students working on contextualized problems improved (from 34 percent to 48 percent); the lower ranked students working on noncontextualized problems improved (from 21 percent to 27 percent); and higher ranked students working on noncontextualized problems improved (from 45 percent to 52 percent). Additional statistical analyzes confirmed that there was a significant difference in performance between pretest and posttest as well as a significant difference between the performance of the students ranked higher and those who ranked lower.

| **Analysis of students' dialogues** | To attempt to determine what might have enabled some students to improve their understanding of decimal fractions more than other students, I analyzed students' dialogues during the intervention. I uncovered a difference in the nature of the dialogue between the pairs that worked on contextualized problems and the pairs that worked on noncontextualized problems but made less progress. In particular, the dialogue of students working on the contexualized problems more often showed greater equity and reciprocity in the arguments. And it was often the lower ranked partner in these pairs who used everyday knowledge to make sense of the problems, whereas the higher ranked partner ignored context in favor of manipulating numbers. Two examples from the *JRME* article are reproduced next so that readers can see how context and the use of everyday knowledge assisted the pairs as they solved the problems. The categories are from the coding scheme used to classify the kinds of student comments (Answer, Explanation, Agreement, Disagreement, Challenge, Question, Consideration of other's view, Request for agreement, Comment, and Incomplete statement). |

Ngata and Mereana

Problem: If you go on a trip and you buy 1 liter of petrol @ 90.9 cents and a meal at McDonald's at $4.95, how much will it cost?

	Dialogue line	**Categories**
M:	Ninety-five dollars about, something cents ... four cents.	Wrong answer
N:	Or five dollars ... 'cus it's the whole thing.	Partial right answer
M:	Yeah, I know, ninety-five dollars and something cents. Probably about ninety-five dollars and four cents.	Wrong answer, repeated
N:	Where's the petrol?	Question to partner
M:	That's how it costs all together, isn't it?	Question to partner
N:	Ninety-five dollars? For McDonald's and petrol?	Challenge
M:	Yeah, *[laugh]* oh.	Agreement (sees error)

. . . [17 lines later]

M:	Oh. See that's five dollars, five, six dollars.	Right answer, expanded
N:	Six dollars.	Agreement
M:	Six dollars point one.	Answer expanded
N:	Five dollars, six dollars ... six dollars point eight, no point, no. Six dollars five ... uh, six dollars five cents point nine.	Answer expanded
M:	Six dollars five cents point nine?	Challenge

N:	Yeah.	Agreement
M:	Point nine? *[pause]* Six dollars and five cents point nine.	Challenge again
N:	Yeah.	Agreement
M:	That means six point five point nine.	Clarifies challenge
N:	Oh man, nah.	Considers partner's view (sees error)

. . . [dialogue continues]

Notice that the lower ranked partner (Ngata) challenged the calculation of the higher ranked partner (Mereana) on the grounds of his everyday knowledge ($95 for a McDonald's and petrol?). On the other hand, Mereana challenged Ngata's answer on mathematical grounds (six dollars five cents point nine?)

Glen and Bruce

Problem: The paper says that one New Zealand dollar equals 0.9309 Australian dollars. Susan said that was 93 point 09 cents. Why do you think she said that? Do you agree?

	Dialogue line	**Categories**
G:	Yeah it would be 93 point 09 dollars. Yeah. *[pause]*	Wrong answer
B:	No, it … *[13-second pause]*	Disagreement
G:	*[unintelligible]*	
B:	Yeah, we're not really sure. *[pause]*	
G:	Maybe that's not cents, 'cus look, over here it says Indonesia 1516 *[pointing to other exchange rates]*. And then it can't be one thousand five hundred sixteen dollars, can it? *[pause]*	Explanation
B:	Umm, no, it has to be cents, 'cus I mean	Answer
G:	Unless that's very poor, something or cents	Explanation
B:	No, because if you took one dollar Indonesia, flip, you just take one dollar and you'd be a millionaire in Indonesia right away. *[laugh]*	Explanation
B:	No, it would have to be cents, so she is probably most likely to be right.	Answer
G:	Yeah.	Agreement

. . . [dialogue continues]

Here, Glen and Bruce—the lower and higher ranked students, respectively—both draw on their everyday knowledge by using a newspaper clipping of exchange rates to give meaning to their discussion (*Glen: over here it say Indonesia 1516 … and then it can't be one thousand five hundred sixteen dollars … ; Bruce: No, because if you took one dollar Indonesia … you'd be a millionaire in Indonesia right away*).

Summary and Suggestions for Practitioners

An important finding from this study was that students who worked on the contextualized problems improved their competence with decimals more than did a comparable group of students who worked only on problems with no referents. There might be two explanations for this result. The first explanation is based on the notion of "scholastic knowledge" (Wardekker, 1998), which refers to those aspects of our culture that are generally learned in school as opposed to those learned in daily life. Cultural or contextual aspects (i.e., everyday knowledge) are not self-evident but are developed over time, usually through dialogue .

The misconceptions held by students in my studies and others like them suggested that the students had not reflected on the concepts involved in decimal fractions. For example, students who believed that "one hundredth" was written as 0.100 or that ¼ could be written either as 0.04 or 0.25 had not reflected on the incompatibility of these notions with principles such as place value. For all students in this investigation, discussing problems that involved some cognitive conflict (for example, a fast-food meal and a liter of gasoline costing $95) gave them an opportunity to address such inconsistencies and reflect on their understanding.

The second explanation for why students who worked on contexualized problems gained in their understanding of decimal fractions derives from Piaget's theory of peer collaboration. Piaget (1932/1965) stated that learning from peers required that partners have a common scale of intellectual values that allowed them to understand both language and a system of ideas in the same way. Despite differences in rankings, those pairs working on contextualized problems did succeed in collaborating fruitfully. This was probably because the lower ranking students not only drew on their everyday knowledge but also held to it, something that their higher ranking peers tended not to do until its value was demonstrated to them.

What do these results mean for teaching? Despite Howard's optimistic statement about the system of decimal fractions being "eminently simple," it still remains a particularly difficult domain for many students. Complete understanding requires multiplicative thinking, which is not natural but requires a reconceptualization of number relationships from that required in additive relationships. For reflection in classroom dialogues to be most effecrive, both teachers and students must be aware of the place that everyday knowledge has in the classroom. However, what is everyday knowledge for one group may not be for another groups. Many of these students came from families that sent money back to relative in Tonga, Samoa, or the Cook Islands and were familiar with exchange rates. That would be true in some U.S. classrooms and not in others. Teachers need to choose contexts wisely.

Thus, teachers of lower income or diverse classrooms need to be aware of students' everyday knowledge and of any misconceptions developed on the way to achieveing knowledge of decimal fractions. They need to pose questions and mediate dialogue to promote reflection. Students who are already used to discussing and arguing about mathematics would likely do well with this type of approach. Presenting pairs of students with the answers and logic of two hypothetical students—one whose answer and logic are correct and the other whose answer is incorrect and contains a common misconception—works for addressing several areas of common confusions. Both students need to be prepared to listen to the contributions of their partner and see what they agree with and what they disagree with and why.

For this method to work for confusions related to decimal fractions, the class needs to first share the situations in which they see decimal points. Money will predominate, but special attention should be paid to other examples, especially those that involve one or three or more decimal places. For example, in New Zealand we have stopped using all coins below 10 cents, and restaurants often list costs of items to only one decimal place, such as $12.5. Yet at the same time you can be charged $2.99 for an item in a store. If you had to add these, would you add 99 and 5? Once you begin to look for such apparent inconsistencies it is not too hard to find them, thus allowing children to use their everyday knowledge to enhance their understanding of decimals.

References

Bell, A., Swan, M., & Taylor, G. (1981). Choice of operation in verbal problems with decimal numbers. *Educational Studies in Mathematics, 12,* 399–420.

Boaler, J. (1998). Open and closed mathematics: Student experiences and understandings. *Journal for Research in Mathematics Education, 29,* 41–62.

Britt, M. S., Irwin, K. C., Ellis, J., & Ritchie, G. (1993). Teachers raising achievement in mathematics: Report to the Ministry of Education. Auckland, NZ: Auckland College of Education.

Irwin, K . C. (1995a). Learning to understand decimals. In R. P. Hunting, G. E. Fitzsimons, P. C. Clarkson, & A. J. Bishop (Eds.), *Regional collaboration in mathematics education* (pp. 19–23). Melbourne, Australia: University of Melbourne.

Irwin, K. C. (1995b). Students' images of decimal fractions. In L. Meira & D. Carraher (Eds.), *Proceedings of the 19th International Conference for the Psychology of Mathematics Education* (Vol. 3, pp. 50–59). Recife, Brazil: PME.

Irwin, K. C. (1996). *Understanding decimals: Report to the Ministry of Education.* Auckland, NZ: Uniservices, University of Auckland.

Irwin, K. C. (1999). *Difficulties with decimals and using everyday knowledge to overcome them, set two.* Auckland: New Zealand Council for Educational Research.

Irwin, K. C. (2001). Using everyday knowledge of decimals to enhance understanding. *Journal for Research in Mathematics Education, 32,* 399–420.

Kerslake, D. (1991). The language of fractions. In K. Durkin & B. Shire (Eds.), *Language in mathematical education: Research and practice* (pp. 85–94). Buckingham, England: Open University Press.

Lubienski, S. T. (2000). Problem solving as a means toward mathematics for all: An exploratory look through a class lens. *Journal for Research in Mathematics Education, 31,* 454–482.

Piaget, J. (1965). *The moral judgement of the child* (M. Gabain, Trans.). New York, NY: Free Press. (Original work published in 1932)

Resnick, L. B., Bill, V. L., Lesgold, S. B., & Leer, M. L. (1991). Thinking in arithmetic class. In B. Means, C. Chelemer, & M. S. Knapp (Eds.), *Teaching advanced skills to at-risk students* (pp. 27–53). San Francisco, CA: Jossey-Bass.

Silver, E. A., Shapiro, L. J., & Deutsch, A. (1993). Sense making and solution of division problems involving remainders: An examination of middle school students' solution processes and their interpretation of solutions. *Journal for Research in Mathematics Education, 24,* 117–135.

Wardekker, W. L. (1998). Scientific concepts and reflection. *Mind, Culture, and Activity, 5,* 143–153.

Chapter 22

Student Reasoning on Probability Tasks with Coins

Laurie H. Rubel
Brooklyn College of the City University of New York

Mathematics teachers often use probabilistic devices like spinners, dice, or coins to conduct experiments with students to teach concepts of probability. The data gathered in probability experiments can be used to contextualize or model concepts, such as randomness, sample space, or independence. A benefit of using familiar devices like spinners, dice, or coins is that students bring informal knowledge about these devices to the mathematics classroom. Lessons can be designed to intentionally draw out, build on, deepen, or formalize students' informal knowledge. However, students' ideas and beliefs about these probabilistic devices can sometimes conflict with mathematics. This chapter presents research results pertaining to student reasoning and sense-making about coins.

A construct that has proven useful in analyzing probabilistic thinking is *representativeness*, in which one assigns a probability to an event "by the degree to which it is (1) similar in essential properties to the parent population and (2) reflects the salient features of the process by which it is generated" (Kahneman & Tversky, 1972, p. 431). Previous studies (e.g., Konold, Pollatsek, Well, Lohmeier, & Lipson, 1993) document the tendency of adults to indicate that a sequence of coin tosses that has the two characteristics of (1) equal numbers of heads and tails and (2) apparent randomness in its distribution of heads and tails is more likely to occur than those sequences without either or both of these characteristics. Similarly, adults often say, according to the gambler's fallacy, that a run of successive heads on a coin is more likely to be followed by a tail than another head.

Four Heads Task

A fair coin is flipped four times, and each time it lands with heads up. What is the most likely outcome if the coin is flipped a fifth time?

Coin Sequences Task

Suppose you toss a fair coin six times, recording the result of each toss. For instance, if you toss a head and then five tails in a row, you would write HTTTTT.

Which is the most likely result?

	Toss 1	Toss 2	Toss 3	Toss 4	Toss 5	Toss 6
a)	H	T	H	T	H	T
b)	H	H	T	H	T	T
c)	H	H	H	T	T	T
d)	T	T	T	H	T	T
e)	All are equally likely

Explain.

Fig. 22.1. Coin tasks

This chapter is adapted from L. H. Rubel (2007), Middle school and high school students' probabilistic reasoning on coin tasks, *Journal for Research in Mathematics Education, 38,* 531–556.

Representativeness thinking does not fully explain findings from previous studies. A more comprehensive way of looking at student thinking about independent trials of a random event is in terms of student beliefs. Students might believe that (1) coins are unpredictable, (2) coin tosses result in random patterns, (3) coins have no memory, (4) a coin has a 50 percent chance of landing heads or tails, or (5) there should be about equal numbers of heads and tails in a string of tosses (Konold et al., 1993). Although the whole set is not consistent, people are known to hold one or more of these beliefs. Various mathematical tasks that involve coins may stimulate different subsets of these beliefs. This chapter summarizes findings from a study (Rubel, 2007) about student thinking in response to the following two coin tasks (see fig. 22.1) using this set of beliefs as a framework. (The tasks were used in previous research by Fischbein and Schnarch [1997] and Konold et al. [1993].)

Methods

Students in grades 5, 7, 9, and 11 ($n = 173$) at a private boys' school participated in the study. As part of the larger study (Rubel, 2002), each student completed a set of ten tasks, in writing, during a regular mathematics period. Within a week of their written work, thirty-three students, from various age groups and classes, were interviewed. Students were selected for interviews on the basis of their written responses. The next sections of this chapter focus on student reasoning about a subset of the results reported in Rubel (2007).

Results

Four Heads task

Nearly three-fourths of the students indicated that there is no most likely outcome on the fifth toss. As shown in table 22.1, representativeness thinking, or the pair of beliefs that coin tosses result in random patterns and about equal numbers of heads and tails, would direct one to respond that tails would be most likely on the fifth toss. However, only thirty students (17 percent) chose tails as the most likely outcome for the fifth toss. Justifications of this incorrect response tended to emphasize, instead, the belief that there should be equal numbers of heads and tails in a sequence of coin tosses.

Table 22.1
Responses to the Four Heads task ("most likely" version)

	Grade 5 (n = 36)	Grade 7 (n = 45)	Grade 9 (n = 50)	Grade 11 (n = 42)	Total (n = 173)
No most likely	39% (14)	89% (40)	70% (35)	83% (35)	72% (124)
Tails	36% (13)	2% (1)	18% (9)	17% (7)	17% (30)
Heads	17% (6)	9% (4)	10% (5)	0	9% (15)
Other or no answer	8% (3)	0	2% (1)	0	2% (4)

The interview with ninth grader Dave presents an example of student thinking that focuses on the belief that coin tosses should result in equal numbers of heads and tails.

Dave: Tails, because the coin has two sides with equal chances. If you flip it four times and get heads all four times on one side, the probability has to even out, to equal out.

Researcher: We flip it four times and get heads all four times. That already happened. Now we take the coin and flip it again. What is the most likely outcome?

D: Heads or tails. Since it did go four times on one side, it will go to the other side.

R: How does that work? Does the coin remember how it has landed?

D: No, but the idea of probability is that it's going to equal out eventually. Let's say you flip it 100 times and get 20 in a row, then there would be a streak of the other side, say 10 in a row, to equal it out.

Dave's responses are indicative of a belief that there should be an equal number of heads and tails in a string of coin tosses. When he says that this should happen "eventually," he seems to be thinking about the long-term trend of a coin but applies it to a short-term example. His response does not address the randomness aspect of the coin-flipping process.

Another fifteen students indicated that another heads would be most likely after a string of four heads. The common written justification for this response was that another heads would be consistent with the pattern. The interview with fifth-grader Bob provides an example of this reasoning.

Bob: Well, heads because it's constant. It's hard to tell in a classroom. Heads, heads, heads, heads, and then heads would probably be right in math and probably right in the real world.

Researcher: In math?

B: In math, on paper, you can look for patterns and whatever and you can still in life. If you're doing a heads and a tails, you're judging he's going to flip it the same way every time.

R: Now, say it's a different person each flip. Four different people flipped the coin and got heads. Now it's your turn for the fifth flip.

B: Heads. Because it's been happening more than tails. You wouldn't bet on a horse that had lost four times in a row unless you had some knowledge.

R: The horse is a little different—it could be sick, or it could be slow.

B: But when you're betting, it's all luck. If you saw people playing cards and you know that one guy had won four times, you'd bet on him to win since he hasn't lost yet.

R: Maybe he is a better player or is using a better strategy. When the coin keeps getting heads, does it have a strategy?

B: It just kinda happens. It's like when you're betting, if you're going to guess, you want to guess the thing that's been happening. When it happens and happens and happens, the way our brain works is that it will happen again. Probably I'd guess heads.

Identifying and extending patterns is a problem-solving approach that is emphasized in school mathematics. The concept behind analyzing probabilities of independent events conflicts with the problem-solving strategy of recognizing and extending a pattern. Although

Bob's written answer seemed incorrect, his reasoning is far from being nonsensical as it calls to question the fairness of the coin based on available frequency data.

Coin Sequences task

Six tosses of a fair coin can result in 2^6, or 64, equally likely sequences. Twenty of those 64 sequences comprise three heads and three tails. The probability that six tosses of the coin results in three heads and three tails in any order is relatively likely ($^{20}/_{64}$), but the probability of each of the possible 64 ordered sequences is $^1/_{64}$. In the Coin Sequences task (see fig. 22.1), 3 sequences with three heads and three tails in various orders were presented: HTHTHT, HHTHTT, and HHHTTT, along with 1 sequence of five tails and one heads. Among these choices, the sequence HHTHTT is often viewed as most likely because it satisfies both characteristics of representativeness, in that it comprises three heads and three tails and has a seemingly random ordering (Kahneman & Tversky 1972; Konold et al., 1993).

Table 22.2 presents the distribution of responses to the Coin Sequences task. In contrast with typical adult responses described in previous studies, few students (ten) chose the seemingly representative sequence HHTHTT as the most likely. Surprisingly, more than twice as many students (twenty-one) said that HTHTHT was the most likely of the given sequences. Although the sequence HTHTHT comprises an equal number of heads and tails, it does not correspond to a belief that coin tosses result in random patterns. Other students (eighteen) resisted choosing a single sequence and instead wrote in a response that any of the sequences with three heads and three tails would be the most likely.

Table 22.2
Responses to the Coin Sequences task ("most likely" version)

	Grade 5 (n = 36)	Grade 7 (n = 45)	Grade 9 (n = 50)	Grade 11 (n = 42)	Total (n = 173)
All equally likely	44% (16)	73% (33)	76% (38)	69% (29)	67% (116)
Any of the sequences with three heads and three tails most likely	11% (4)	11% (5)	4% (2)	17% (7)	10% (18)
HTHTHT most likely (choice A)	22% (8)	11% (5)	6% (3)	12% (5)	12% (21)
HHTHTT most likely (choice B)	11% (4)	0	12% (6)	0	6% (10)
HHHTTT most likely (choice C)	0	4% (2)	0	0	1% (2)
Other or no answer	11% (4)	0	2% (1)	2% (1)	4% (7)

About two-thirds of the students correctly indicated that each of the coin sequences listed in the Coin Sequences task is equally likely. Most of the high school students justified their answer using an approach that highlighted the independence of the each individual coin toss or even that indicated directly that each sequence has probability $^1/_{64}$.

Middle school students, on the other hand, more often used other approaches to justify their answers. For example, one of these approaches is the "50-50 approach," which employs the belief that a coin has a 50 percent chance of landing heads or tails. Twenty-seven students, mostly in middle school, justified their answer that all of the coin sequences are equally likely using the 50-50 approach. For example, a seventh grader explained that "all coins have a 50 percent chance of heads or tail." Similarly, a fifth grader explained, "Because there are two sides and a 50-50 chance of getting any pattern."

A second approach used by students to arrive at the correct answer to the Coin Sequences task is known as "outcome approach" (Konold et al., 1993), which centers on the belief that coin tosses are unpredictable. For example, a seventh grader explained, "The result of a coin toss is purely chance." Similarly, a ninth grader justified, "You cannot predict exactly what a coin is going to do." Both the outcome and 50-50 approaches yield correct answers to this particular task but would lead to incorrect answers to other tasks. For instance, these students might indicate that a sequence of three tosses resulting in HHT would be as likely as a sequence of four tosses resulting in HTTH, using either the 50-50 or outcome approaches, even though the shorter sequence is twice as likely. The distribution of student justifications to the correct answer, that all of the sequences are equally likely, is shown in table 22.3.

Table 22.3
Justifications for "all equally likely" on the Coin Sequences task

Justification	Grade 5 (16 of 36)	Grade 7 (33 of 45)	Grade 9 (38 of 50)	Grade 11 (29 of 42)	Total (116 of 173)
Each has probability $1/64$		6% (2)	5% (2)	21% (6)	9% (10)
Independent trials	6% (1)	36% (12)	61% (23)	66% (19)	47% (55)
50-50 approach	56% (9)	36% (12)	13% (5)	3% (1)	23% (27)
Outcome approach	25% (4)	14% (5)	21% (8)	7% (2)	16% (19)
Other or no justification	13% (2)	6% (2)	0	3% (1)	4% (5)

Inconsistencies across tasks

The Four Heads and Coin Sequences tasks address the concept of compound and independent events, and it is interesting to consider the consistency of students' responses across these two tasks. Four Heads is a predictive task, prompting the students to indicate what will happen on the very next toss; the Coin Sequences task gives subjects four events that could have already happened, asking them to judge if any is more or less likely than any other. The information in table 22.4 focuses specifically on the 143 students who answered the Coin Sequences task correctly by indicating that the sequences are equally likely with justification of exact probability (10 students), independence of trials (55 students), 50-50 approach (27 students) or who gave an incorrect answer that any of the sequences with three heads and three tails would be most likely (total of 51 students).

Nearly all of the students who answered the Coin Sequences task correctly with the justification of exact probability, independence of trials, or the 50-50 approach were consistent with the response that there was no most likely outcome on the Four Heads task, as shown in the first three rows of table 22.4. The exact probability or independence justifications yield correct responses to both tasks. The 50-50 approach generates correct answers to the Four Heads and Coin Sequences tasks but does not imply understanding of the concept of independence or of sample space.

Table 22.4
Classification of responses to Coin Sequences task

Response	Frequency (n = 143 of 173 students)	Equally Likely response to Four Heads task (n = 110 of 173 students)
All sequences equally likely justified by exact probability	10	10
All sequences equally likely justified by independence of trials	55	52
All sequences equally likely, using 50-50 approach	27	26
Sequence(s) with three tails/three heads chosen as most likely	51	22

The most frequent inconsistency across student thinking in response to these two tasks was to indicate that a sequence with three heads and three tails would be most likely on the Coin Sequences task but that there is no most likely outcome on the Four Heads task. As shown in the last row of table 22.4, 51 students answered the Coin Sequences task by indicating that some sequence containing three heads and three tails would be the most likely (either HHTHTT, HTHTHT, or a written-in response that any order of three heads and three tails would be most likely). However, about half of those students (22 of 51) responded that there is no most likely outcome on the Four Heads task. These students seem to be struggling with competing beliefs, namely, that coins have no memory and that coin tosses ought to result in equal numbers of heads and tails.

An interview with a seventh grader illustrates an example of this inconsistency. Kendall explained his answer to the Four Heads task, saying that there was neither a most likely nor a least likely outcome on the fifth toss. I then directed his attention toward the Coin Sequences task and his written response that any of the sequences with three heads and three tails is most likely.

Researcher: Which is the most likely or are they all equally likely?

Kendall: OK, so . . . um so I would think that, well, choice A [HTHTHT] has three heads and three tails, which, I don't know, it also seems that all of them could. You don't know exactly what's going to happen. Probability wise, I'd say that A [HTHTHT] and B [HHTHTT] and C [HHHTTT] all have three heads and three tails so I think that those are the most likely.

R: Which is the least likely?

K: Probably D [TTTHTT] because you're getting one heads up.

Because Kendall's earlier response regarding the Four Heads task was indicative of his belief that coin tosses have no memory, my intention was to activate this belief in the context of the Coin Sequences task. One way to accomplish this was to attune Kendall toward each individual toss.

R: *[I use my hand to cover up all tosses but the first toss.]* Which is the most likely or are they equally likely?

K: Heads or tails? *[Researcher nods yes.]* Both are equally likely. I think they're the same, if it's a fair coin.

R: *[I use my hand to cover up all tosses but the second toss.]* Which is the most likely or are they equally likely?

K: My answer is the same.

R: *[I use my hand to cover up all tosses but the third toss.]* Which is the most likely or are they equally likely?

K: I would have to answer the same.

R: How about the whole question?

K: Oooh!

R: What's that?

K: If you look at it that way, they're equally likely because, yeah, if you were to put, yeah. If you were to toss these all up, since there's no way to know, there's not one that's more than the others. They're all equally likely.

The inconsistency seems to have been resolved, as evidenced by Kendall's exclamation ("Oooh!") and his new answer. However, the following dialogue suggest otherwise:

R: You have two ways of answering this question. What's your final answer?

K: Probably, maybe, mathematically wise, maybe my first answer, three heads and three tails. The second way makes more sense overall.

R: OK. Which one are you going to go with?

K: Overall sense.

Although my work to activate the belief that coins have no memory seems to have resulted in Kendall's new idea that all of the sequences are equally likely, he seems to have resolved this inconsistency by compartmentalizing (see Vinner, 1990) one answer, in this case, the incorrect answer, as a "math answer" and the other as the answer making "more sense overall."

An interview with an eleventh grader led to similar findings. Will wrote that there was no most likely outcome for the next toss on the Four Heads task. Will also wrote that the alternating HTHTHT was the most likely of the coin sequences. During the interview, when I asked him to explain his answer to the Coin Sequences task, Will made a distinction between his sense of a real-world scenario and what is true mathematically.

Researcher: Which is the most likely or are they equally likely?

Will: I tried to address this from a common sense point of view. It definitely wasn't D—you're more likely to get three heads and three tails, but I realize that it's an independent event every time. Mathematically, it's equally likely to get all of the, every time it's 50-50, but the way I see it, in a real-world scenario, because three heads and three tails. I chose A [HTHTHT] because it's an alternation.

R: If I asked you, "Are you more likely to get 3 heads, 3 tails in any order or 5 heads and 1 tail"—that's a different question, because here I wrote out specific orders.

W: Strictly mathematically, A, B, and C all have equal numbers of heads and tails.

R: What about D [TTTHTT]?

W: Also, because it's independent. In the real world, it's more likely to get three heads and three tails. The correct math answer is E [All are equally likely].

In the first interview example, Kendall referred to the belief that coin tosses should result in an equal number of heads and tails as the "math answer," while indicating that independence of coin trials makes more "overall sense." In the second interview example, Will's ascribed the "math answer" to the correct mathematical answer, and his "real world" answer corresponded to the belief that coin tosses should result in an equal number of heads and tails. These two examples demonstrate that when students are confronted with their own inconsistencies, some may reconcile the inconsistency by accepting the conflicting answers and compartmentalizing them according to an in-school or out-of-school distinction.

Implications for Instruction

This study highlights the general importance of student explanations and justifications. Explanations or justifications of answers can allow teachers to analyze students' responses deeply, beyond whether an answer is correct. It is not enough for teachers to know that students perform poorly, or perform well, on a given task: A knowledge of the common errors or of the different strategies being used can help a teacher prepare activities that confront errors, build on student thinking, or direct students' attention toward specificity of language.

Organizing student responses around the structure of beliefs about coins appears to be effective; their incompatibility as a set provides a way to think further about the implications of this work for classroom teaching. One suggestion is to create situations in which students solve a variety of problems and then are led toward a confrontation by inconsistencies. For example, as this study shows, a student could reply that there are no most or least likely outcomes on the fifth toss of a fair coin that has come up heads four times in a row, according to the belief that coins have no memory. That same student could then indicate that the sequences of coin tosses possessing an equal number of heads and tails are more likely than those that do not, according to the belief that coin tosses result in an equal number of heads and tails. The teacher could attempt to direct the student's attention toward this inconsistency with the goal of creating cognitive conflict, which might lead to the student's resolution of the inconsistency and, perhaps, to new or more developed understandings. Of course, an additional finding of this study is that a teacher's intention of creating cognitive conflict might not result in a conflict for the student or change the student's thinking in a way that is expected.

Finally, the extent of students' belief in the acceptability of multiple, conflicting answers to the same mathematics question is worthy of further research. How might this be unique to probability, or are there other mathematical content areas in which this occurs? What are the effects of different types of probability instruction or curriculum on this phenomenon? Earlier, I described students' multiple, conflicting answers in terms of a tension between students' beliefs about mathematical thinking in school or out of school. An alternative is to view the multiple and conflicting answers in terms of a distinction between theoretical and empirical probabilities. Students' attributions of "math answers" or perhaps theoretical probabilities as being different from "real-world answers" or empirical probabilities could be interpreted as a call to better integrate empirical and theoretical probability into classroom teaching and learning.

References

Fischbein, E., & Schnarch, D. (1997). The evolution with age of probabilistic, intuitively based misconceptions. *Journal for Research in Mathematics Education, 28,* 96–105.

Kahneman, D., & Tversky, A. (1972). Subjective probability: A judgment of representativeness. *Cognitive Psychology, 3,* 430–54.

Konold, C., Pollatsek, A., Well, A. Lohmeier, J., & Lipson, A. (1993). Inconsistencies in students' reasoning about probability. *Journal for Research in Mathematics Education, 24,* 392–414.

Rubel, L. (2002). *Probabilistic misconceptions: Middle and high school students' judgments under uncertainty* (Unpublished doctoral dissertation). Teachers College, Columbia University, New York, NY.

Rubel, L. H. (2007). Middle school and high school students' probabilistic reasoning on coin tasks. *Journal for Research in Mathematics Education, 38,* 531–556.

Vinner, S. (1990). Inconsistencies: Their causes and function in learning mathematics. *Focus on Learning Problems in Mathematics, 12,* 85–98.

Chapter 23

Dice: Fair or Not Fair? That Is the Question

Jane M. Watson
University of Tasmania, Australia

The issues that students face when rolling dice today are not new but similar to those faced by others throughout history. The ancient astragalus, which comes from the heel bone of a hoofed animal and was probably the earliest chance device, was definitely not fair in the sense that each of its four sides was not equally likely to occur topmost when tossed. This "fairness" was not important at the time, however, because it was believed that a god or fate determined the outcome, not chance based on the shape of the astragalus. Whatever the belief in the mechanism for determining outcomes, the outcome of winning has been an important concept since the advent of gambling in ancient times. The desire to win led to the early "loading" of dice. By the time of Cardano and Galileo, the idea of fairness was firmly entrenched, with Cardano qualifying his analysis of probabilities with "if the die be honest" (Bennett, 1998, p. 77) and Galileo describing a fair die as one with "six faces and when it is thrown it can equally well fall on any one of these" (p. 47). Using trials to test hypotheses about fairness was an idea known, at least theoretically, to Cicero and Cardano.

The issue of fairness is often dismissed in classrooms with questions such as "What is the chance any side will come up when this die is tossed?" Hearing a chorus of "$\frac{1}{6}$" from students, teachers are likely to move on to issues they consider to be more sophisticated and more likely to challenge their students, such as what happens when two dice are tossed and the outcomes summed, or the fairness of games whose rules are determined based on the outcomes of dice tosses. Even when trials are performed for a single die, the purpose is usually to verify the fairness of the die rather than put fairness to the test, and any empirical deviations from an even distribution are likely to be dismissed as random variation. Researchers have found, however, that the beliefs about dice that students bring with them to school include beliefs that God, fate, or mental powers determine dice outcomes and that such beliefs also include understandings about dice developed through experiences when playing games, such as the importance of rolling technique, experiences of losing games, and the difficulty of obtaining a 6 to start in a game. For many students, these intuitions about dice and probability are resistant to instruction and do not improve with age.

Summary of the Study and Its Results

In this chapter, I report on a study that sought to explore the development of students' beliefs about, and strategies in dealing with, the fairness of dice. The following questions provided a starting point:

1. What do students believe about the fairness of dice? What experiences or understandings do students state to support their beliefs? Do these show a hierarchical progression? Do beliefs differ for students of different grades?

This chapter is adapted from J. M. Watson & J. B. Moritz (2003), Fairness of dice: A longitudinal study of students' strategies and beliefs for making judgments, *Journal for Research in Mathematics Education, 34,* 270–304. The author wishes to acknowledge the Australian Research Council for funding the research reported in the *JRME* article.

2. What are students' dominant strategies for assessing the fairness of dice? What are the qualitative differences among these strategies, and do they fit within a framework of increasing sophistication and statistical appropriateness? Do strategies differ for students of different grades?

3. Is there an association between students' beliefs in the fairness of dice and their strategies for assessing the fairness of dice?

Interviews were conducted with 108 Australian students in grades 3, 5, 6, 7, and 9 (ages eight to fifteen years), representing rural and urban, and public and private schools. Students were selected for interviews on the basis of the variety and sometimes the unusual nature of their responses to survey items about chance and data. Teachers confirmed that the students selected by the researchers were articulate and willing to be interviewed. There were 32 students in grade 3, 42 in grades 5 to 7 (combined middle school group), and 34 in grade 9.

Students were given three wooden dice with 3-cm edges. The dice were typical in that they had "dots" on each face representing numbers (e.g., one dot representing the number 1). A red die was theoretically fair and with the numbers 1 through 6 represented; a white one had only the numbers 1, 2, and 3 each represented twice (opposite faces had the same number of dots); and a blue one, again with the numbers 1 to 6 represented, had been weighted on the side with the five dots, and hence the side opposite 5 was more likely to come up. The blue die was configured with opposite faces summing to 7, and the trials ($n = 200$) conducted before the study yielded the following distribution of outcomes: 1 (16 percent), 2 (33 percent), 3 (20 percent), 4 (16 percent), 5 (7 percent), and 6 (9 percent).

The interview protocol began with the placing of the dice on the table in front of the student and asking an introductory question such as "Do you play games with dice?" Next, a question was asked to elicit students' beliefs about fairness of dice. Sometimes a frequency form of the question was used, similar to "Do some numbers come up more often than others?" and sometimes a chance form was used, such as "Do all numbers have the same chance of coming up?" The interviewer then asked for clarification, either of unfair beliefs by asking which numbers occur more often, or of fair beliefs by asking for confirmation that all numbers come up equally often *and* that they have the same chance.

As a transition into considering strategies for determining fairness, the question "Do you know what it means for dice to be fair?" was used. This was for clarification only; if the students were unclear, a statement such as the following was made: "We say a die is fair if all six numbers have the same chance of coming up." The interviewer then asked students how they would work out which of the dice on the table were fair and which were unfair. Students who failed to inspect or throw the dice were told that they were permitted to do so. Students who continued to assert beliefs without engaging the dice in front of them were further asked what they might do to be *sure* or to demonstrate they were right to be skeptical. In interviews where students used observation alone to conclude that the white die was unfair and the other dice fair, students were sometimes told that one of the remaining dice in front of them was fair and one was unfair, and were asked to judge their fairness. The wide variation in initial responses necessitated flexibility on the part of the interviewer, and the excerpts that follow include interviewer prompts to clarify the part these may have played in the students' responses. Because of ethics requirements, at the end of the protocol, students who confirmed the blue die to be unfair had the process of creating the bias described; they were assured that such bias was not normal.

The identification of different beliefs and strategies was informed by a combination of (a) the statistical appropriateness of the students' comments and (b) their structural complexity. Statistical appropriateness included appreciation that fair dice should have equally likely outcomes, subject to a degree of random variation when trialed, and knowledge of strategies for examining and empirically trialing dice outcomes. Structural complexity was informed by the cognitive development model of Biggs and Collis (1982), incorporating various modes of thinking similar to Piagetian stages, including the *ikonic* mode, involving intuitions or storytelling about experiences, and the *concrete symbolic* mode, involving symbols and propositions referring to concrete objects, as often taught in schools. Within the concrete symbolic mode, three hierarchical levels were identified based on the structure of the observed response:

- *Unistructural* responses that employed single elements

- *Multistructural* responses that employed multiple elements, usually in sequence, sometimes recognizing but not resolving conflicts among the elements

- *Relational* responses that related elements and created closure for the question

These three levels of response plus ikonic responses were identified for the first two research questions, reflecting increasingly more appropriate statistical understanding.

Research Question 1: Beliefs about fairness

The four levels of response for students' *beliefs* about fairness of dice are summarized in table 23.1, along with the number of students in grades 3, 5 to 7 (middle school), or 9, as well as the total, giving each level of response. At the ikonic level students believed dice are unfair with idiosyncratic reasoning. At the unistructural level they had a single-minded belief that all dice are fair, whereas at the multistructural level this was qualified with other conditions. At the relational level there was recognition of the more complex relationship of short-term and long-term variation.

Table 23.1
Frequency of level of belief in fairness of dice by grade

Level of belief	Explanation	Grade			Totals
		3	5–7	9	
Ikonic—Unfair	Beliefs that dice are unfair, involving idiosyncratic or inconsistent stories about experiences, often with games	20	17	7	44
Unistructural—Fair	Single belief in fairness or equal chance in a propositional form	9	20	21	50
Multistructural—Fair qualified	Belief that dice are fair subject to the rolling condition or to the physical condition of the dice	3	2	5	10
Relational—Short-term vs long-term	Belief that outcomes are fair in the long-term, but that short-term experience may suggest otherwise	0	3	1	4
Totals		32	42	34	108

Overall, 87 percent of the students expressed firm beliefs that dice are either fair or unfair, with 41 percent believing unfair. Responses shown from the various grades at the four levels illustrate the comments teachers are likely to hear and need to be prepared to respond to in the classroom. In the responses, " . . . " denotes a pause in the student's response, and "[. . .]" denotes dialogue that has been edited, which does not affect the meaning of the extract. "*I*" denotes the interviewer comments, which have in some cases been edited for brevity and appear in italics within the student comments. Students are uniquely identified (e.g., S1) because occasional multiple responses are presented from a single student.

Ikonic—Unfair beliefs

Responses classified as Unfair beliefs were either idiosyncratic or inconsistent. Thirty-eight students stated idiosyncratic beliefs that some specific numbers occurred more often than others; that is, they had "more chance." Some responses were apparently based on remembered personal experience or other ideas, such as "Yeah, 3s, 4s, and 2s. . . . Or the ones that you don't want to come up" (S1, grade 6). Other responses were based on specific experiences (e.g., games) from rolling dice with a focus on getting the number 6. Focusing on this outcome appears to have distorted the students' perceptions of the outcomes that occurred for *all* rolls, as illustrated in these comments:

S2:   6s don't come up as often as smaller numbers. *[I: So that's when you are playing games?]* Yes, especially when you want a 6 to start. (grade 9)

S3:   Yes. My mum always gets 6s and I always get 2s or 1s. [. . .] My hand just must be not very good at throwing the dice. (grade 5)

One student appeared to show confusion concerning "which number is most likely to occur" and "which number is the highest number":

S4:   6. *[I: Do you think a 6 comes up more often?]* That's the most in the dice. Well, it's the most number in the dice. (grade 3)

Six students responded inconsistently with some evidence of fairness, but contradicted theses responses with other comments suggesting certain numbers were favored. The following response reflected both fair and unfair views, as well as a conflict between fairy-tale beliefs and personal experience:

S5:   Yes, number 3 because in fairy-tales there's 3 wishes and there's 3 fairy godmothers and there's 3 wishes and there's all sorts of 3 things. *[I: . . . do you find that 3 comes up a lot, or do all of the numbers come up?]* Well 6 comes up not the most because it's the biggest, and for me 2 comes up usually. *[I: Do you think that they all have the same chance of coming up?]* Yes, I think that's just because you turn the dice differently every time. (grade 3)

Unistructural—Fair beliefs

Fifty students simply stated, or agreed to the statement, that no numbers were more likely to come up than others and hence all numbers have the same chance on fair dice. The basis of this understanding was often not stated, that is, whether it was based on personal experience or on a theoretical belief. These responses are typical of the concrete symbolic mode in which propositions are stated referring to concrete situations and not restricted to telling of isolated stories as at the ikonic level:

S6:   *[5-second pause] [I: Or do you think they come up about the same?]* They all come up the same. (grade 3)

S7:   No, I think that they all have the same chance of getting rolled than all of the rest of the numbers. (grade 9)

Multistructural—Fair Qualified beliefs

Ten responses were more complex structurally as students volunteered qualifications about the fairness of dice. Responses included requirements for fair dice outcomes, such as how the dice are rolled or physical characteristics of the dice. Four students stated that no numbers were more likely to come up than others, but that this depended on an appropriate rolling technique. One student, for example, commented that minimal thrust when rolling might introduce bias in the outcomes, and that it is possible to exploit some aspects of rolling technique to bias outcomes:

S8:   Not really. *[I: They all have about the same chance, do you think?]* Yes. It depends how you roll them. *[I: Right, how does it depend on that?]* *[S8 manipulates red die.]* If you roll it so that you roll it that way, it could land on any number along there, but if you rolled it along that way, it could land on any of those numbers. (grade 5)

Six students stated that no numbers were more likely to come up than others, provided the dice were manufactured appropriately, but they also suggested that imperfections in shape or weight distribution would affect outcomes. A few students were very sensitive to these conditions, believing that dice manufactured with grooves for dots to denote the number on the face would result in an uneven weight distribution, as illustrated by this comment: "Well, if they had huge holes *[manipulates blue die while talking]*, then 6 would probably come up the most because 1 is set heaviest and it's got the least numbers, but this looks like it won't happen that much . . . and that they're both random" (S9, grade 7).

Relational—Short-Term Variation beliefs

Four students suggested that dice are fair but that it may appear that some numbers come up more often when dice are tossed. These contrasting ideas were resolved in favor of belief in fairness, but were related to acknowledgment of apparent unfairness. This unfairness was attributed to short-term observation or selective recall. These students dealt with a conflict of sometimes-observed short-term results and the understanding of long-term trends, and related the ideas in an appropriate fashion.

S10:   No. Umm . . . it sometimes seems that the other person always gets a 6, but it's just the luck of the draw really. It's just that you might get a row of 6s, but then probably in the next game you'll get a row of 1s *[laughs]*. And, but usually it just depends, sometimes it might come up more often, but then the next time it won't. It's just the luck of the draw. (grade 7)

S11:   Well, if it was, if you had to say, sort of like the chance, you would say that all numbers have the same chance, but sometimes it doesn't turn out that way. Because sometimes we do, in our class, we have things that we, we make our own dice, and then we have to roll them 60 times, and see which comes up the most, out of all the numbers. (grade 6)

Research Question 2: Strategies for judging fairness

When asked for *strategies* to judge fairness of specific dice, students engaged with the dice at four levels, which were similar structurally to responses for beliefs. Ikonic responses incorporated intuitive beliefs, including anthropomorphism and luck, in idiosyncratic answers. Unistructural responses just asserted the single idea that dice are fair, without further suggestions. Multistructural responses considered physical conditions or rolling conditions on a die to determine fairness, whereas at the Relational level students suggested systematic trials, recording results and comparing frequencies, to test for fairness. As shown in table 23.2, most third-grade students responded with either idiosyncratic strategies or observational strategies, whereas for students in grades 5 to 7 and grade 9, the modal response level involved observational strategies. Very few students in any grade level used empirical strategies.

Table 23.2
Frequency of level of strategy for determining fairness by grade

Level of strategy	Explanation	Grade			Totals
		3	5–7	9	
Ikonic—Idiosyncratic	Incorporation of nonstatistical issues	15	7	2	24
Unistructural—Untestable	Assertion that dice are fair; no test is necessary	5	4	5	14
Multistructural—Observational	Reference to physical features, rolling technique, or a few unsystematic trials	12	30	21	63
Relational—Empirical	Systematic trials of the dice	0	1	6	7
Totals		32	42	34	108

Ikonic–Idiosyncratic strategies

Twenty-four students attached special significance to individual number outcomes or to combinations of outcomes across two or more dice. In a few cases, the presentation of the dice with certain numbers facing up was the basis on which students judged fairness, as was the case for the following student:

S12:   *[White die shows 2, others show higher numbers.]* This one [white] is a bit unfair because you don't get much like these two [red and blue]. *[Interviewer turns dice so all show 2.]* [. . .] *[I: Does that make any difference?]* Yes. That is fair if they were all on 2. (grade 5)

This student did not perform any trials, as was the case for the following student who used the numbers facing up on each die compared to her beliefs about the numbers:

S13:   *[dice showing red 6, white 2, and blue 5]* Well, I reckon number 6 isn't fair, because it just doesn't really come up as much as the others. I just think that it doesn't come up. I reckon a 2 does. I reckon it sort of lands on it most of the times. And the 5, well, it's in the middle really. (grade 6)

When prompted, this student acknowledged the white die would be unfair because of the repeated numbers but, when further prompted, did not consider trialing the others:

S13:   *[I: Suppose that someone came along and they said, "I think that one of these two is unfair, it seems to keep coming up on one of the numbers a lot." Could you test out if they were right?]* No, not really, you couldn't really say that there wasn't. It's probably just a lucky chance that it came up on that one all the time.

Unistructural–Untestable strategies

For fourteen students, beliefs strongly influenced their responses, and they relied solely on these beliefs for judging fairness. Of the following two students, the first (S7) had earlier stated a unistructural belief that dice would be fair, whereas the second (S14) had stated an ikonic belief:

S7:   *[holds all dice, casually turning; 10-second pause]* I don't know.
[I: Suppose someone came along and said, "I think one of these might be pretty unfair, it just seems to keep coming up on just a few of the numbers."] *[picks up all dice, 7-second pause]* They all look pretty fine to me. (grade 9)

S14:   *[I: . . . which ones might be fair, coming out evenly, and which ones might be unfair?] [doesn't pick up dice]* They would all be the same. (grade 9)

When prompted about checking over the faces of the dice, these two students could identify the repeated numbers on the white die but used only this strategy provided by the interviewer for judging fairness of the other dice, initiating no strategies of their own.

Multistructural—Observational strategies

Sixty-three students focused on physical characteristics of the dice in front of them, either by observation or by manipulation. The three strategies identified at this level involved (a) possible outcomes—that is, observing that each of the six numbers appeared on one and only one face, (b) feeling if the dice were weighted symmetrically or not, or (c) rolling the dice to observe rolling irregularity rather than the outcome of the roll. None suggested on their own initiative to trial the dice and record results to draw conclusions about fairness, although some did so after prompting.

Twenty-one students judged fairness according to whether each of the six numbers appeared on one and only one face. Students thus judged the white die with repeated numbers as unfair, whereas the other dice were judged as fair without any other considerations. Following closure on this decision, further questioning prompted twelve students to perform some trials, whereas nine students continued to be uninterested in doing so: "This [white] isn't a fair die. [. . .] Because it doesn't have 4, 5, or 6 like these two [blue and red]. [. . .] Those two are fair and that one isn't" (S15, grade 9). A few students suggested that the white die with two faces each of 1, 2, and 3 was still fair because it had two of each number.

Twenty students focused on physical symmetric characteristics of the dice instead of the numbers on the faces, such as the importance of the cubic shape being exact or the weight distribution being even. After judging that closure had occurred with regard to the fairness issue, the interviewer prompted the students further, asking for other ways to demonstrate the conclusion to people who might disagree. Based on this prompt, only four students trialed the dice in any systematic way, one student only after repeated prompting. Ten students could not be prompted to trial the dice, with only a few rolling the dice to demonstrate the uneven weight by the bias in the roll. The following dialogue illustrates the belief that weight is the only possible strategy for judging fairness:

S16:   *[manipulates each die]* Those two [white and red] are fair, and that one [blue] is unfair. [. . .] By putting them on an angle and seeing if they swung around. You could feel the weight, that it was at the top, like that *[demonstrates swinging]*. With the other ones, they didn't swing or anything. [. . .] *[I: Is there any other way that you could do it other than letting them feel it?]* Not really, because it's just chance. With any dice, it's chance if it lands on one number anyway, so that could happen with a normal dice as it would with a weighted dice. *[I: Do you think it's more likely to come up on some numbers than others?]* Yes. *[I: Would you be able to work out which ones? Is there a way of doing that?]* Yes, by the opposite side of the weight. (grade 6)

Six students combined ideas of possible outcomes and symmetry to cover both types of physical characteristics. A ninth-grade student, for example, first noted the difference in weight, but then in manipulating the dice stumbled across the repeated numbers. The fact that many students identified one or other physical aspect and that few students considered both probably reflects in part the motivation of most students, who, having identified one strategy, felt that this was sufficient for the task. Twenty-two students rolled the dice to demonstrate that a variety of outcomes occurred but did not employ any systematic testing and recording to conclude that these outcomes were equally frequent. Eight students rolled the dice together, as if the task were to decide whether all three dice would together yield a fair distribution of outcomes, rather than whether fairness applied to each individual die:

S17:   *[Rolls each die once, then once again, and considers all dice together]* 6 came up more times than all the others. *[I: Did it? How many times did you throw them?]* I threw them twice, and 6 came up twice, and the other numbers only came up once. *[I: Right. Could that help you decide whether they were fair or not?]* They're unfair. (grade 3)

Fourteen students rolled the dice and considered each independently, sometimes with a few rolls, but had no system for collecting data, recording them, or summarizing the results. Some of these students appeared to be observing if a variety of outcomes occurred rather than documenting the frequency distribution. The following student, for example, first commented about the results across the three dice, but then appeared to draw conclusions across the three trials of each die, based on which had repeated outcomes:

S18:   If you roll them each 3 times. *[rolls red die once, then blue and white]* Well, I suppose that's a kind of . . . they're all different, so they've all got a chance of . . . *[rolls each once more]* They're all still different again, except 1 keeps coming up *[points to white die]. [rolls each once more]* That one [red] is the fairest because all different ones come up, and these two have had the same [blue and white]. (grade 6)

Other students were also unsystematic in trialing, simply rolling and observing outcomes without planning how many rolls to do or recording the results. One sixth-grade student, for example, also drew a conclusion from a few trials based on repeated outcomes; she concluded that the blue die "seems to come up on the same numbers most of the time" and that the white and red dice "usually come up on different ones," and hence that white and red "are more fair." Other students using unsystematic trialing recorded results but appeared to have no way of using them to draw a conclusion about the fairness.

Relational—Empirical strategies

Seven students suggested trialing the dice, and they did so with appropriate recording methods to use empirical data concerning the frequency distribution of outcomes for each die to judge fairness. It is interesting to note that of the seven students who responded in this level, none checked the numbers on the dice before trialing, and only two checked them after their trials, whereas the other five required prompting by the interviewer to check the dice and notice the repeated digits on the white die. The physical device hence mattered less than the outcomes, which were the basis for the decision. Of the seven students, four used only a small number of trials (e.g., fewer than eighteen rolls of each die) and three used more trials.

Four students who performed trials only used six to twelve rolls of each die. This may have been a function of the perceived time available in the interview, because some students asked how much time they could have. Some had simple recording strategies based on memory or a sequential list of outcomes. The following student, for example, decided to do twelve trials, wrote the numbers 1 to 12 in a column, wrote the outcomes for each roll of the blue die in an adjacent untitled column, and wrote the outcomes for the red and white dice in columns titled by color. She also hinted at an idea about short-term variation in anticipating results, though her belief quoted earlier concerned physical conditions for fairness:

S9:   Well you could roll it say 12 times and then expect around each number to come up twice, but then if . . . it's okay to be not, to be unfair, because it's just chance like, maybe a 1 might not turn up all the time, that time. But then maybe it would turn up a lot another time. [. . .] Right, so that one was a 6 *[records blue results]* 3, 2, 3, 6, 2, 3, 2, 5, 4, 4, 5. It could have been the way that I rolled it that made it not turn up on the 1, or else the indents were too large and it was very heavy on the bottom side [. . .] But by that it looks quite fair. At the start it looked a bit unfair, but then it looked quite fair. (grade 7)

Three students trialed the dice more than eighteen times and systematically recorded results to consider whether the distributions of outcomes were even. The recording strategy used involved tallying frequencies for each of the six outcomes, thus readily permitting interpretation against the hypothesis of an even distribution of outcomes:

S19:   *[without comment, rolls each die 20 times and records in a tally table (numerals that follow refer to the number of dots on the die): white: four 1s, nine 2s, seven 3s; blue: one 1, three 2s, eight 3s, six 4s, one 5, one 6; red: two 1s, three 2s, four 3s, four 4s, three 5s, four 6s]* I definitely think the blue one's weighted. You can feel it's heavier, and the white one possibly, but I don't think the red one is at all because it's fairly even across . . . *[points to results]* [. . .] The white one could be, I don't know. It could be just freaky that it came up with those 3s all the time but then again who knows. (grade 9)

This student only noticed the repeated numbers on the white dice after excessive prompting, as the continuation of the dialogue shows:

S19:   I think the red one is normal because it seems to have fairly even distribution with the numbers. *[I: . . . the white one?]* I think it is [unfair] because it seemed to always roll 1, 2 or 3. [. . .] *[I: prompts to inspect]* It doesn't seem to look like it is or anything. It could have been like a freak that it was coming up like that. [. . .] *[I: Well just as it happens this one has a 2 and a 2 here . . .]* [. . .] Oh, no wonder! So it's not weighted, it's just like a cheat dice. I didn't even notice that.

Research Question 3: Association of beliefs and strategies

Research Question 3 concerned the association between students' beliefs and their strategies for judging fairness of dice. Table 23.3 shows that there was not a strong association between the levels of response for beliefs about fairness of dice and the levels of response for strategies for determining fairness. Of those students with Unfair or Fair beliefs about dice, about a quarter used idiosyncratic strategies for judging fairness, over half used observational strategies, and only four students performed empirical trials. The ten students who expressed Fair Qualified beliefs generally used observational or empirical strategies (five students and three students, respectively). Of the four students who believed in fair dice subject to short-term variation, none used an empirical strategy for judging fairness; all preferred observational strategies.

Overall, forty-three out of forty-four students who believed dice were unfair and forty-seven out of fifty students who believed dice were fair suggested idiosyncratic, untestable, or observational strategies, suggesting they felt no need to test what they *knew* to be true. On the other hand, only three of fourteen students who had a fair qualified or short-term variation view of fairness used an empirical strategy to test fairness, whereas nine suggested weaker observational strategies and two gave idiosyncratic or untestable strategies.

Table 23.3
Frequency of level of strategy by level of belief

Level of strategy	Level of belief				
	Unfair	Fair	Fair qualified	Short-term vs long-term	Totals
Idiosyncratic	14	9	1	0	24
Untestable	7	6	1	0	14
Observational	22	32	5	4	63
Empirical	1	3	3	0	7
TOTALS	**44**	**50**	**10**	**4**	**108**

Implications for Teaching

The inconsistencies between beliefs about and strategies for determining fairness point to a dilemma for teachers. If students do not believe dice are fair to start with, will a lesson trialing real dice help them to change their beliefs or not? Perhaps students "learn" to repeat what the teachers want them to say—that dice are fair. This is the "in-school" belief. What they believe outside the mathematics classroom may be another matter. This point requires up-front discussion in the classroom, perhaps even with examples from the media in which claims of luck or foresight are made. In the overall context of the probability part of the mathematics curriculum, the topics addressed here provide an excellent opportunity to address all three perspectives on probability: subjective, frequency, and theoretical. If this approach is taken from the beginning, perhaps students will develop a more balanced and circumspect belief structure. There is also the possibility to compare and contrast fairness in probability with fairness in sampling (e.g., Watson & Moritz, 2002). This relationship becomes an important issue in the senior years of schooling.

The outcomes of the study suggest the importance of two instructional strategies when introducing random generators to students, at whatever grade the introduction is made. One is the use of concrete materials. Computer simulations of dice outcomes were not included as part of the protocol in this study, and it will be interesting to follow the development of student understanding as student-friendly software becomes more common in classrooms (e.g., English & Watson, 2016; Watson & English, 2015). The abstract nature of simulation may be far removed from the hands-on approach using physical objects as a starting point to build understanding about outcomes of rolling dice. This is not to say that handling dice is without its difficulties. Observations in this study showed that many students found it very difficult to roll dice in an unbiased fashion. I realized that this was not a trivial exercise when conducting trials of the loaded die.

The other important instructional strategy is the focused use of language, particularly related to the word *chance* itself. One aspect of this has been mentioned in relation to the outcomes for students' beliefs. Specific discussion should take place about frequencies as observed phenomena and chances as theoretical entities. It would appear that the expectation of frequencies over increased numbers of trials approaching theoretical chances (probabilities) needs much explicit discussion. There is also the aspect of "chance" as the mechanism that provides the variation from theoretical expectations. One student (S16), for example, in explaining the lack of need to test her belief in fairness, attributed deviations to "lucky chance." Although random behavior is an appropriate alternative description, the colloquial use of terms like *chance* in different contexts means that teachers must always clarify their own use of the terminology.

In the days when only theoretical probability based on sample spaces was taught, with no experimentation in the classroom, curriculum documents and textbooks assumed that dice were fair as a starting point for work in probability. This assumption was adequate and fit the mathematics curriculum. Today, however, given the rise in the prominence of statistics in the curriculum, it may be more appropriate to put *questioning the fairness of the random generator* on the curriculum agenda. The only curriculum framework to address this important issue explicitly is the GAISE Report (Franklin et al., 2007). In illustrating probability as a tool for statistics rather than just a mathematical model, two problems are set as shown in figure 23.1.

Problem 1: A mathematical probability problem

Assume a coin is "fair."
Question: If we toss the coin five times, how many heads will we get?

Problem 2: A statistics problem that can use the mathematical probability model from Problem 1 as a tool to seek a solution

You pick up a coin.
Question: Is this a fair coin?

Figure 23.1. Two problems from GAISE (Franklin et al., 2007, p. 8)

This study was about belief in the assumption in problem 1 in the figure and the need for that assumption in problem 2. The results for these students, if typical, would indicate that much discussion needs to take place in the classroom before understanding of fairness and its application in problem solving can be assured. Students' contact with probability can then be linked to other parts of the statistics and probability curriculum: performing trials and using them to make decisions to justify the physical model they are going to use. Having developed empirical techniques for handling data and confidence that dice are fair based on the results of trials using single random generators, students can then transfer these strategies and beliefs in modeling and describing compound events, such as the sum of two dice, and more complex phenomena, such as the fairness of games.

References

Bennett, D. J. (1998). *Randomness*. Cambridge, MA: Harvard University Press.

Biggs, J. B., & Collis, K. F. (1982). *Evaluating the quality of learning: The SOLO taxonomy (structure of the observed learning outcome)*. New York, NY: Academic Press.

English, L. D., & Watson, J. M. (2016). Development of probabilistic understanding in fourth grade. *Journal for Research in Mathematics Education, 47*, 28–62.

Franklin, C., Kader, G., Mewborn, D., Moreno, J., Peck, R., Perry, M., & Scheaffer, R. (2007). *Guidelines for assessment and instruction in statistics education (GAISE) report: A preK–12 curriculum framework*. Alexandria, VA: American Statistical Association.

Watson, J., & English, L. (2015). Expectation and variation with a virtual die. *Australian Mathematics Teacher, 71*(3), 3-9.

Watson, J. M., & Moritz, J. B. (2002). Developing concepts of sampling for statistical literacy. In J. Sowder & B. Schappelle (Eds.), *Lessons learned from research* (pp. 117–124). Reston, VA: National Council of Teachers of Mathematics.

Watson, J. M., & Moritz, J. B. (2003). Fairness of dice: A longitudinal study of students' strategies and beliefs for making judgments. *Journal for Research in Mathematics Education, 34*, 270–304.

Chapter 24

Using a Computer Microworld to Make Sense of the Total of Two Dice

Dave Pratt
Institute of Education, University of London

Many studies have shown that strategies used in making judgments about chance are subject to systematic bias. In the case of randomness and chance in the domain of probability, I found that little was known about the relationship between the external structuring resources made available in a pedagogical (teaching) environment and the construction of new internal resources on the part of learners—especially learners as young as ten or eleven years of age. In my *Journal for Research in Mathematics Education* article (Pratt, 2000) I used a computer microworld environment to investigate the interplay between children's informal intuition and computer-based resources as the children constructed new internal resources for making sense of chance. For this chapter, my focus is on the section of the *JRME* article that details the work on two children (to whom I assigned the pseudonyms Anne and Rebecca). I begin the chapter with a brief overview of the theoretical basis for my research. (For a more complete consideration of the theoretical considerations, I refer the reader to the *JRME* article.)

Theoretical Considerations

Piaget and Inhelder (1951) have argued that a sophisticated knowledge of probability—in particular, in the case of the total of two dice—depends upon mental schemas that have not yet developed at, say, age ten or eleven years. A proper grasp of the probability of obtaining a total of 7 with two dice involves knowing how many configurations of two dice results in that total (e.g., 6 + 1, 1 + 6, 2 + 5, 5 + 2, etc.). The probability of obtaining a 7 is directly proportional to this number. Piaget and Inhelder proposed that the schema for understanding chance like this emerge at a later age.

A bias relevant to the research reported here has been called the *equiprobability* bias, a tendency to assume that different outcomes are equally likely. In the case of throwing dice, this bias might be articulated as "different totals are equally easy to obtain—it is just a matter of chance." In a series of studies Lecoutre (1992) found that the equiprobability bias was resistant to modification but that masking the chance element of the problem could induce a correct response. Also, Fischbein and Schnarch (1997) sought to establish at what age one's probabilistic intuitions stabilize, and found that the equiprobability bias was remarkably stable across all ages. However, Tarr (1998) found that after instruction fifth-grade students exhibited the equiprobability bias far less frequently. Thus, there may be learning environments that cue more probabilistic ways of thinking.

In my research I created such a learning environment, which is described next, and then used it in a case study of sixteen children (ages ten and eleven years). Based on an analysis of the case studies, I found that the children used two kinds of internal resources for making sense of stochastic situations: local and global.

This chapter is adapted from D. Pratt (2000), Making sense of the total of two dice, *Journal for Research in Mathematics Education*, *31*, 602–625.

Local resources are those that focus on trial-by-trial variation and are immediately accessible. I identified four local resources and describe them briefly below (for additional information on the local resources see Pratt, 2000; Pratt & Noss, 2002, 2010):

1. *Unpredictability*—The next outcome is not predictable, though some apparent success in the short term may be experienced.

2. *Irregularity*—No patterned sequence in prior results is evident.

3. *Unsteerability*—The observer is unable to exert physical control over the outcome of the phenomenon.

4. *Fairness*—There exists a rough symmetry in appearance of the events.

Global resources focus on an aggregated view of the stochastic. I identified three global resources, and characterize them briefly as—

1. *Probability*—the proportion of outcomes for each possibility is predictable;

2. *Large numbers*—the proportion of prior results for each possibility in the sample size will stabilize as an increasing number of results is considered; and

3. *Distribution*—the observer is able to exert control over these proportions through the sample space.

There is an interesting correspondence between the two sets of resources in that local resources tend to be inverted in relation to their global counterparts. For example, unpredictability as a local resource is inverted in comparison to the global resource of predictability. And control cannot be exerted locally, but there is a global resource for control through manipulation of the distribution. I refer to examples of local and global resources in the next sections of the chapter that detail Anne's and Rebecca's work in the Chance-Maker microworld activities.[1]

The Chance-Maker Microworld

At the time of my research, the Boxer environment had evolved out of Logo, and it provided a powerful environment in which children were able to express their mathematical ideas in computational terms. The central objects of the Chance-Maker microworld[1] were a series of *gadgets*, computational tools that behaved like their everyday counterparts. The gadgets were designed as familiar devices so as to maximize the likelihood that their behavior and appearance would cue intuitions and expectations based on their real-world equivalents. A variety of gadgets were available—for example, coins that can be tossed and coins that roll, spinners, dice, and a Frisbee—and each gadget contained a variety of tools. The focus of this chapter is on two of the gadgets: one involving two spinners and another involving two dice.

Figure 24.1 shows the tools inside the two-spinners gadget. The outcome resulting from activating this gadget is controlled by the *workings* box (see upper right graphic in fig. 24.1). As shown in the figure, the workings box is in the default mode and only contains six of the nine possibilities (e.g., 2 + 3 and 3 + 2 are two of the missing possibilities). As children work through the gadget, they can add other sums to the workings box. There is also a repeat box (see lower left graphic in the figure), into which a child can enter the number of trials to be carried out (e.g., 10, 100, 1,000, etc.). The results box (see lower right graphic in the figure) allows results from each trial to be displayed either as a pie chart or as a pictogram. The gadget for the two dice was similar to that for the two spinners, but with two dice shown and with the outcomes box containing six of thirty-six possible outcomes. The two-spinners gadget served as a way forward to the work that Anne and Rebecca did on the two-dice gadget.

---

[1] You can download a free version of ChanceMaker that does not require Boxer and will run on a PC from http://people.ioe.ac.uk/dave_pratt/Dave_Pratt/Software.html.

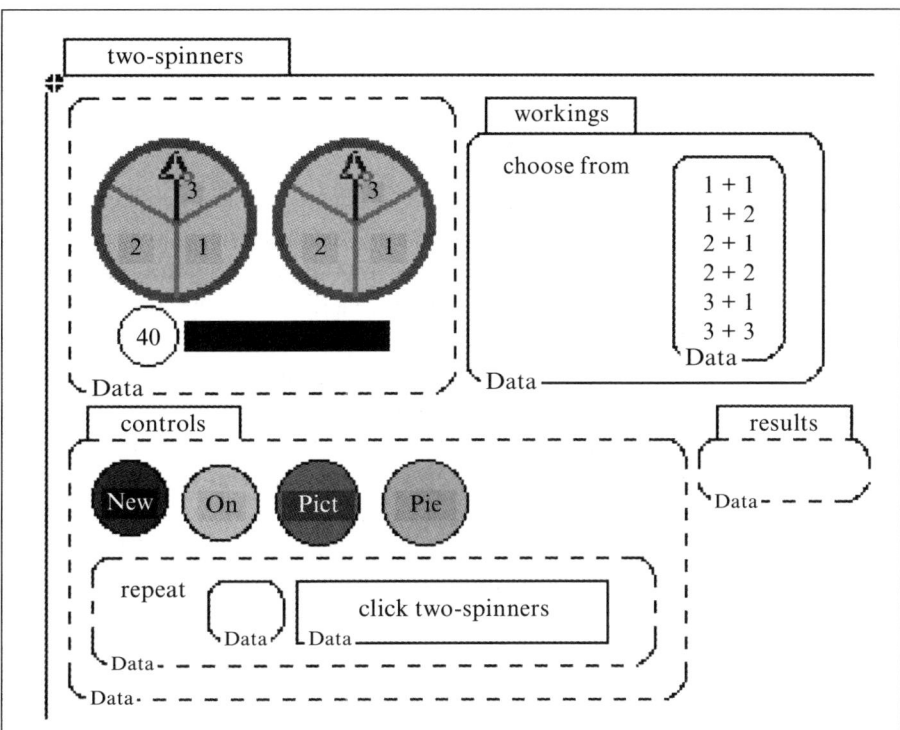

Fig. 24.1. Tools inside the two-spinners gadget

How Anne and Rebecca Made Sense of Chance Using the Chance-Maker Gadgets

In this section, I present results from interactions with Anne and Rebecca as they worked through the gadgets for two spinners and two dice. Before Anne and Rebecca worked on the gadgets, I interviewed them in order to identify their prior notions about chance. With respect to the two-dice situation, the girls declared that no total was harder to obtain than any other (e.g., a total of 2 had the same "chance" of occurring as a total of 7). For example, Anne intuited that the totals must be equally likely because the dice were (individually) fair and so the combination of them must be fair.

1. *Anne:* Because you can't estimate what number you'll get because they're all fair, both the numbers are fair.

Rebecca also saw the total of two dice as equiprobable, but for a different reason:

2. *Rebecca:* 'Cos it's random; you can't control which number it lands on.

The girls' comments were a direct affirmation of equiprobability bias and showed evidence that Anne and Rebecca exhibited the local resource of fairness. Further, in line 2 Rebecca applied the unsteerability local resource to her likelihood comment.

Equiprobability was also apparent in their early interactions with the two-spinners gadget below:

3. *Researcher:* If these were real spinners and we can get any total between 2 and 6, do you think there is any total which is harder to get than the others, any total that is easier to get than any others?

4. *Anne:* What it was in real life? *[Researcher confirms that that is what he means.]* No.

5. *Rebecca:* There's a 50/50 chance of getting any total.

6. *Researcher:* So you think all totals are equally easy or hard to get?

7. *Both:* Yes.

In the next sections, I report on how Anne and Rebecca used the two-spinners and the two-dice gadgets and focus on how they developed global resources from the activities.

Using the two-spinners gadget

Leading up to the following exchange, Anne and Rebecca had already worked with three other gadgets, so they were familiar with how to use the tools such as the workings box and the results box. Figure 24.2 is a schematic representation of the girls' early interactions with the two-spinners gadget. Box A depicts how they began with 50 trials of the default version of the gadget (i.e., only six of nine possible outcomes shown). They generated the pie chart (box B), and felt there were too many 3s and 4s. I could tell by this stage that Anne and Rebecca expected the four sections of the pie chart to be equal, an interpretation of their equiprobability bias. Rebecca commented, "Maybe if we do it more times, it [the size of the pie wedges] might be more even" (comment in box C), referring to need for more trials. They then set the trials to 1,000. The pictogram in fact showed that the total of 5 had not appeared (box E). Anne incorrectly suggested that there were no numbers on the spinners that made 5, but Rebecca pointed out that one could have a 2 and a 3 but that this combination was missing from the default workings chart. They edited the workings to include both 2 + 3 and 3 + 2, though they failed to consider that 1 + 3 was also missing. Rebecca explained their dual inclusion.

8. *Rebecca:* Because the first number is representing the first spinner, so you have to have it both ways.

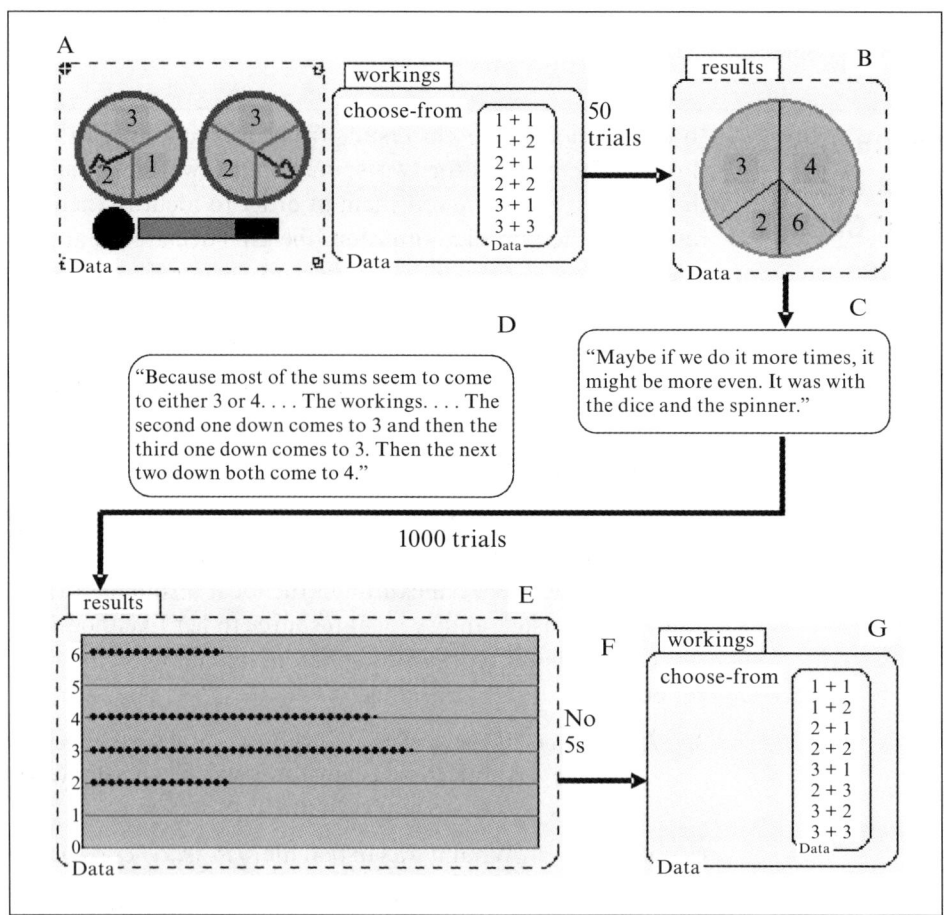

Fig. 24.2. Anne's and Rebecca's early interactions with the two-spinners gadget

Figure 24.3 contains a representation of the girls' subsequent interactions with this gadget, with the story continuing from box H in the top left-hand corner of the figure. Anne and Rebecca repeated 1,000 new trials with eight sums in the workings box, and Anne predicted that the pie chart would now be "even." However, the resulting pie chart (box I) showed unequal sections, with 2s appearing less frequently than other numbers.

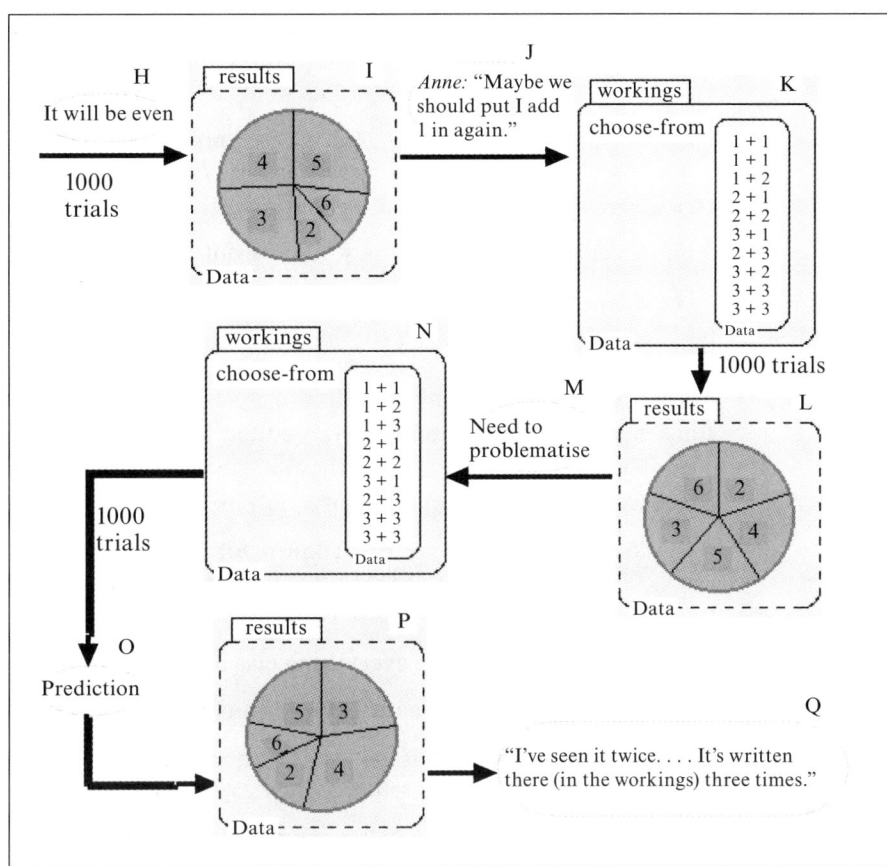

Fig 24.3. Anne's and Rebecca's further interactions with the two-spinners gadget

The next interchange revealed the girls' thinking. Anne wanted to add another 1 + 1 as a way to equalize the sections of the pie chart (make it fair), prompting the following exchange:

9. *Rebecca:* No, because it's already done, so it would just be the same.

10. *Anne:* Yes, because some of them have to be the same again.

11. *Rebecca:* Yes, maybe we should actually put some of them in again, because then there's more of a chance of them coming out more even.

Anne saw the insertion of another 1 + 1 as a way to equalize the sections of the pie chart. Rebecca was torn because of her belief that the different totals should be equally likely (line 11) and her analysis that the doubles, unlike the nondoubles, did not have to be repeated in the workings (line 9). It was now clear that both girls believed strongly that the different totals should be equally likely, confirming previous evidence; this belief was so strong that they sought to fix the workings to make the pie chart more uniform. I decided to intervene to offer support for Rebecca's view regarding doubles, asking Anne what she meant when she said that some of them were the same (line 10).

12. *Anne:* Well, 1 and 2 and 2 and 1 are the same . . . ; they come to the same number.

13. *Researcher:* They come to the same total, but are they the same as far as the spinners are concerned?

14. *Rebecca:* No they are not. Because, the second one down, that number *[pointing to the 1 of 1 + 2 in the workings box]* refers to that spinner *[pointing to the first spinner]*, and that number *[pointing to the 2 of the 1 + 2]* refers to that spinner *[pointing to the second spinner]*. So, say, if that one [the first spinner] lands on 1 and that one [the second spinner] lands on 2, it would be 3. And if that one [the second spinner] lands on 1 and that one [the first spinner] lands on 2, it would be 3 as well.

15. *Anne:* Exactly. . . . I think we should add that one *[pointing to the 1 + 1]* and that one *[pointing to the 2 + 2]* again, because then we get more of a chance of getting them.

16. *Researcher:* But then you would be putting in 1 plus 1 twice.

17. *Anne:* Yes, because 2 doesn't come up much, does it?

18. *Rebecca:* So maybe if we do that.

Anne, who was still convinced that the pie-chart sections should be uniform in size, rejected my intervention; she and Rebecca edited the workings again (box K). After another 1,000 trials with the new workings, the pie chart was fairly uniform (box L). So that Anne and Rebecca might question this new solution, I made the following intervention:

19. *Researcher:* What we don't know for sure is whether that is how real spinners would behave. I think what you need to try and do is justify why you should have 1 plus 1 in there twice over.

20. *Rebecca:* Because everything else has two ways of coming, except maybe 2 plus 2.

21. *Researcher:* But in reality, does 1 plus 1 have two different ways of coming?

22. *Rebecca:* I think it is more fair because the pie chart looks roughly even and before there were barely any 2s and barely any 4s. . . .

23. *Researcher:* I think you have certainly made it more fair. What I am not convinced about is that you have made it more like real spinners would be . . . ; maybe with real spinners that would not be the case.

24. *Anne:* Oh, yes, mmm.

25. *Researcher:* You see, I am not sure you are being fair by putting 1 + 1 in twice.

26. *Anne:* We don't want it to be even. We want it to work like a real spinner.

My intervention was intended to suggest an alternative way of thinking about fairness by suggesting that it might be regarded as unfair to represent the same outcome twice over, especially inasmuch as the aim was to make the two-spinners gadget behave like real spinners. In Anne's comment in line 26, it is important to know whether "working like a real spinner" for her then meant reviewing the ordered pairs and making them fair or whether she was still wedded to the idea that the totals should be represented equally.

Anne and Rebecca immediately and spontaneously removed one 1 + 1 and one 3 + 3 from the workings, suggesting they recognized that the unfairness of including 1 + 1 twice extended to the other doubles. They recognized the need to add 1 + 3 only after I intervened with this suggestion. The girls then generated another 1,000 trials, and I asked them to predict what the pie chart would look like (box O):

27. *Anne:* A bit uneven . . . because 1 and 1 had only got once, because that is what a real spinner would be like.

Using a Computer Microworld to Make Sense of the Total of Two Dice

28. *Rebecca:*   I think maybe . . . 2 won't come up as much, and 6.

The pie chart showed most 4s and fewest 2s and 6s (box P). For the first time in this activity, Anne and Rebecca suggested that the two-spinners gadget should not generate a uniform-looking pie chart. I also looked for an explanation from the students for the size of the 4 sector in this pie chart. Rebecca referred directly to the frequency of the total 4 in the workings box (box Q):

29. *Anne:*   I've seen it twice. . . . It's written there three times. . . . Because it's got more of the numbers. It's got like three different numbers, so it's coming up much more.

30. *Researcher:*   So do you think the different totals, 2, 3, 4, 5, 6, are all just as easy to get or all just as hard to get? Or is there one that's easy to get?

31. *Anne & Rebecca:*   *[overlapping]* Four is easier to get.

32. *Researcher:*   Is there one that is hard?

33. *Anne & Rebecca:*   Six.

34. *Researcher:*   Just the 6?

35. *Anne & Rebecca:*   Two.

I took these responses as evidence that Anne and Rebecca were articulating a global resource—that is, "The more often a total is represented in the workings box, the larger will be its sector in the pie chart." But as the discussion continued, the girls still depended on the pie chart (box P) and concluded that because the sector for 6 is smaller than that for 2, 6 is harder to get. I suggested they repeat the 1,000 trials to see what would happen, and the following discussion ensued while we waited:

39. *Rebecca:*   There seems to be less ways of getting 6 and 2 than there are of 4, 5. *[The pie chart appears on screen with a larger sector for 6s and 2s.]*

40. *Rebecca:*   The 2 is smaller this time than last time. Because last time 6 was smaller.

41. *Researcher:*   So do you think 2 is easier to get, just the same, or harder to get than the 6?

42. *Rebecca:*   Just the same, I think, because last time 6 was slightly bigger than 2 and this time . . . the other way round; 2 was bigger than 6 and this time 6 is bigger than 2.

43. *Anne:*   Yes, I agree.

44. *Researcher:*   So, if we were going to play a game in which we had two spinners like this [i.e., like the ones on screen] and you are going to bet a pound and I am going to bet a pound and you're going to bet that a total of 4 comes up and I'm going to bet that a total of 6 comes up, would you take that bet?

45. *Anne:*   No . . . well, sort of . . . I wouldn't bet as much as a pound.

46. *Researcher:*   How much would you bet?

47. *Anne:*   Twenty p.

48. *Rebecca:*   Two p.

49. *Researcher:*   If we were to bet 2p the other way round that I'm betting on a 4 and you're betting on a 6, would you take that bet?

50. *Rebecca:*   No.

51: *Anne:*   Definitely not.

265

The discussion indicates that the girls were now placing quite a high priority on the global resource based on frequency of the totals in the workings box rather than on the appearance of the pie chart.

Using the two-dice gadget

At this point, Anne and Rebecca turned their attention to the two-dice gadget. Figure 24.4 traces their interactions. It was clear to me from their first discussion that they believed there were missing data in the workings.

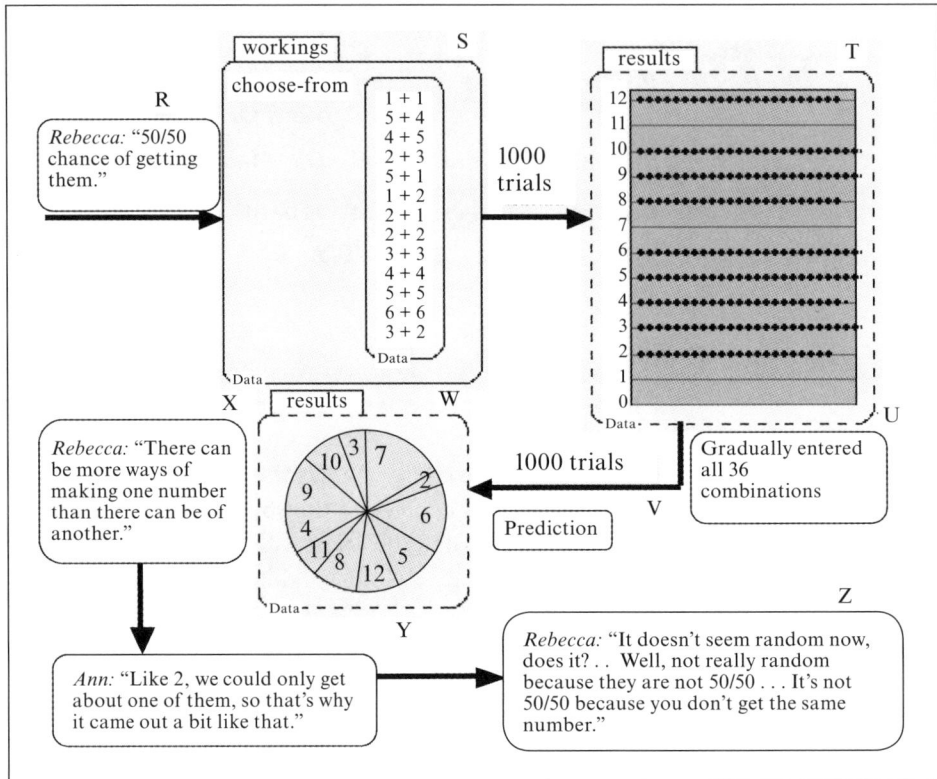

Fig. 24.4. Anne's and Rebecca's interactions with the two-dice gadget

52. *Rebecca:* There's lots of different ways . . . there's probably more ways of making 6 . . .

53. *Anne:* Ah, 5 add 1 is missing to make 6.

This discussion resulted in some modifications to the workings box. At this point, I wanted to find out whether their thinking about the totals of the two dice had changed in light of insights gained from exploration of the two-spinners gadget just concluded.

54. *Researcher:* If we were shaking two real dice, do you think all the totals you could get are just as easy, just as hard, or do you think some totals are easier than others, harder than others?

55. *Rebecca:* Fifty-fifty chance of getting them. *[Anne agreed.]*

56. *Researcher:* So you think they are all about the same chance.

57. *Both:* Yes

Anne and Rebecca clearly articulated the equiprobability bias with respect to the total of two dice, a surprising turn, given the way they had ended the discussion about the total of two spinners. The global resources constructed from the two-spinners work were not cued by this new situation. Instead, local resources articulated in the preinterviews (e.g., fairness) were cued.

Anne and Rebecca then noticed other outcomes missing and added them to the workings box (box S), but after 1,000 trials, they generated a pictogram showing no 7s or 11s (box T). With some encouragement from me, they incorporated all thirty-six ordered pairs into the working box (box U) and tried the new situation with 1,000 trials. Anne made this prediction about the pie chart:

58. *Anne:* I think some will be a bit less because they haven't got as much as the others . . . because some of the numbers will not be the same, will be less, because we didn't find enough sums for them . . . like 1 and 1.

59. *Researcher:* Can you give me an example of one that had a lot of different ways of getting it?

60. *Anne:* Seven.

Through their interactions with the workings box, Anne recognized that some totals were represented more often than others, although Rebecca still thought that the pie chart might turn out to be uniform. Because of their previous experience with the two-spinners gadget, they considered the possibility that the pie chart for two dice might be uneven. There were then two types of internal resources available to the girls: (a) the global resource that the sectors in the pie chart were determined by the frequencies of the corresponding totals in the workings box and (b) the local resource that each individual die was unsteerable (i.e., uncontrollable) and so the various totals were equally likely.

The pie chart (box W) showed most 7s and fewest 2s and 12s. Rebecca's first reaction was that there might still be some missing outcomes in the workings box, but then she made the following observation:

61. *Rebecca:* Ah, I bet there are various ways of making a number. There can be more ways of making one number than there can be of another.

By inspecting the workings box, the girls identified that the 12s, 2s, and 3s could not be obtained many ways whereas the 7 resulted for the most possible pairs. I asked why the pie chart had that appearance, and Anne replied:

62. *Anne:* Because some of the sums we put as more. Like 2, we could only get about one of them, so that's why it came out a bit like that.

Rebecca's final comment ("It doesn't seem random now, does it? . . . Well, not really random because they are not 50/50 . . . because you don't get the same number.") shows the extent to which she was constructing new global resources for the behavior of the two-dice gadget.

Conclusions and Suggestions for Instruction

Based on the results from my work with Anne and Rebecca (and the other fourteen children), I found that:

1. The children came to this activity with a variety of local resources based on symmetry and experience of short-term behavior, through which they supposed that the totals would be uniformly distributed. The equiprobability bias was strongly articulated and was associated with local resources such as unsteerability and fairness (see lines 1, 2, and 5; lines 42 and 43).

2. In making sense of the total of two dice, Anne and Rebecca (as an example for all subjects) constructed new global resources, such as "The frequency of representations of a total in the workings box of the two-spinners gadget controls the size of its sector in the

pie chart (box Q, fig. 24.3; and lines 27–28). A similar new global resource was also constructed for the two-dice gadget (boxes X, Y, and Z, fig. 24.4; and lines 37–38).

3. Newly acquired global resources are not automatically cued in different circumstances. Although Anne and Rebecca had constructed the global resource ("The frequency of representations of a total in the workings box controls the size of its sector in the pie chart"), in the two-dice explorations the girls reverted to the equiprobability bias.

The arguments about the possible benefits of the use of technology to support teaching and learning of mathematics are often over simplified. Chance-Maker seems to be beneficial in this content because of certain features of its design:

1. It allows children space to confirm or disconfirm their prior knowledge about the total of two dice (or spinners).

2. The workings box allows children to set up conjectures that can then be tested by running the simulation.

3. The repeat control facilitates the capture of many results quickly, though it was in practice not self-evident to the students that it was necessary to do so.

Nevertheless, the interventions of the researcher were critical. In a sense the microworld merely opened up the possibility for these interventions to be made at opportune moments. It was the researcher's actions that prompted the notion of fairness to be redirected from the totals to the individual outcomes of each throw. Teachers may also need to engage in this kind of intervention in order to redirect their students' thinking.

Even so, children did not re-use in a trivial way what they had learned about the total of two spinners to make sense of the total of two dice. The researcher needed to wait patiently while the children discovered that what they had learned about two spinners very recently carried more explanatory power for the total of two dice than did their long-held beliefs. Again, this kind of patience on the part of teachers can be of great benefit as students engage in discovery learning.

References

Fischbein, E., & Schnarch, D. (1997). The evolution with age of probabilistic, intuitively based misconceptions. *Journal for Research in Mathematics Education, 28*, 96–105.

Lecoutre, M. P. (1992). Cognitive models and problem spaces in "purely random" situations. *Educational Studies in Mathematics, 23*, 557–568.

Piaget, J., & Inhelder, B. (1951). *The origin of the idea of chance in children.* New York, NY: Norton.

Pratt, D. (2000). Making sense of the total of two dice. *Journal for Research in Mathematics Education, 31*, 602–625.

Pratt, D., & Noss, R. (2002). The micro-evolution of mathematical knowledge: The case of randomness. *Journal of the Learning Sciences, 11*, 453–488.

Pratt, D., & Noss, R. (2010). Designing for mathematical abstraction. *International Journal of Computers for Mathematical Learning, 15*, 81–97.

Tarr, J. E. (1998). Middle school students' misuse of the phrase "50-50 chance" in probability instruction. In S. Berenson, K. Dawkins, M. Blanton, W. Coulombe, J. Kolbe, K. Norwood, & L. Stiff (Eds.), *Proceedings of the 20th annual meeting of the North American Chapter of the International Group for the Psychology of Mathematics Education* (pp. 401–406). Columbus, OH: ERIC Clearinghouse for Science, Mathematics, and Environmental Education.